Encyclopedia of
Human Genetics and Disease

Encyclopedia of
Human Genetics and Disease

Evelyn B. Kelly

VOLUME 2
L–Z

GREENWOOD

AN IMPRINT OF ABC-CLIO, LLC
Santa Barbara, California • Denver, Colorado • Oxford, England

Library of Congress Cataloging-in-Publication Data

Kelly, Evelyn B.
  Encyclopedia of human genetics and disease / Evelyn B. Kelly.
    p. cm.
  Includes bibliographical references and index.
  ISBN 978–0–313–38713–5 (hardback) — ISBN 978–0–313–38714–2 (ebook)
  1. Human genetics—Encyclopedias. 2. Genetic disorders—Encyclopedias. I. Title.
QH431.K32   2013
616′.04203—dc23   2012018368

ISBN: 978–0–313–38713–5
EISBN: 978–0–313–38714–2

17 16 15 14 13   2 3 4 5

This book is also available on the World Wide Web as an eBook.
Visit www.abc-clio.com for details.

Greenwood
An Imprint of ABC-CLIO, LLC

ABC-CLIO, LLC
130 Cremona Drive, P.O. Box 1911
Santa Barbara, California 93116-1911

This book is printed on acid-free paper ∞

Manufactured in the United States of America

Contents

Contents

Contents

Contents

L

L1 Syndrome

Prevalence   1 in 25,000 to 60,000 males; females rarely affected

Other Names   CRASH syndrome; MASA syndrome; SPG1; X-linked
complicated hereditary spastic paraplegia type 1; X-linked
corpus callosum agenesis; X-linked hydrocephalus with
stenosis of the aqueduct of Sylvius (HSAS)

Naming syndromes and disorders presents a challenge to investigators. This syndrome
has had many names, with the acronym CRASH designated in 1995. However, some
scientists thought this name was not appropriate and could possibly be offensive. In
1998, the name for the syndrome was changed to L1, now used as the preferred name.

What Is L1 Syndrome?

L1 syndrome is a disorder of the nervous system. At one time, because of the variety of features, researchers thought they were dealing with several disorders. Now
they believe these different symptoms are part of the same syndrome. Features of
the syndrome include muscles stiffness of the lower limbs or spasticity, mental
retardation, fluid collection in the brain or hydrocephalus, and thumbs bent toward
the palm called adducted thumbs.

Each symptom presents a distinct side to the disorder and has been labeled in
the following manner:

- *SPG1 or X-linked complicated spastic paraplegia type 1*: Affected males
  have spastic paraplasia or stiffness in the lower limbs, mild to moderate
  intellectual disability, but a normal MRI of the brain.
- *MASA syndrome*: MASA is an acronym for: M—mental retardation; A—
  aphasia or delayed speech; S—spastic paraplasia; and A—adducted thumbs.
  Males with this condition have mild to moderate intellectual disability,

delayed onset of speech, muscle weakness that leads to spasticity, adducted or clasped thumbs, and an enlarged third ventricle area of the brain.

- *X-linked complicated corpus callosum agenesis*: The corpus callosum is the tough band of fibers that joins the two halves of the brain. In this condition, this part of the brain did not develop properly, leading to paraplegia and intellectual disability.

- *X-linked hydrocephalus with stenosis of the aqueduct of Sylvius (HSAS)*: These males have severe hydrocephalus even before birth and increased pressure and fluid collecting in the ventricles of the brain. They also have adducted thumbs, severe spasticity, and severe intellectual disability.

How can these seemingly disparate conditions be linked as one syndrome? They all are mutations of the same gene, *L1CAM*.

What Is the Genetic Cause of L1 Syndrome?

Mutations in the *L1CAM* gene, officially called the "L1 cell adhesion molecule" gene, cause L1 syndrome. Normally, *L1CAM* instructs for the L1 protein found on the surfaces of neurons throughout the nervous system. Similar to many proteins, L1 has part of the protein inside the cell and part of it protruding out of the cell. Such a position enables the protein to activate chemical signals within the cell and to communicate to other cells. The function of L1 appears to be developing and organizing neurons, forming the protective myelin sheath, and making the connections between cells called the synapses. All of these functions are essential to normal development.

Over 200 mutations in *L1CAM* cause L1 syndrome. Any small change in a building block or amino acid can disrupt the normal functioning of the molecule and cause serious malfunction of the neurons. This impairment of the function of the neurons is the hallmark of this condition and leads to the diverse characteristics of L1 syndrome. *L1CAM* is inherited in an X-linked recessive pattern and is located on the long arm (q) of chromosome X at position 28.

Because of the inheritance in an X-linked recessive manner, women who are carriers have a 50% chance of passing the gene to offspring. Daughters who inherit the mutation will be carriers; sons will have the condition. Affected males do not reproduce. If a woman knows of L1 syndrome in her background, she should seek genetic counseling.

What Is the Treatment for L1 Syndrome?

This serious genetic condition has no cure. Management of the symptoms must involve a team of experts in pediatrics, child neurology, neurosurgery, rehabilitation, and medical genetics. Depending upon the type and symptoms, care should be given. If the child has fluid on the brain, surgery may be necessary to relieve pressure. If the thumb is adducted, a splint may be used to improved function. Physical and occupational therapy should be used to help the child learn to function to their optimal capacity.

Further Reading

"L1 Syndrome." 2011. RightDiagnosis.com. http://www.rightdiagnosis.com/l/l1 _syndrome/intro.htm. Accessed 5/21/12.

"L1 Syndrome." 2011. Health.com. http://www.health.com/health/library/mdp/0,, nord1097,00.html. Accessed 5/21/12.

Schrander-Stumpel, Connie, and Yvonne Vos. 2010."L1 Syndrome." *GeneReviews*. http://www.ncbi.nlm.nih.gov/books/NBK1484. Accessed 5/21/12.

Lactose Intolerance

Prevalence 30 to 50 million Americans are lactose intolerant; congenital intolerance rare but most common in Finland, where it affects about 1 in 60,000 infants; lactase non-persistence about 65% prevalence in some populations, such as those of West African, Jewish, Greek, and Italian heritage; only 5% in people of northern European descent

Other Names alactasia; dairy product intolerance; hypolactasia; lactose malabsorption; milk sugar intolerance

What Is Lactose Intolerance?

Lactose intolerance is the inability to digest lactose, a natural sugar found in milk and other dairy products. People who are lactose intolerant have a shortage of an enzyme called lactase, which is produced in the lining of the small intestine. Lactase breaks down lactose into two simpler sugars, glucose and galactose. These simple sugars can then be absorbed in the bloodstream for use in the body to make energy.

Following are kinds of lactose intolerance:

- Congenital alactasia: The infant is born without the ability to break down lactose in both breast milk and formula. When milk is drunk, the child will have cramps and severe diarrhea and must be given a lactose-free diet immediately, or he will dehydrate and lose weight.

- Adult lactose intolerance: This form, called lactase nonpersistence, develops later in life. People note that about 30 minutes to two hours after eating dairy products, they develop abdominal pain, bloating, gas, nausea, or diarrhea. These signs indicate lactose intolerance.

- Secondary lactase deficiency: This form may result from injury to the small intestine. Certain diseases, such as severe diarrhea, Crohn disease, celiac disease, or chemotherapy, may affect the production of the lactase enzyme. This type can occur at any age.

• Some tolerance: Some people with lactase nonpersistence can eat some dairy products without the unpleasant symptoms. They may still have difficulty digesting fresh milk but can eat products such as yogurt or cheese, which are made from fermentation processes that break down lactose.

NOTE: Lactose intolerance is not an allergy to milk. A lactose allergy is a response that the immune system triggers. People with milk allergy must avoid any milk products altogether, or it can be life-threatening. Some people with food intolerances may eat a small amount of the food with milk products with minimal symptoms.

What Is the Genetic Cause of Lactose Intolerance?

Mutations in two genes—*LCT* and *MCM6*—cause lactose intolerance.

LCT

Mutations in the *LCT* gene or the "lactase" gene cause lactose intolerance. Normally, *LCT* encodes the enzyme lactase, an essential protein for digesting lactose. In the small intestine are cells known as epithelial cells that have small microvilli or fingerlike projections. These projections absorb nutrients from food as it passes through the small intestine. The villi are called the brush border, with the function to break down lactose into glucose and galactose, which the body can use.

Nine mutations in *LCT* cause the congenital form of congenital alactasia. The mutations change a single amino acid in the lactase enzyme and result in an abnormally short protein. This short enzyme disrupts the function of the digestive process, leading to undigested lactose in the small intestine and the unpleasant symptoms. Lactose intolerance in adulthood occurs from decreasing activity of the *LCT* gene. *LCT* is inherited in an autosomal recessive pattern and is located on the long arm (q) of chromosome 2 at position 21.

MCM6

Mutations in the *MCM6* gene, or the "minichromosome maintenance complex component 6" gene, are also related to lactose intolerance. Normally, *MCM6* encodes for making a segment of the MCM complex, which is a group of proteins that works as a helicase. Helicases are enzymes that attach to certain regions of DNA and unwind the two spirals so a cell can divide. The DNA is then copied. Helicases are also involved in the production of RNA and regulation of other genes. MCM6 protein regulates the activity of *LCT*.

At least four changes in the MCM6 protein have been involved in a regulatory element of the *LCT* gene that makes the enzyme lactase. The changes involve only one building block in the regulatory area. Each change results in constant production of lactase in the small intestine and disrupting the ability to digest lactose. This process affects people as they get older. *MCM6* is inherited in an autosomal recessive pattern and is located on the long arm (q) of chromosome 2 at position 21.

What Is the Treatment for Lactose Intolerance?

No cure for lactose intolerance exists, but it can be controlled through diet and education about nutrients. Individuals may not have to avoid lactose completely or may be able to tolerate small amounts. Knowing about foods with lactose is essential. Following is a list of items that contain lactose:

- Milks and creams, including canned evaporated and condensed milk. There is a lactose-free product that may be substituted for milk
- Ice creams, sherbets, and yogurts.
- Some cheese including cottage cheese.
- Butters.
- Prepared foods such as breads, cereals, mixes for cakes and cookies, and lunch meats other than kosher meats.
- Instant potatoes, soups, and breakfast drinks.
- Frozen dinners.
- Salad dressings.
- Candies and other snacks.
- Nondairy creamers and whipped toppings. These foods have sodium casein-ate that may contain low levels of lactose.
- Some medicines. This unexpected source includes items such as birth control pills and over-the-counter medicines to treat stomach acid and gas. These medications may cause symptoms in people with severe intolerance.

Learning to check food labels in important. Take care if the label says the product contains milk or milk by-products, whey, curds, or nonfat milk powder.

Further Reading

"Lactose Intolerance." 2011. International Foundation for Gastrointestinal Disorders. http://iffgd.org/site/gi-disorders/other/lactose. Accessed 11/13/11.

"Lactose Intolerance." 2011. MedlinePlus. National Library of Medicine (U.S.). http://www.nlm.nih.gov/medlineplus/lactoseintolerance.html. Accessed 11/13/11.

"Lactose Intolerance." 2011. National Institute of Diabetes Digestive and Kidney Diseases (U.S.). http://digestive.niddk.nih.gov/ddiseases/pubs/lactoseintolerance. Accessed 11/13/11.

Lafora Progressive Myoclonus Epilepsy

Prevalence   Unknown, but occurs worldwide; common in countries around the Mediterranean, Central Asia, India, Pakistan, North Africa, and Middle East.

Other Names   epilepsy, progressive myoclonic, Lafora; Lafora body disease; Lafora disease; Lafora progressive myoclonic epilepsy; Lafora type progressive myoclonic epilepsy; myoclonic epilepsy of Lafora; progressive myoclonic epilepsy type 2; progressive myoclonus epilepsy, Lafora type

Lafora progressive myoclonus epilepsy is only one of a very large group of diseases characterized by myoclonic seizures. A myoclonic seizure is known as the fall-down, jerking-all-over seizure that most people associate with epilepsy; this seizure is also called a grand mal seizure. This sudden involuntary jerking can affect any part of the body. The term "progressive myoclonus epilepsy" or PME is used to describe a group of epilepsies that appear to worsen over time. Lafora is one of these types of epilepsy.

What Is Lafora Progressive Myoclonus Epilepsy?

Lafora progressive myoclonus epilepsy is a brain disorder. The following characteristics of the disorder distinguish Lafora from other types of epilepsies:

- Person is healthy until first seizures, which may begin between 12 and 17.
- Recurring myoclonus seizures can begin when the person is at rest and is made worse by motion or excitement of flashing lights.
- Person may have grand mal seizures that cause muscle rigidity over the entire body, along with loss of consciousness.
- Individuals may have occipital seizures that cause temporary blindness and hallucinations.
- Seizures worsen over time and become continuous.
- Progressive neurological deterioration includes both intellectual decline and behavioral disorders.
- Emotional disturbance and confusion are common after onset of seizures.

Deterioration of the nervous system causes most individuals to die within 10 years after the onset of seizures.

What Are the Genetic Causes of Lafora Progressive Myoclonus Epilepsy?

Two genes are responsible for Lafora Progressive Myoclonus Epilepsy. They are *EPM2A* and *NHLRC1*.

EPM2A

The first gene is *EPM2A* or "epilepsy, progressive myoclonus type 2A, Lafora disease (laforin)." *EPM2A* instructs for the protein laforin, a critical player in the survival of neurons in the brain. Laforin wears many hats. It is present in all cells in the body and interacts with other proteins, especially malin, produced by the *NHLRC1* gene. These two proteins create a network to transmit signals to break down abnormal proteins. Another important function of laforin is to act as a tumor suppressor, keeping cells from unnecessary dividing.

In addition, laforin and malin regulate the production a sugar called glycogen. This sugar, which is stored in muscles and liver, is a source of fuel energy when needed by the body. Cells of the nervous system do not need these molecules like the muscles must have to produce energy.

More than 50 mutations in *EPM2A* cause Lafora progressive myoclonus epilepsy. Most of the mutations result from a change in only a single building block in the laforin protein. A few mutations are related to the deletion or insertion of genetic material. The change in the gene disrupts the action of laforin and prevents the cells from regulating the production of glycogen. The production of an abnormal form of glycogen causes clumps of Lafora bodies to form within the cells. These clumps cannot be used for fuel or energy and instead begin to damage cells, especially neurons. The degenerating neurons then cause seizures, abnormal movements, and intellectual and behavioral decline. *EPM2A* is inherited in an autosomal recessive pattern and is located on the long arm (q) of chromosome 6 at position 24.

NHLRC1

The second gene involved is *NHLRC1* or "NHL repeat containing 1." Normally, *NHLRC1* instructs for the protein malin, which like its companion *EPM2A* plays an important role in the survival of nerve cells in the brain. Malin helps to break down unwanted proteins in the cells in a unique way. It tags damaged and unwanted proteins with a signal molecule called ubiquitin. This system becomes the cell's quality control agent by disposing of unwanted proteins. The interaction between malin and laforin is important in regulating glycogen. The two together protect the nervous system.

About 45 mutations in the *NHLRC1* gene may cause Lafora progressive myoclonus epilepsy. Many of the mutations are caused by only one change in a single building block. One of the most common is the substitution of alanine for the amino acid proline. This particular mutation is found in people with Portuguese, Spanish, and Italian heritage. These changes lead to a nonworking version of the protein. Any disruption causes a protein that does not work properly and leads to the symptoms of Lafora progressive myoclonus epilepsy. *NHLRC1* is inherited in an autosomal recessive pattern and is located on the short arm (p) of chromosome 6 at position 22.3.

What Is the Treatment for Lafora Progressive Myoclonus Epilepsy?

Antiepileptic drugs, called AEDs, are effective against general seizures. However, terrible side effects may occur with these drugs if the person is overmedicated. As the person progresses with seizures, they become more difficult to control with medications.

Further Reading

"Lafora Progressive Myoclonus Epilepsy." 2011. Genetics Home Reference. National Library of Medicine (U.S.). http://ghr.nlm.nih.gov/condition/lafora-progressive-myoclonus-epilepsy. Accessed 5/21/12.

Jansen, Anna, and Eva Andermann. 2011. "Progressive Myoclonus Epilepsy, Lafora Type." *GeneReviews*. http://www.ncbi.nlm.nih.gov/books/NBK1389. Accessed 5/21/12.

Madisons Foundation. 2011. http://www.madisonsfoundation.org/index.php/component/option,com_mpower/Itemid,49/diseaseID,624. Accessed 5/21/12.

Laing Distal Myopathy

Prevalence   Unknown; several families worldwide identified

Other Names   distal myopathy 1; Laing early-onset distal myopathy; MPD1

Nigel Laing, a professor at Western Australian Institute for Medical Research, noted a condition that usually begins before the age of five and starts with the areas farthest from the body core. The term "distal" is used to describe these areas, generally the hands and feet; the term "promixal" is used to describe the body core of the heart, intestine, and other such areas. The term "myopathy" comes from two Greek terms: *myo*, meaning "muscle," and *path*, meaning "disease." Thus, Laing distal myopathy is a muscle condition that affects the areas that are away from the body core, at least in the early stages.

What Is Laing Distal Myopathy?

Laing distal myopathy is a disorder of the skeletal muscles. The condition is noted when the child develops weakness in the muscles of the feet and ankles. First affected is the Achilles tendon, the band that connects the heel to the muscles of the calf of the leg. The child also cannot lift the big toe and appears to have a high-stepping walk. Over a period of months or perhaps years, the disorder is noted in the hands and wrists. The child cannot lift the fingers, especially the third and fourth digits. In addition to hands and feet, the muscles that flex the neck may begin to show weakness along with muscles of the face.

The muscle weakness in the hands and feet becomes progressively worse. After about 10 years, other skeletal muscles in the legs, hips, and shoulders are affected. However, the person has a normal life expectancy and can remain mobile throughout life.

What Is the Genetic Cause of Laing Distal Myopathy?

Mutations in the *MYH7* gene, also called the "myosin, heavy chain 7, cardiac muscle, beta" gene, cause Laing distal myopathy. Normally, *MYH7* instructs for a protein called the cardiac β-myosin heavy chain. This protein is found in two major areas: heart muscle and type I skeletal muscle fibers.

Type I skeletal muscle fibers are called slow-twitch fibers and are one of the two types of fibers that make up the muscles of the skeletal system. Type I fibers provide strength to the muscles. They are primarily responsible for keeping the head and neck steady and for helping the muscles resist fatigue.

In both heart and skeletal muscle cells, the β-myosin heavy chain is part of a larger protein complex called type II myosin. Type II myosin is the protein that enables the skeletal muscles to contract and move. The gene *MYH7* instructs for the two heavy chains that are part of type II myosin protein; two light chains also that are part of Type II myosin are controlled by other genes. The two heavy chains have a head area or a beginning and a tail area. The head area interacts with actin, a protein essential for cell movement. The tail region interacts with other proteins.

About five mutations in the *MYH7* genes cause Laing distal myopathy. The genetic changes are found in the tail area of the β-myosin heavy chain. Most of the mutations involve only the exchange of one amino acid from the heavy chain. This exchange is enough to disturb the normal function of Type II myosin. The altered tail region cannot interact with other proteins, leading to the progressive symptoms of Laing distal myopathy. *MYH7* is inherited in an autosomal dominant pattern and is located on the long arm (q) of chromosome 14 at position 12.

What Is the Treatment for Laing Distal Myopathy?

When the weakness in the Achilles tendon is first noticed, the child should receive physical therapy to prevent progressive tightening. A lightweight splint can possibly help. The heart should be carefully monitored to make sure that no problems occur in this area.

Further Reading

"Laing Distal Myopathy." 2011. Genetics Home Reference. National Library of Medicine (U.S.). http://ghr.nlm.nih.gov/condition/laing-distal-myopathy. Accessed 5/21/12.

"Laing Distal Myopathy." 2011. RightDiagnosis.com. http://www.rightdiagnosis.com/l/laing_distal_myopathy/intro.htm. Accessed 5/21/12.

Lamont, Phillipa; William Wallefeld; and Nigel G. Laing. 2010. "Laing Distal Myopathy." *GeneReviews*. http://www.ncbi.nlm.nih.gov/books/NBK1433. Accessed 5/21/12.

LAL Deficiency. *See* Wolman Disease

LAMB—Lentigines, Atrial Myxoma, Mucocutaneous Myoma, Blue Nevus Syndrome. *See* Carney Complex

Langer-Giedion Syndrome

Prevalence   Rare; incidence unknown

Other Names   Giedion-Langer syndrome; trichorhinophalangeal syndrome type II; tricho-rhino-phalangeal syndrome type II; TRPS II

Leonard Langer, a professor at the University of Minnesota, and Andreas Giedion, a professor at the University of Zurich, researched a condition called trichorhinophalangeal syndrome with exostoses in the late 1960s. Both men were active in efforts to help people with this rare syndrome. To honor these two doctors, the condition was named Langer-Giedion syndrome.

What Is Langer-Giedion Syndrome?

Langer-Giedion syndrome is a condition that affects the skeletal system and produces the distinctive facial features. Following are the characteristics of the syndrome:

- Short stature.
- Cone-shaped end of bones.
- Distinct facial appearance: Individuals have a rounded nose with a long flat area between the nose and the thin upper lip.
- Benign bone tumors called exostoses: Although one may not be too annoying, many exostoses can result in pain. Exostoses limit joint movement putting pressure on nerves, the spinal cord and surrounding tissues.
- Sparse scalp hair.
- Loose skin in childhood, which may improve with age.
- Some intellectual disability.

What Is the Genetic Cause of Langer-Giedion Syndrome?

Two genes located on chromosome 8 are related to this syndrome. The genes are *EXT1* and *TRPS1*.

EXT1

Mutations in the *EXT1*, or the "exostosin 1" gene, cause Langer-Giedion syndrome. Normally, *EXT1* instructs for the protein exotosis-1, which is found in the Golgi apparatus. The function of the Golgi apparatus is to modify enzymes and other proteins. Here, exotosin-1 binds to another protein called exotosin-2 to form a supermolecule that modifies heparan sulfate. Heparan sulfate, a group of sugar molecules called a polysaccharide, is then added to other proteins to form proteoglycans. Heparan sulfate is a regulator of body process and plays a major role in spreading of cancer cells.

Mutations in *EXT1* and some other genes on chromosome 8 cause Langer-Giedion syndrome. The mutations keep exotosin-1 from being produced, and thus the multiple noncancerous bone tumors or exotoses occur. Most cases are not inherited but occur during the formation of egg or sperm. It appears as inherited in a dominant pattern because one copy of an altered gene may cause the condition. *EXT1* is located on the long arm (q) of chromosome 8 at position 24.11.

TRPS1

Mutations in the *TRPS1* gene, or the "trichorhinophalangeal syndrome I" gene, cause Langer-Giedion syndrome. Normally, *TRPS1* instructs for making a protein that controls the activity of many other genes. This activity is called a transcription factor. Much of the embryonic tissue is regulated by this protein.

Along with other genes on chromosome 8, the mutations disrupt the TRPS1 protein from the normal functioning of normal bone growth, which leads to short stature and distinctive facial features. *TRPS1* is inherited in an autosomal dominant pattern and is located on the long arm (q) of chromosome 8 at position 24.12.

Chromosome 8

This chromosome has between 700 and 1,100 genes and represents about 4.5% to 5% of total DNA in cells. Researchers have determined that a section of this chromosome that has the two genes is missing and causes several disorders relating to growth and tumors.

What Is the Treatment for Langer-Giedion Syndrome?

Treatment for this disorder is symptomatic. Some of the bone tumors that are removed tend to grow back. Physical and occupational therapy may improve the person's self-esteem.

Further Reading

"Langer-Giedion Syndrome." 2011. Genetics Home Reference. National Library of Medicine (U.S.). http://ghr.nlm.nih.gov/condition/langer-giedion-syndrome. Accessed 11/5/11.

"Langer-Giedion Syndrome." 2011. University of Kansas Medical Center. http://www.kumc.edu/gec/support/langer_g.html. Accessed 11/5/11.

"Trichorhinophalangeal Syndrome Type 2." 2011. Genetics and Rare Disease Information Center. National Institutes of Health (U.S.). http://rarediseases.info.nih.gov/GARD/Disease.aspx?PageID=4&diseaseID=7801. Accessed 11/5/11.

Larsen Syndrome

Prevalence   Affects about 1 in 100,000 newborns
Other Names   LRS

In 1950, L. J. Larsen and a team of researchers wrote about a condition that had many congenital dislocations, in addition to distinct facial abnormalities. On the basis of studies of six families, the team published their study in the *Journal of Pediatrics*. The syndrome was named in honor of Dr. Larsen—Larsen syndrome.

What Is Larsen Syndrome?

Larsen syndrome is a condition that affects the entire skeletal system. It is classified as a skeletal dysplasia because of the abnormalities of many bone structures. Following are some of the symptoms of the disorder, which may vary widely even within a family:

- Short stature.
- Clubfeet. The infant is born with inward and upward turning feet.
- Distinct facial appearance. The person has a large high forehead, flat middle face and nose, and wide-set eyes.
- Dislocation of the hips, knees, and elbows.
- Small extra bones in the wrists and ankles. These bones can be seen on X-rays.
- Blunt tips of fingers. The long fingers look much like a square-shaped shovel.
- Joints have a large range of motion, called hypermobility.
- Abnormal curvature of the spine.
- Weakness in the limbs because of pressure on the spinal cord.
- Cleft palate. Some children have a cleft palate.
- Hearing loss because of malformation of bones in the ear.

- Respiratory problems. The person may have many infections and short pauses in breathing.
- Heart defects.
- Cataracts.

Individuals may live into adulthood. Intelligence is normal.

What Is the Genetic Cause of Larsen Syndrome?

Mutations in the *FLNB* gene, known also as the "filamin B, beta" gene, cause Larsen syndrome. Normally, *FLNB* encodes for a protein called filamin B. This protein is essential in building the network of cells called the cytoskeleton, which is the lattice-like structure that supports cells and allows them to move and change shape. Filamin B binds to another protein called actin, an important substance that makes up the branching network of the cytoskeleton. Filamin B is active in the embryonic development of the skeletal system. It is active in many cells of the body, especially the cartilage-forming cells called chondrocytes. Most of the cartilage later becomes bone.

At least 13 mutations of *FLNB* cause Larsen syndrome. The mutations change only one amino acid in the filamin B protein or delete a small portion of genetic material to result in an abnormal protein. This abnormal structure disrupts the normal differentiation and growth of the chondrocytes, keeping the bones from forming properly and causing the many deformities of Larsen syndrome. *FLNB* is generally inherited in an autosomal dominant pattern, although a few mutations have recessive patterns. It is located on the short arm (p) of chromosome 3 at position 14.3.

What Is the Treatment for Larsen Syndrome?

Most people with Larsen syndrome have mild symptoms that can be treated. However, a few children with severe symptoms may die at birth or soon after. A team of specialists must determine the nature of the symptoms and recommend a course of treatment.

Further Reading

"Larsen Syndrome." 2005. *Gale Encyclopedia of Public Health*. http://www.healthline.com/galecontent/larsen-syndrome-1. Accessed 11/15/11.

"Larsen Syndrome." 2011. Genetics Home Reference. National Library of Medicine (U.S.). http://ghr.nlm.nih.gov/condition/larsen-syndrome. Accessed 11/15/11.

"Larsen Syndrome Research Project." 2011. Cedars Sinai Hospital. http://www.cedars-sinai.edu/Patients/Programs-and-Services/Skeletal-Dysplasia/Larsen-Syndrome-Research-Project.aspx. Accessed 11/15/11.

Laurence-Moon Syndrome. *See* Bardet-Biedl Syndrome

Learning Disorders. *See* 47,XYY Syndrome; Aarskog-Scott Syndrome (AS); Triple X Syndrome

---

Leber Hereditary Optic Neuropathy (LHON)

Prevalence   Estimated 1 in 30,000 to 50,000, especially in northern Europe and Finland; generally begins in early adulthood with mean age of onset between 18 and 35.

Other Names   hereditary optic neuroretinopathy; Leber hereditary optic atrophy; Leber optic atrophy; Leber's optic atrophy; Leber's optic neuropathy; LHON

In 1871, the German ophthalmologist Theodore Leber noted four families with a number of young men who suddenly became blind. In a paper he described these young men, who lost vision in either or both eyes at the same time or one at a time. At first, this disorder was thought to be linked to something carried by the mother, later described as an X-linked disorder. However, in 1988, D. C. Wallace and a team published findings in the journal *Science* that LHON is one of the disorders inherited only through the mother's mitochondria. Leber contributed his name to the disorder.

What Is Leber Hereditary Optic Neuropathy (LHON)?

Leber hereditary optic neuropathy, or LHON, is a disorder of the eyes and vision. The word "optic" comes from the Latin word *opt* that means "eye." "Neuropathy" comes from two Greek words: *neuro*, meaning "nerve," and *path*, meaning "disease." The actual cause of LHON is death of the nerve cells in the optic nerve. The optic nerve carries the messages from the eyes to the brain.

The condition usually begins in the teens with blurring or clouding of vision. The problem may begin with one eye and then progress to the other within several weeks to several months, or it may happen in both eyes at the same time. Over time, vision becomes worse, and the person cannot see to read, drive, or even recognize faces. This happening is called loss of central vision. LHON is usually painless, but vision loss is serious and permanent.

Some forms of LHON do have pain and other related symptoms. This condition is called LHON plus. In addition to vision loss, families with LHON plus may display the following symptoms:

- Movement disorders
- Tremors
- Heart electrical system defects or arrhythmias
- Features similar to multiple sclerosis
- Muscle weakness
- Numbness
- Other medical problems

Some people with the gene mutation do not develop symptoms of the disorder. Researchers believe that environmental factors such as smoking or alcohol may trigger the disorder.

What Are the Genetic Causes of Leber Hereditary Optic Neuropathy (LHON)?

Mutations in genes in the mitochondria of the mother cause LHON. Although most genetic conditions are caused by genes on chromosomes, a small number are related to the genes found in the mitochondria of the mother, called mitochondrial DNA or mtDNA. Normally, mitochondria are bean-shaped structures that are the powerhouses of the cells. The mtDNA have only 37 genes, but these genes are essential to life processes. Thirteen of the genes instruct for the enzymes relating to oxidative phosphorylation, a process that converts oxygen and simple sugars from food to energy. The remaining 24 genes instruct for making transfer RNA or tRNA and ribosomal RNA or rDNA. These two molecules are important in assembling the amino acids for proteins.

Mutations in the mitochondrial genes—*MT-ND1*, *MT-ND4*, *MT-ND4L*, and *MT-ND6*—are related to Leber hereditary optic neuropathy. The mutations disrupt the normal production of the enzyme complex that is responsible for oxidative phosphorylation. Usually only one change in the amino acids is involved, but one exchange is enough to affect the creation of energy in the cells. Exactly why the visual information from eye to optic nerve is damaged is unknown. Other genetic and environmental factors may be involved.

The four genes that are involved—*MT-ND1*, *MT-ND4*, *MT-ND4L*, and *MT-ND6*—all disrupt the creation of the enzyme. The enzyme group named Complex I is the first step in transporting electrons from the molecule NADH to another molecule called ubiquinone. These electrons then go through several other enzyme complexes to provide energy. The workings of the mitochondrial genes are similar with small variation. The genes are named according to the NADH molecule affected. For example, the *MT-ND1* gene is also called "mitochondrially encoded

NADH dehydrogenase 1." Although researchers know the location and function of the genes, they are continuing to work on the basic cause of the disorder.

What Is the Treatment for Leber Hereditary Optic Neuropathy (LHON)?

No cure for LHON exists. Treatment of the manifestations is according to symptoms. If the condition is LHON plus, life-threatening conditions such as heart arrhythmias must be attended to first. According to the International Foundation for Optic Nerve Disease, no treatment has proven effective in controlled trials of LHON. There is a long list of things that people with LHON-causing mtDNA should avoid. Prominent among these is cigarette smoke and alcohol consumption. The website for the foundation is http://www.ifond.org/lhon.php3.

Further Reading

International Foundation for Optic Nerve Disease. 2011. http://www.ifond.org/lhon.php3. Accessed 5/21/12.

"Leber Hereditary Optic Neuropathy." 2011. Genetics Home Reference. National Library of Medicine (U.S.). http://ghr.nlm.nih.gov/condition/leber-hereditary-optic-neuropathy. Accessed 5/21/12.

"Leber's Hereditary Optic Neuropathy." 2011. Orphanet. http://www.orpha.net/data/patho/GB/uk-LHON.pdf. Accessed 5/21/12.

Legal and Ethical Issues: A Special Topic

All people with genetic disorders have certain relevant legal rights. This discussion will focus on two aspects: the laws and the acts that have established these laws; and the ethics of treatment, especially when dealing with experimental drugs or trials.

Legal Rights of People with Disabilities

In past years, people with both mental and physical disabilities were considered "throwaways." Before the thirteenth century, a person with a mental disability in Europe, for example, would be banned from society to roam in forests and live in caves. Around the 1300s in England, people with disabilities began to be placed in an asylum of some sort. The most infamous was Bethlem, which became pronounced "Bedlam" and led to the creation of that word to mean a state of wild disorder and confusion. There was no medical treatment, and "inmates" were treated cruelly and inhumanely. People who had Alzheimer disease or Parkinson disease were placed with those who had syphilis, sexual deviation, and serious infections. It has been reported that it was Sunday entertainment for people to go to the asylum and look at the patients through peepholes. They would poke sticks through the holes to get the inmates to fight.

Although there were some who began to think that those with mental and physical disabilities had rights, in the nineteenth century, people with all types of mental and physical disorders were still placed in chains, cribs, or locked rooms. By the end of the century, however some advocates began to talk and act. Thomas Gallaudet founded a school in 1817 in Hartford, Connecticut, called the American School for the Deaf and Dumb. In 1832, New York established a school for blind children, and in 1852, New York, Pennsylvania, and Massachusetts gave money for children called at the time "retarded."

Yet, most children with disabilities were denied an education, and the law did not help. Consider the following cases:

- 1893: A Massachusetts court ruled that imbecility was grounds for expulsion from school.

- 1919: A Wisconsin court ruled that a student with a disability was capable but he could be excluded from school because "his disability had a depressing and nauseating effect on the teachers and school children."

Thus, children with cerebral palsy or poliomyelitis were excluded from learning, even though they may have been very capable.

Four Important Events

Four things happened to make a difference and could be considered turning points:

- 1918–1919: World War I soldiers returned from the war with disabilities. The Soldiers' Rehabilitation Act and the Smith-Bankhead Act offered vocational services in form of counseling and training.

- 1944 (World War II): The Veterans' Preference Act was amended to include people who had service-related disabilities, including developmental disabilities.

- 1954: The *Brown v. Board of Education* desegregation order set the scene to include access to all children, including those with disabilities.

- 1971: A Pennsylvania court determined that children with intellectual disabilities in Pennsylvania have a right to a free, public education. When possible, the children should be in a regular classroom and not segregated in an institution. A 1972 decision, *Mills v. Board of Education of the District of Columbia*, included all children with disabilities. And the following services were mandated: a free appropriate public education; an individualized education program; and due process procedures to follow if there is a question.

Development of Legal Rights for Children and People with Disabilities

Movement for rights of people with disabilities began to develop at a fast and furious pace. Five major acts have determined the action and are now law:

1. Section 504 of the Rehabilitation Act of 1973
2. Education of All Handicapped Children Act (PL 94-142) of 1975
3. Americans with Disabilities Act (ADA) of 1990
4. Individuals with Disabilities Act (IDEA) of 2004.
5. Family Education Rights and Privacy Act (FERPA) of 1974

Each of these laws relates to all students with disabilities to assist in their learning and to avoid discrimination.

Section 504 of the Rehabilitation Act of 1973

This act authorized federal funds to be paid to institutions when they comply with regulations for educating students with disabilities. The act protects schools from discriminating against students who meet one of the following criteria:

- Has a physical or mental impairment that limits one or more of a person's major life activities. These activities include caring for oneself, performing manual tasks, walking, seeing, hearing, speaking, breathing, learning, and working. The law protects against both intentional and unintentional discrimination.
- Has a record of such impairment. People with a communicable disease, diabetes, ADHD, asthma, or allergies may be considered here. Most genetic defects fall in this category.
- Is regarded as having such an impairment.

Education for All Handicapped Children's Act of 1975 (PL 94-142)

This landmark legislation captured an attitude in the country to help those who had disabilities, especially in the area of education. The language of the *Mills* case was adopted expressing three main areas:

- Children with disabilities have a right to a free appropriate public education (FAPE).
- An individualized education plan would be developed for each child (IEP).
- Due process procedures for determining compliance were implemented.

The law must be revisited every five years.

Americans with Disabilities Act of 1990 (ADA)

Although the Vocational Rehabilitation Act was passed in 1973, people with disabilities still maintained an inferior status and were disadvantaged in many ways.

Congress saw a need to strengthen the law and did so in 1990 by passing the Americans with Disabilities Act (PL 101-336). The two laws were important to the lives of children with disabilities because they provide legal venues for resolution of complaints. If the complaints are validated, the school district can lose federal funds. ADA strives for equality of opportunity, full participation, independent living, and economic self-sufficiency. The main purpose was to extend civil rights to 43 million Americans with disabilities who have been unable to access communities and necessary services.

Individuals with Disabilities Education Act of 2004 (IDEA)

In this law, the terminology was changed from "handicapped children" to "individuals with disabilities." Seven areas were addressed in this law:

- All children with disabilities are entitled to a free and appropriate education.
- Students are to be educated in the least restricted environment. In other words, they will be in a regular classroom rather than pulled out (the old model) for special classes. If it is determined that the child is best educated in a pull-out classroom, it must be in his or her best interest.
- Parents are expected to participate in all decision making.
- Related services must be provided if needed.
- Placement must be unbiased.
- Due process procedures must be followed.
- Transition plans must be developed for when the person leaves school.

The No Child Left Behind Act (NCLB) was part of the reauthorization of an original funding act of 1965 as part of the civil rights legislation. Some of the provisions of the act are complex and controversial. Each state must implement an accountability system that ensures all students, including students with disabilities, make Adequate Yearly Progress (AYP). For the first time, rigid consequences were placed for schools that did not make progress.

Family Education Rights and Privacy Act (FERPA) of 1974

Also known as the Buckley Amendment, this act defines who may and may not see student records. Parents have access to all information gathered about students with disabilities and can challenge information placed there. After the age of 18, only the student can access the records.

These five laws have made a great deal of difference in the lives of people with disabilities.

Ethics Issues Involving People with Disabilities

For many of the disorders in this encyclopedia, treatments or management are related to pharmaceuticals, surgery, or procedures. Most treatments are just to

the individual patient must be the beneficiary. When Jesse Gelsinger was recruited for the gene therapy experiment for ornithine transcarbamylase (OTC) deficiency, his disease was being controlled with drugs and food. He was not actively ill and probably had a less severe version of OTC. Because the experiment was designed to look at treating children with the condition, he would never have benefited from the results.

Principle 3: Every Participant in an Experiment Must Be Fully Informed

Informed consent is an outgrowth of respect for people. Many of the great discoveries in medicine came from the abuse of persons who were used as guinea pigs. Most of the people had no idea that they were the subject of an experiment. During World War II, Hitler encouraged Nazi doctors to experiment on Jews, gypsies, persons with mental disabilities, homosexuals, and others. All this was done without their consent, and with no concern for pain, so that it often became torture for the people who were experimented on, and they often died or were executed after the experiments. From the trials at Nuremberg after World War II, doctors and researchers established the Nuremberg Code in 1947, which said that researchers engaged in medical trials must first do not harm and then ensure that persons involved in research sign a form saying that they voluntarily give their consent.

Now, in order to participate in medical trials, the researcher is bound to fully inform the recipient of the nature of the test, what will happen, what the risks are, and what to do if there are side effects. The people in the trials are carefully monitored. They may give their consent for experimentation, but they must know what they are doing. Even with consent, some ethical problems and several unanswered questions remain. Can the researcher make sure that a person who does not understand the language or who cannot read understand what they are about to do? Can a parent give consent for a child to be involved in a medical experiment? Can a guardian give consent for a person with mental illness to participate in a trial that will not benefit him or her specifically? Is research involving animals, which can feel pain and fear, proper because they cannot understand what is happening to them? Is it ethical to use animals solely for the benefit and greater good of mankind?

Principle 4: Researchers Must Guard for Malpractice and Misuse

Malpractice and misuse come about when doctors do things that they should not do, and when they try to cover up. For example, if many of the animals die in preclinical trials, it is imperative that a human experiment not be initiated. That is what happened in the case of Jesse Gelsinger. The researchers fraudulently covered up and falsified many of the reports in preclinical trials. In addition, the vector that was being used was created by a company owned by one of the researchers.

However, as medical research advances, many other issues will arise. In practice, if a doctor knows that a child will be born with a genetic disorder and does not explain the consequences to the parents, is this ground for a malpractice suit? If medicine advances enough for gene therapy to be an accepted treatment, can a doctor be sued for not recommending it to the parents?

The legal and ethical issues that are involved in the practice of medicine, especially medicine affecting people who have severe genetic disorders, are still being refined. Legal rights of people with disabilities have come a long way in the last next few decades; accepting people with disabilities has also made great headway among the general populations. Ethics, especially when dealing with human medical experiments, will continue to be an issue. Although the promise of technology is very great, it can also be very dangerous. However, there is one obvious conclusion: regulations and watchfulness are imperative.

Further Reading

Kelly, Evelyn B. 2007. *Gene Therapy*. Westport, CT: Greenwood Press.

Kelly, Evelyn B. 2007. *Stem Cells*. Westport, CT: Greenwood Press.

"Medical Ethics." 2011. MedlinePlus. National Library of Medicine (U.S.). http://www.nlm.nih.gov/medlineplus/medicalethics.html. Accessed 5/21/12.

Zoloft, Laurie. 2006. "The Ethics of Human Stem Cell Research: Immortal Cells, Moral Selves." In Robert Lanza (Ed.), *Essentials of Stem Cell Biology* (pp. 479–488). Burlington, MA: Elsevier Academic Press.

Legius Syndrome

Prevalence   Unknown; misdiagnosed because of similarity to neurofibromatosis type 1.

Other Names neurofibromatosis type 1-like syndrome; NFLS

This condition is fairly new in its designation. Doctors began to notice symptoms of a disorder that was similar to neurofibromatosis 1 (NF1), but later in life the people did not develop the characteristic tumors of NF1. They began to call the condition NF1-like syndrome, which became confusing. So attendees at the 13th European Neurofibromatosis Meeting in 2007 suggested the disorder be named Legius syndrome after Eric Legius, who first wrote about the syndrome.

What Is Legius Syndrome?

Legius syndrome is similar to a condition called neurofibromatosis 1 or NF1 in which the skin develops colored pigments. The flat patches of skin are darker than the surrounding area and are called café-au-lait spots. Other pigment changes include freckles in the groin and armpit areas. In addition to the dark spots and freckles, the following symptoms may occur:

- Large head
- Distinct facial characteristics
- Learning disabilities
- Attention deficit disorder (ADD) or attention deficit hyperactivity disorder (ADHD).
- Normal intelligence

The signs are the same as NF1 are very difficult to distinguish during the early years. However, as the person ages, he does not get the tumors that form at the nerve endings. The two disorders are very different in later life.

What Is the Genetic Cause of Legius Syndrome?

Mutations in the *SPRED1* gene, also called the "sprouty-related, EVH1 domain containing 1" gene, cause Legius syndrome. Normally, *SPRED1* provides instructions for making the Spred-1 protein that regulates an important signaling pathway called Ras/MAPK. This pathway is essential for the growth and division of cells. Also, the pathway works specifically in cell differentiation, cell movement, and the process of cell destruction, called apoptosis. The Spred-1 protein binds and activates another protein called Raf, which stops the Ras/MAPK signals at the time for proper development.

Several mutations in the *SPRED1* gene cause Legius syndrome. Most of the mutations cause a shortened and nonfunctional protein. The short protein is not able then to bind and block activation of Raf, leaving the Ras/MAPK pathway to act continually, leading to the symptoms of Legius syndrome. *SPRED1* is inherited in an autosomal dominant pattern and is found on the long arm (q) of chromosome 15 at position 14.

What Is the Treatment for Legius Syndrome?

Because the tumors of NF1 do not develop, most of the issues relating to Legius syndrome relate to behavior and education. Both behavioral modification and drugs may control ADHD. Speech, physical, and occupational therapy can help those with educational disabilities.

Further Reading

Neurofibromatosis Network Support Community. https://www.inspire.com/groups/neurofibromatosis-inc/discussion/nf1-like-syndrome-or-legius-syndrome. Accessed 11/14/11.

Stevenson, David; David Viskochil; Rong Mao; and Talia Muram-Zborovski. 2011.

"Legius Syndrome." *GeneReviews*. http://www.ncbi.nlm.nih.gov/books/NBK47312. Accessed 11/14/11.

"Syndrome Misdiagnosed as Neurofibromatosis." 2011. Perelman School of Medicine. University of Pennsylvania. http://www.medpagetoday.com/HematologyOncology/OtherCancers/17075. Accessed 11/14/11.

Leigh Syndrome

Prevalence   Affects about 1 in 40,000 newborns; about 1 in 2,000 in the Saguenay Lac-Saint-Jean region of Quebec, Canada.

Other Names   infantile subacute necrotizing encephalopathy; juvenile subacute necrotizing encephalopathy; Leigh disease; Leigh's disease; subacute necrotizing encephalomyelopathy

In 1951, Denis Archibald Leigh, a British psychiatrist, noted a rare, progressive condition in which young patients lost mental and physical abilities and died within a few years. He called the condition subacute necrotizing encephalomyelopathy (SNEM). Later, the name was changed to Leigh syndrome after the doctor who first recognized it.

What Is Leigh Syndrome?

Leigh syndrome is a rare metabolic disorder that affects the central nervous system. The infant appears normal at birth, but at some time between the ages of three months and two years, he begins to lose mental and physical abilities that he has acquired. Death will usually occur in a few years. Following are the symptoms of Leigh syndrome:

- Eating problems: The infant may vomit and have difficulty swallowing. Poor sucking ability is one of the first symptoms.
- Diarrhea.
- Failure to thrive: Because of the eating problems, the child loses weight and does not grow.
- Muscle and movement problems: Muscles become weak with no muscle tone. The child develops problems with balance and movement. He may also develop peripheral neuropathy, which is the loss of feeling in the limbs.
- Paralysis of muscles of eyes: This leads to involuntary eye movements and degeneration of the optic nerve.
- Severe breathing problems: These problems are a leading cause of death in people with this disorder.
- Heart problems: The child may develop cardiomyopathy, a condition in which the heart muscles thicken.
- Lactate buildup in the blood and spinal fluid: This error in metabolism causes episodes of lactic acidosis, which can lead to kidney failure.

The symptoms appear to be caused by developing brain lesions in people with the condition. Magnetic resonance imaging, or MRI, reveals growths in following regions of the brain:

- The basal ganglia, an area that controls movement
- Cerebellum, the area related to balance and movement
- Brainstem, the area that controls swallowing, breathing, hearing, and seeing
- Myelin sheaths of nerves; the myelin sheath is the protective coating around nerves that enables the neuron to relay sensory information

A small number of people with Leigh syndrome may develop more slowly, and some may even live to adulthood.

What Are the Genetic Causes of Leigh Syndrome?

The genetic causes of Leigh syndrome are quite complex. It can be caused by mutations in one of over 30 different genes. Most of the mutations are found in the DNA of the nucleus, but 20% to 25% have a mutation in the mitochondrial DNA. Another gene responsible for Leigh syndrome is located on the X chromosome.

Nearly all of the genes related to this syndrome involve the complex process of energy production. The mitochondria are the powerhouses of the cell, and the place where energy is produced. The workings of the mitochondria involve many chemical reactions that involve the way oxygen is used to convert the energy from food into a form of energy that cells can use. The process is called oxidative phosphorylation and involves five protein complexes: called complex I through complex V. As oxidative phosphorylation proceeds, the end product of the five protein complexes is adenosine triphosphate, or ATP.

Most of the mutations related to Leigh syndrome affect one or more proteins in complexes I, II, IV, or V or disrupt their assembly. Most hereditary DNA is packaged in chromosomes in the nucleus, but the mitochondrial DNA is an exception. This DNA has 37 genes, all of which are essential to the production of energy, and spans 16,500 base pairs. Thirteen genes instruct for the enzymes essential to oxidative phosphorylation. The remaining genes give instructions for making transfer RNA or tRNA and ribosomal RNA or rRNA, the substances that are essential for assembling working amino acids. Mitochondrial DNA is designated mtDNA.

Mutations in several different mitochondrial genes cause Leigh syndrome.

MT-ATP6

Mutations in the *MT-ATP6*, officially named the "mitochondrially encoded ATP synthase 6" gene, cause Leigh syndrome. Normally, *MT-ATP6* instructs for a protein that is critical for normal mitochondrial action. The protein makes up one part of a very large enzyme called complex V, which is the final step of oxidative phosphorylation. One part of the ATP synthase enzyme lets protons or positive charges cross an important membrane inside the mitochondria. Another part uses the energy from the proton flow to change a molecule called adenosine diphosphate or ADP into ATP, which is then used for energy.

Mutations in *MT-ATP6* affect about 10% to 20% of people with Leigh syndrome. The mutations are the result of changes in only one building block. The most common mutation occurs when guanine replaces thymine. The mutations disrupt the ATP synthase complex, thereby stopping ATP production and affecting oxidative phosphorylation. Researchers think that this impairment to the process causes cell death because energy to the cell is decreased. The body tissues that need the greatest amounts of energy are the brain, heart, and muscles. Cell death in these areas causes the symptoms of Leigh syndrome. Mitochondrial DNA is inherited from the mother; *MT-ATP6* is located in the mitochondrial DNA.

SURF1

Mutations in the *SURF1* gene, officially called the "surfeit 1" gene, are the most common cause of Leigh syndrome. Normally, *SURF1* encodes for a protein that is essential in the oxidative phosphorylation process. The SURF1 protein affects the process step in complex IV, which is also known as cytochrome *c* oxidase or COX. This step in the process takes the electrons from earlier steps, along with protons from inside the mitochondrion, and uses them to convert oxygen to water. The process will eventually generate ATP for energy.

Over 40 mutations in *SURF1* cause Leigh syndrome. Most of the mutations result in a very short protein or replace one building block with another. In the cell, the mutated proteins are broken down, leading to the absence of the protein and disrupting the complete process. *SURF1* is inherited in an autosomal recessive pattern and is located on the long arm (q) of chromosome 9 at position 34.2.

DNA Mutations in the Nucleus

Some mutations that cause Leigh syndrome are the result of other proteins that affect oxidative phosphorylation. Mutations can lead to a shortage in the proteins that form a complex called pyruvate dehydrogenase, an enzyme essential for energy production.

Other Patterns

A small number of cases of Leigh syndrome appear to have mutations in the nuclear DNA, which are X-linked recessive. Also, mitochondrial DNA appears to be prone to somatic non-inherited mutations that are not passed on to future generations.

What Is the Treatment for Leigh Syndrome?

The prognosis for people with Leigh syndrome is poor. Those who have a deficiency in complex IV and with the pyruvate dehydrogenase deficiency have the worst prognosis and usually die within a few years. Other forms are treated with thiamine or vitamin B1. Those with X-linked form may be put on a high-fat, low-carbohydrate diet. Individuals with partial deficiencies may live to ages 6 or 7 or into mid-teenage years.

Further Reading

"Leigh Syndrome." 2011. Genetics Home Reference. National Library of Medicine (U.S.). http://ghr.nlm.nih.gov/condition/leigh-syndrome. Accessed 11/14/11.

"NINDS Leigh's Disease Information Page." 2011. National Institute of Neurological Disorders and Stroke (U.S.). http://www.ninds.nih.gov/disorders/leighsdisease/leighsdisease.htm. Accessed 11/14/11.

United Mitochondrial Disease Foundation. 2011. http://www.umdf.org/site/c.otJVJ7MMIqE/b.5692881/k.4B7B/Types_of_Mitochondrial_Disease.htm. Accessed 11/14/11.

Lenz Microphthalmia Syndrome

Prevalence   Rare; incidence unknown; identified in only a few families

Other Names   Lenz dysmorphogenic syndrome; Lenz dysplasia; Lenz syndrome; MAA; MCOPS1; microphthalmia or anophthalmos

The term "microphthalmia" comes from two Greek words: *micro*, meaning "small," and *ophthalmos*, meaning "eye." This syndrome is a condition affecting the eyes and is found exclusively in males.

What Is Lenz Microphthalmia Syndrome?

Lenz microphthalmia syndrome is primarily an eye disorder but also can affect other body systems. The person is born with very small eyeballs, or with no eyeballs, a condition called anophthalmia. Following are some of the symptoms of the disorder:

- Eyes: Small eyes or no eyes lead to vision loss and blindness. Problems may also include cataracts; involuntary eye movements; split in the structures that make the eyes, called a coloboma; and risk for glaucoma.
- Skeletal abnormalities, such as hands.
- Problems with ears.
- Issues with urinary system.
- Heart defects.
- Intellectual disability ranging from mild to severe.

What Is the Genetic Cause of Lenz Microphthalmia Syndrome?

Mutations in the *BCOR* gene, officially called the "BCL6 corepressor" gene, causes Lenz syndrome. Normally, *BCOR* programs for the BCL6 corepressor, a protein that cannot function by itself but must interact with several other proteins produced from the *BCL6* gene. *BCOR* is not only active in the immune system; it appears to play a role in embryonic development in tissues throughout the body, especially the eyes. The BCL6 corepressor may be involved in determining the right and left sides of the body.

A mutation in *BCOR* causes Lenz syndrome. The change occurs when leucine replaces proline in the amino acid building block. The defective version of the BCL6 corepressor disrupts the normal formation of eyes and other embryonic tissues. *BCOR* is inherited in an X-linked recessive pattern and is located on the short arm (p) of the X chromosome at position 11.4.

What Is the Treatment for Lenz Microphthalmia Syndrome?

The eye condition is difficult to treat. For a small eye or no eyeballs, an ocularist may replace the ball with an artificial orbital expander to help the facial bones develop normally. Early intervention with physical and occupational therapy can help children with visual impairments avoid motor delays. Other treatments are symptomatic depending on the severity and nature of the disorder.

Further Reading

"Lenz Microphthalmia Syndrome." 2010. Columbus Clinical Research. http://www.columbusclinical.com/health-topics/L/LENZ-MICROPHTHALMIA-SYNDROME.html. Accessed 11/16/11.

National Organization for Rare Disorders. 2009. "Lenz Microphthalmia Syndrome." http://www.rarediseases.org/rare-disease-information/rare-diseases/byID/1057/view Abstract. Accessed 11/16/11.

Ng, David. 2010. "Lenz Microphthalmia Syndrome." *GeneReviews*. http://www.ncbi.nlm.nih.gov/books/NBK1521. Accessed 11/16/11.

LEOPARD Syndrome

Prevalence   Rare; about 200 cases worldwide

Other Names   cardio-cutaneous syndrome; diffuse lentiginosis; lentiginosis profuse; Moynahan syndrome; multiple lentigines syndrome; progressive cardiomyopathic lentiginosis

The name LEOPARD syndrome is actually a mnemonic or memory device to help people remember the many symptoms of the disorder. In 1936, Zeisle and Becker first encountered this disorder and noted the person had many spots on the skin along with several other physical problems. Later, in 1962, Moynahan and Walther studied the condition and wrote about it in a peer-reviewed journal.

However, Gorlin introduced the name LEOPARD syndrome in 1969. The symptoms that are related to the syndrome along with the spots or lentigines on the skin could be associated with a leopard cat that is covered with spots.

What Is LEOPARD Syndrome?

LEOPARD syndrome affects many parts of the body. The following symptoms create the acronym:

- *L or lentigines*: The individual has over 10,000 brown skin lesions that cover about 80% of the body. The spots range in size from that of a very small freckle to large café-au-lait spots of many centimeters. The spots may be surrounded by an area with no pigment that appears very light. Lentigine spots may occur inside the mouth and in the eye. Lentigines are the most common symptom of the disorder.

- *E or electrocardiographic abnormalities*: Disorders of the heart are seen in a test called an electrocardiograph. The electrical patterns of the heart are disrupted and display a bundle pattern on the test. Eighty percent of the people have an enlarged heart or hypertrophic cardiomyopathy, which forces the heart to work very hard. The aorta, the large artery that pushes blood to the other parts of the body, may be narrow, a condition called aortic stenosis. The valves of the heart may not close and open properly.

- *O or ocular disorder*: The word ocular means eye. The person has wide-set eyes, a condition called hypertelorism. The individual also has a broad nose, droopy eyelids, protruding lower jaw, and low-set, rotated ears. The distinct look and facial abnormalities are the second-highest occurring symptom after lentigines.

- *P or pulmonary stenosis*: A stenosis occurs with an abnormal narrowing of an important passageway. The pulmonary artery leads from the heart to the lungs. In the person with LEOPARD syndrome, the pulmonary artery may be very narrow as it exits the heart, which causes an obstruction in the blood flowing to the lungs.

- *A or abnormal reproductive organs*: Males may have undescended testicles or a single testicle. The opening of the urethra may be on the underside of the penis. Males may be infertile. Females may have a missing ovary and delayed puberty.

- *R or retarded growth*: Most newborns are normal at birth, but as they progress into the first year, growth becomes slow or stunted.

- *D or deafness*: The auditory nerve may not function. Deafness occurs in about 20% of persons with this disorder. Deafness may be present at birth or may occur later in life.

Not everyone with the syndrome will have all the symptoms, and the features may vary even within people with the disorder in the same family. Learning disabilities and developmental delay may be present in some children.

What Are the Genetic Causes of LEOPARD Syndrome?

Mutations in three genes cause LEOPARD syndrome. The genes are *PTPN11, RAF1*, and *BRAF*.

PTPN11

The first gene, *PTPN11* or "protein tyrosine phosphatase, non-receptor type 11," is related to about 80% of cases of LEOPARD syndrome. Normally, *PTPN11* instructs for the protein SHP-2. This protein activates the Ras/MAPK signaling pathway which controls the following cell processes:

- Growth and division of cells
- Differentiation of cells as they mature to carry out specialized functions
- Movement of cells
- Division of cells
- Self-destruction of cells or apoptosis

SHP-2 plays a special role in the differentiation of stem cells that form the heart, blood cells, bone, and other tissues.

About 10 mutations in *PTPN11* cause Leopard syndrome. *PTPN11* is a member of a group of genes called oncogenes, meaning that the mutations can cause normal cells to become cancerous. Most mutations involve only one change in an amino acid. For example, one mutation occurs when cysteine replaces tyrosine. When the gene changes, the structure of the SHP-2 protein is also changed. This decrease in the SHP-2 protein disrupts the signaling pathway of Ras/MAPK, interfering with cell division and growth and causing the other symptoms of LEOPARD syndrome. *PTPN11* is inherited in an autosomal dominant pattern and is located on the long arm (q) of chromosome 12 at position 24.

RAF1

The second gene associated with LEOPARD syndrome is the *RAF1* gene or "v-raf-1 murine leukemia viral oncogene homolog 1" gene. Normally, *RAF1* instructs for a protein that is also part of the Ras/MAPK pathway. Sending signals from the outside to the nucleus, the protein carries out all important cell functions. *RAF1* is also an oncogene.

At least two mutations in *RAF1* cause LEOPARD syndrome. In the most common mutation, leucine replaces serine. In another mutation, valine replaces leucine. The mutations make an abnormal RAF1 protein and disrupt the signaling pathway of Ras/MAPK. The symptoms of LEOPARD syndrome are the result. *RAF1* is inherited in an autosomal dominant pattern and is located on the short arm (p) of chromosome 3 at position 25.

irritable and, at around 4 to 6 months, will show a lack of muscle tone including the ability to lift the head. Then developmental problems are noted. The child arches his back and is unable to crawl, stand, or walk. By the age of one year, the child will have spasms in the limbs.

The hallmark symptoms of biting the fingers and lips appear as the teeth develop. As the children grow, self-injury becomes more severe, and eventually various mechanical constraints must be used to keep the child from serious injury. Older children may become violent and aggressive toward others.

What Is the Genetic Cause of Lesch-Nyhan Syndrome (LNS)?

Mutations in the *HPRT1* gene or "hypoxanthine phosphoribosyltransferase 1" cause Lesch-Nyhan syndrome. Normally, *HPRT1* instructs for the enzyme hypoxanthine phosphoribosyltransferase 1, which enables cells to recycle purines. When purines, the building blocks of DNA and RNA, are first manufactured, it takes a lot of energy. As the process develops and recycling of the purines occurs, the process becomes much more refined and efficient. There must be plenty of building blocks for this pathway known as the purine salvage pathway. Because the process is precise and efficient, anything that goes wrong in the smallest way can disrupt the production.

The more than 200 mutations in the *HPRT1* gene cause LNS. Most of the mutations are from the exchange of only one building block or nucleotide; a few mutations have insertions or deletions of small amounts in the gene. This small disruption causes the enzyme hypoxanthine phosphoribosyltransferase 1 to either shut down or malfunction. Then dangerous waste products build up and accumulate in the joints, kidneys, and bladder. The exact relationship between the buildup and the hallmark neurological and behavioral symptoms of self-mutilation is not known. *HPRT1* is inherited in an X-linked pattern and located on the long arm (q) of the X gene at position 26.1.

What Is the Treatment for Lesch-Nyhan Syndrome (LNS)?

No treatment exists for Lesch-Nyhan syndrome. However, medication may help. Allopurinol may combat the high levels of uric acid in the body. Some of the problem behaviors may be controlled by medications such as diazepam or haloperidol. With careful psychological and medical care, the person may live into the 20s or beyond. No prevention exists for LNS. If an individual has a family history of the condition, a genetic counselor may help in deciding to have children.

Further Reading

"*HPRT1*." 2011. Genetics Home Reference. National Library of Medicine (U.S.). http://ghr.nlm.nih.gov/gene/HPRT1. Accessed 5/22/12.

Jinnah, H. A., and T. Friedmann. 2000. "Lesch-Nyhan Disease and Its Variants." In *The Metabolic and Molecular Bases of Inherited Diseases*, 8th ed. (pp. 2537–2570). http://emedicine.medscape.com/article/1181356-overview. Accessed 5/22/12.

Lesch-Nyhan Disease International Study Group. http://www.lesch-nyhan.org. Accessed 5/22/12.

Leukemia, Acute Promyelocytic. *See* Acute Promyelocytic Leukemia

---

Leukodystrophy with Rosenthal Fibers. *See* Alexander Disease

---

Leukodystrophy, Spongiform. *See* Canavan Disease

Lewy Body Disease. *See* Parkinson Disease (PD)

Li-Fraumeni Syndrome (LFS)

Prevalence   Unknown; in the United States, about 400 people from 64 families

Other Names   LFS; sarcoma, breast, leukemia, and adrenal gland (SBLA) syndrome; sarcoma family syndrome of Li and Fraumeni; SBLA syndrome

In 1969, Frederick Pei Li and Joseph F. Fraumeni described a syndrome in which children and young adults have an increased risk for cancer. They reported their findings in the *Annals of Internal Medicine*. Previously, the condition had been known as SBLA syndrome because of cancers in several areas including breast, blood, and adrenal glands. When Li and Fraumeni published their studies, the medical community named the condition for the two American physicians.

What Is Li-Fraumeni Syndrome (LFS)?

Li-Fraumeni syndrome is a rare disorder in which the children and young adults develop several types of cancer. The types of cancer include the following:

- Breast cancer
- A form of bone cancer called osteosarcoma
- Soft tissue cancers called sarcoma
- Brain tumors
- Blood cancers such as leukemia
- Cancer of the outer layer of the adrenal gland called adrenocortical carcinoma
- Several other types of cancer

There are three things that make LFS so unusual:

- Cancers appear in children and young adults.
- Cancers are several different types and locations.
- Cancers may be cured but reappear several times throughout the life of the person.

What Are the Genetic Cause of Li-Fraumeni Syndrome (LFS)?

Two genes appear to be involved in LFS: *CHEK2* and *TP53*.

CHEK2

Mutations in the gene *CHEK2*, also known as the "CHK2 checkpoint homolog (S. pombe)," are associated with LFS. Normally, *CHEK2* instructs for the protein called checkpoint kinase 2 (CK2), which acts as a tumor suppressor. A tumor suppressor works to regulate cells to keep them from growing wildly and dividing rapidly. CHK2 proteins detect when DNA in cells is broken or damaged. Several things can cause this DNA damage: sunlight ultraviolet rays, other radiation, or toxic chemicals. When the damage occurs, CHK2 rushes to the site and interacts with other proteins. One such protein is the tumor protein 53, is produced by the *TP53* gene. These two work together to keep cell division in check and to determine if the cell will be repaired or self-destruct, a process called apoptosis. Thus abnormal cells are kept from developing into tumors.

Mutations in the *CHEK2* gene have been found in several families with this condition. Researchers are exploring the actual relationship of *CHEK2* with LFS. They are trying to determine if the mutations in the gene cause the condition or simply increase risk. *CHEK2* is inherited in an autosomal dominant pattern and is located on the long arm (q) of chromosome 22 at position 11.

TP53

The second gene related to LFS is the *TP53* gene or the "tumor protein p53" gene. Normally, *TP53* instructs for the tumor protein p53. This protein is a tumor suppressor, which regulates division of cells by keeping them from growing too rapidly. The protein has earned the nickname "The Guardian of the Genome." The protein binds to DNA in cells throughout the body. When some type of damage occurs, tumor protein p53 calls on other genes to fix the damage. If the damage cannot be repaired, the protein then keeps the cell from dividing. The process keeps the development of wildly growing cells from forming cancers.

More than 60 mutations have been found in people with LFS. Many of the mutations are a change in only one amino acid; others delete DNA from the gene. The mutations in the gene cause the protein to be faulty and unable to fulfill its function as the guardian of the genome. Damage then accumulates in the cells, and the cells then continue to divide in an uncontrolled way that causes cancers or tumors. *TP53* is inherited in an autosomal dominant pattern and is located on the short arm (p) of chromosome 17 at position 13.1.

What Is the Treatment for Li-Fraumeni Syndrome (LFS)?

Treatment for this condition is difficult. If the person has early screening and is identified with LFS, the oncology team can treat cancers early by removing them and administering radiation or chemotherapy. However, people with LFS will develop another malignancy at a future time.

Further Reading

"Li-Fraumeni Syndrome." 2006. The M. D. Anderson Medical Center. http://www2 .mdanderson.org/app/pe/index.cfm?pageName=opendoc&docid=2293. Accessed 5/22/12.

"Li-Fraumeni Syndrome." 2010. Medscape. http://emedicine.medscape.com/article/ 987356-overview. Accessed 5/22/12.

"Li-Fraumeni Syndrome." 2011. The Ohio State Medical Center. http://medicalcenter.osu.edu/ patientcare/healthcare_services/breast_health/lifraumeni_syndrome/Pages/index.aspx. Accessed 5/22/12.

LIPA Deficiency. *See* Wolman Disease

Lip-Pit Syndrome. *See* Van Der Woude Syndrome

Lissencephaly

Prevalence   About 1 in 85,000 people; males more prevalent because this is an X-linked condition

Other Names   classic lissencephaly; lissencephaly and agenesis of corpus callosum; lissencephaly type 1; lissencephaly, X-linked; LISX; XLIS

The term "lissencephaly" comes from two Greek words: *liss*, meaning "smooth," and *encephal*, meaning "brain." This condition is named from the outside appearance of the brain. Normally, the brain has folds and crevices, but the brain of a child with lissencephaly has a rare malformation in which the normal brain structures do not develop. It all happens in the 12th to 14th week of embryonic development when normal neuron migration did not occur at the proper time. Terms such as *agyria*, meaning "no folds," or *pachygyria*, meaning "broad folds," may be used to describe the brain's surface.

What Is Lissencephaly?

Lissencephaly is a serious developmental disorder of the brain that affects males. If one looks at a normal brain or a picture of the brain, one immediately notes the outside of the brain, especially the front and upper part, is covered with what appears like many mountains and canyons. These folds and grooves are common in the cerebral cortex, the thinking part of the brain. People with lissencephaly do not have these folds and grooves in a normal configuration. Individuals with this condition have the following symptoms:

- Unusual facial appearance
- Difficulty swallowing and eating
- Failure to thrive
- Hands, fingers, and toes deformed
- Severe intellectual and developmental delay
- Seizures
- Abnormal muscle stiffness or spasticity
- Weak muscle tone

Infants born with no folds or agyria have the most severe symptoms and probably will not live for more than a few days.

Two forms of lissencephaly exist. Both forms are liked to the X chromosome.

- Type 1: Children with this form have only the smooth brain of lissencephaly. Only the brain is affected, and other parts of the body are not involved.
- Type 2: Children with this form have not only smooth brain, but also abnormal genitalia. This form is extremely serious, with seizures beginning before

birth and continuing after birth. Males have very small sex organs, undescended testicles, or genitals that look like they are not male or female. This condition is called ambiguous genitalia. The corpus callosum, the tough band of fibers that connect the two hemispheres of the brain is missing. The infant may experience very low body temperature or hypothermia and chronic diarrhea.

What Are the Genetic Causes for Lissencephaly?

Two genes appear to be responsible for lissencephaly. They are the *ARX* gene and the *DCX* gene.

ARX

Mutations in the *ARX* gene, also called the "aristaless related homeobox" gene are related to lissencephaly. Normally, *ARX* instructs for a protein that controls other genes. This protein is called a transcription factor. *ARX* is part of a huge family of genes called homeobox genes, which are active during embryonic development to turn on the formation of many organs. *ARX* appears to activate development of testes, pancreas, intestines, and especially the brain.

More than 20 mutations in the *ARX* gene are related to lissencephaly with abnormal genitals. These mutations cause the *ARX* protein not to work properly to control the action of specific genes. The shortage of the *ARX* protein disrupts the function of the other organs, such as pancreas and testes, and causes the severe symptoms of lissencephaly. *ARX* is inherited in an autosomal dominant pattern and is located on the X chromosome at position 21.

DCX

Mutations in the *DCX* gene, also called the "doublecortin" gene, cause the milder from of lissencephaly without abnormal genitalia. Normally, *DCX* instructs for the protein doublecortin. This protein is essential for helping neurons get to this proper place in the developing embryonic brain. Doublecortin binds to structures called microtubules, which are rigid hollow fibers that direct neurons to form networks around cells. This protein is essential for the proper organization of neuron organization that forms the folds of the cerebral cortex part of the brain.

More than 60 mutations in *DCX* cause lissencephaly. The mutations cause doublecortin not to function properly. The folds and grooves that are part of the cerebral cortex are not normal and lead to the abnormal characteristic of lissencephaly—the smooth brain. *DCX* is inherited in a dominant pattern and is located on the long arm (q) of chromosome X at position 22.3-q23.

What Is the Treatment for Lissencephaly?

The severe malformations of the brain do not respond to any treatment. Treating the symptoms such as diarrhea or seizures with medication is possible. If fluid collects on the brain causing hydrocephalus, a shunt may be required. If the child cannot eat, a feeding tube may be necessary. The prognosis for this condition is not hopeful but depends on the degree of the malformation. Most children die before the age of two because of respiratory problems from breathing small bits of food or fluid. However, some children may have near-normal development and intelligence.

Further Reading

"Lissencephaly." 2011. Cleveland Clinic. http://my.clevelandclinic.org/disorders/lissencephaly/hic_lissencephaly.aspx. Accessed 5/22/12.

"Lissencephaly." 2011. Genetics Home Reference. National Library of Medicine (U.S.). http://ghr.nlm.nih.gov/condition/x-linked-lissencephaly. Accessed 5/22/12.

LNS. *See* Lesch-Nyhan Syndrome (LNS)

Lou Gehrig Disease. *See* Amyotrophic Lateral Sclerosis (ALS)

Lowe Syndrome

Prevalence	Uncommon; about 1 in 500,000 people
Other Names	Lowe's disease; Lowe's oculocerebrorenal disease/syndrome; OCRL; oculocerebrorenal dystrophy; oculocerebrorenal syndrome, OCR; phosphatidylinositol-4,5-bisphosphate-5-phosphatase deficiency

In 1952, Charles Upton Lowe and colleagues observed a disorder that had many unique symptoms. The individuals were mentally retarded and had bulging eyes. The urine had a very high acid content and ammonia, a usual product found in urine, was decreased. Two years later, a serious renal syndrome called Fanconi syndrome was added to the list of characteristics of the symptoms. It was not until 1965 that the connection was made that this syndrome was an X-linked gene carried by the mother.

What Is Lowe Syndrome?

Lowe syndrome is a disorder that affects many systems. Sometimes, the condition is referred to as the oculocerebrorenal syndrome. Breaking down the word "oculocerebrorenal," one can see three systems in three root words: *oculo*, meaning "eye"; *cerebro*, meaning "brain" and relating to the nervous system; and *renal*, meaning "kidney."

The following abnormalities relate to the three symptoms:

- Eyes: The child is born with congenital clouding of the lenses of the eye, called congenital cataracts. Several other eye defects may affect vision. About half of the children have glaucoma, a condition in which pressure builds up on the retina of the eyes.

- Brain and central nervous system: Intellectual disabilities may range from mild to profound. Many of the children have seizures. Motor skills may also be impaired and delayed development of sitting, standing, and walking is usual.

- Kidney or renal system: A condition called Fanconi syndrome develops as children with Lowe syndrome get older. Kidneys have several roles in the body: maintaining the proper amounts of minerals, salts, and water; processing important nutrients to be reabsorbed for use in the body; processing wastes; and then excreting the waste products in the urine. In Fanconi syndrome, the kidneys do not reabsorb the important nutrients into the bloodstream and consequently valuable food nutrients are excreted. Several things then happen: excess urination, dehydration, and very acidic blood. The loss of nutrients leads to soft bones and bowed legs and other metabolic conditions. As the child ages, Lowe syndrome can lead to end-stage renal disease.

What Is the Genetic Cause of Lowe Syndrome?

Mutations in the *OCRL* gene, also known as the "oculocerebrorenal syndrome of Lowe" gene, cause Lowe syndrome. Normally, *OCRL* instructs for an enzyme that is found in all body cells. This enzyme breaks down phospholipids or fat (lipid) molecules. The OCRL enzyme acts on the chemicals that are in the cell membrane called phosphoinositides. When the enzyme gets into the cells, it forms a network of membranes called the trans-Golgi network. The network sorts through proteins and molecules to get them to the proper place either within the cell or outside the cell. Controlling the phosphoinositides, the ORCL enzyme then can transport the materials from the cell membrane. The enzyme may also regulate actin, a network of threads that make the cell's framework. Actin is critical for maintaining the proper shape of the cell and helping cells move.

More than 120 mutations in *OCRL* cause Lowe syndrome. Some of the mutations disrupt the process so severely that no OCRL enzyme is made. Other mutations prevent the enzyme's interaction with other proteins and substances in the cell. Disruptions in the transport of critical proteins cause the eye, brain, and kidney problems of Lowe syndrome. *OCRL* is inherited in an autosomal recessive pattern and is located on the long arm (q) of the X chromosome at position 25.

What Is the Treatment for Lowe Syndrome?

Because this is a genetic disorder, no cure for Lowe syndrome exists. Treatment focuses on the three major problems. Cataracts that are present in the eyes should be removed. Other strategies to assist vision loss, such as glass or contact lenses, are usually essential. Glaucoma is treated with anti-glaucoma medication. Early rehabilitation may assist in treatment of motor disorders connected with the brain. Seizures require treatment specific for the symptoms. Drugs for behavior problems such as clomipramine may also be effective. Renal acidosis needs to be diagnosed and treated with alkali supplements, such as citrates. Other disorders should be treated according to the symptoms.

Further Reading

Loi, Mario. 2006. "Lowe Syndrome." *Orphanet Journal of Rare Diseases.* http://www.ncbi.nlm.nih.gov/pmc/articles/PMC1526415. Accessed 5/22/12.

"Lowe Syndrome." 2007. National Organization for Rare Disorders. http://www.rarediseases.org/rare-disease-information/rare-diseases/byID/109/viewAbstract. Accessed 5/22/12.

"Lowe Syndrome." 2011. Genetics Home Reference. National Library of Medicine (U.S.). http://ghr.nlm.nih.gov/condition/lowe-syndrome. Accessed 5/22/12.

Lujan-Fryns Syndrome

Prevalence   Cases in general population unknown; rare X-linked dominant syndrome

Other Names   Luhan syndrome; X-linked mental retardation with marfanoid habitus

Marfan syndrome is a group of features called a habitus that are present in persons designated with this disorder. In 1984, J. Enrique Lujan noted patients who had the head and facial features and high-arched palate of Marfan syndrome but also had intellectual and behavioral disabilities. He suspected it was linked to the X chromosome because of the appearance in boys. In 1987, Jean-Pierre Fryns provided additional investigations of other families. At one time the disorder was named X-linked mental retardation with Marfanoid habitus but was later changed to Lujan-Fryns syndrome.

What Is Lujan-Fryns Syndrome?

Lujan-Fryns syndrome (LFS) is a condition in which the individual has mild to moderate retardation and features similar to Marfan syndrome. The features are called "marfanoid," using the Greek ending *oid*, meaning "like." Following are the features of Lujan-Fryns syndrome:

- Marfanoid habitus: These features include a tall, slim stature with long limbs and fingers.
- Abnormalities of brain: An error of embryonic development causes the corpus callosum, the nerve structure that allows communication between the right and left hemisphere of the brain does not develop. This error is called agenesis.
- Abnormalities of the heart.
- Distinct facial appearance: Long, thin face with an underdeveloped lower jaw and receding chin. The upper jaw bone is also underdeveloped, leading to crowding and misalignment of teeth. The roof of the mouth has a high-arched palate. Also present are a long nose with a high nasal bridge, enlarged skull with prominent forehead, and low-set ears.
- Voice high and nasal.
- Poor muscle tone.
- Malformed chest.
- Mild to moderate mental retardation.
- Psychiatric disorders: Some children have autistic-like behavior, schizophrenia, oppositional defiant disorder (ODD), aggression, or ADHD.
- Seizures.

The marfanoid habitus with mental disabilities characterizes Lujan-Fryns syndrome.

What Is the Genetic Cause of Lujan-Fryns Syndrome?

Mutations in the gene *MED12*, officially called the "mediator complex subunit 12" gene, causes LFS. Normally, *MED12* provides instructions for the protein mediator complex subunit 12. This protein, named number 12, is one of 25 proteins that are essential for gene activity. A mediator complex links transcription factors with the enzyme RNA polymerase II, leading to the storage of information in a gene's DNA. *MED12* is essential for early embryonic development, including nerve cells and chemical pathways to cell differentiation.

A mutation in *MED12* causes Lujan-Fryns syndrome. This mutation occurs when the amino acid serine replaces asparagine. The change disrupts the MED12 protein, which is involved in so many signaling pathways, and results in the varied symptoms of Lujan-Fryns syndrome. *MED12* is inherited in an X-linked dominant pattern and is located on the long arm (q) of the X chromosome at position 13.

What Is the Treatment for Lujan-Fryns Syndrome?

No specific treatment exists for the genetic causes of LFS. However, some measures can be taken to make life of the person more comfortable. Attention to heart defects and seizures or other life-threatening conditions is essential. A variety of corrective, preventive, and intervention therapies may be prescribed.

Further Reading

"Lujan-Fryns Syndrome." 2011. RightDiagnosis.com. http://www.rightdiagnosis.com/l/lujan_fryns_syndrome/intro.htm. Accessed 11/17/11.

Van Buggenhout, Griet, and Jean-Pierre Fryns. 2006. "Lujan-Fryns Syndrome (X-Linked, Mental Retardation, Marfanoid Habitus)." *Orphanet Journal of Rare Diseases*. http://www.ncbi.nlm.nih.gov/pmc/articles/PMC1538574. Accessed 11/17/11.

Lymphedema-Distichiasis Syndrome

Prevalence   Unknown; extra eyelashes overlooked because it is not part of a general medical exam

Other Names   lymphedema with distichiasis

Swelling in the lower limbs and an extra set of eyelashes appear to be an unusual combination, but in 1954, J. V. Neel and W. J. Scull wrote of such a possible condition in their book *Human Heredity*. About 10 years later in 1964, H. S. Falls and E. D. Kertesz reported on four siblings with an uneven puffiness in the legs and a double set of eyelashes. The father, one of his brothers, and a paternal grandmother reportedly had similar conditions. However, it was not until 2002 that G. Brice and colleagues described in detail 74 affected subjects in 18 families and in 6 isolated cases. All showed linkage to the same gene.

What Is Lymphedema-Distichiasis Syndrome?

Lymphedema-distichiasis syndrome is a disorder of the lymphatic system, combined with abnormal eyelashes. The lymphatic system is a system of vessels that moves lymph, the fluid part of the blood, and immune cells throughout the body. People with a disorder in this system develop a condition called lymphedema, in which fluid collects and causes swelling. Lymphedema generally appears in late childhood or early puberty and is confined to the legs. The puffiness or swelling often affects one limb and can vary in families. Males appear to develop edema at an earlier age and have more complications than females.

The unusual eyelashes are present at birth but may not be noticed until later because it is generally not part of a health examination. Both superior (upper) and lower (inferior) eyelids exist to protect the eye. The inner part of the eyelids has sebaceous glands called the meibomian glands, which secrete fluid to keep the eyelids from sticking together. Distichiasis occurs when wild or aberrant eyelashes grow from these glands. There may be a full set of lashes or a single lash. About 75% of the cases have difficulties arising from the extra eyelashes. The individual may have an irritation of the cornea, the clear part of the eye; recurring conjunctivitis; and extreme sensitivity to light, a condition called photophobia.

Other abnormalities may be present with lymphedema-distichiasis. The following symptoms may occur:

- Varicose veins: About 50% of people have varicose veins. These are present in both deep and superficial veins and suggest association with a mutation in the gene.
- Ptosis: About 30% of cases have ptosis or droopy eyelids.
- Congenital heart defects.
- Cleft palate.
- Scoliosis.
- Neck webbing.
- Kidney disorders.

What Is the Genetic Cause of Lymphedema-Distichiasis Syndrome?

Mutations in the *FOXC2* gene, also called the "forkhead box C2 (MFH-1 forkhead 1)" gene, cause lymphedema-distichiasis syndrome. Normally, *FOXC2* instructs for a protein that forms many organs and tissues during embryonic development. This protein, called a transcription factor, is responsible for regulating the activity of other genes. It binds to certain regions of DNA to give these directions. The FOXC2 protein directs the genes that form the lungs, eyes, kidneys, urinary tract, heart and blood vessels, and the lymphatic vessels that hold immune cells.

More than 50 mutations in the *FOXC2* gene are associated with lymphedema-distichiasis syndrome. Most mutations either insert or delete a few building blocks which disrupt the function of the protein. The mutated FOXC2 protein is very small and cannot bind to DNA. Thus, the changed protein cannot regulate the activity of important developmental genes, causing symptoms of the syndrome. *FOXC2* is inherited in an autosomal dominant pattern and is found on the long arm (q) of chromosome 16 at position 24.1.

What Is the Treatment for Lymphedema-Distichiasis Syndrome?

To treat the aberrant eyelids, the ophthalmologist may pluck, lubricate, use cryo-surgery (cold) on, or split the eyelids to remove the extra lashes. To treat lymphedema, fitted stockings and bandages may improve the swelling. If the individual has ptosis, surgery may correct. A conservative approach to varicose veins is recommended because surgery may aggravate the edema. Diuretics generally used to reduce fluid collection are not effective with this condition.

Further Reading

"Lymphedema." 2011. Mayo Clinic. http://www.mayoclinic.com/health/lymphedema/DS00609. Accessed 5/22/12.

"Lymphedema-Distichiasis Syndrome." 2011. Genetics Home Reference. National Library of Medicine (U.S.). http://ghr.nlm.nih.gov/condition/lymphedema-distichiasis-syndrome. Accessed 5/22/12.

Lymphedema People. http://www.lymphedemapeople.com/thesite/lymphedema_distichiasis.htm. Accessed 5/22/12.

National Lymphedema Network. 2011. http://www.lymphnet.org. Accessed 5/22/12.

Lynch Syndrome/Hereditary Nonpolyposis Colorectal Cancer

Prevalence   In the United States, 160,000 new colorectal cancers cases diagnosed each year; 2% to 7% caused by Lynch syndrome

Other Names   cancer family syndrome; COCA 1; familial nonpolyposis colon cancer; hereditary nonpolyposis colorectal cancer; hereditary nonpolyposis colorectal neoplasms; HNPCC; Lynch syndrome I; Lynch syndrome II

According to the American Cancer Society, colorectal cancer is one of the leading causes of cancer-related deaths in the United States. Most colon cancers begin as benign polyps in the glands of the lining of the colon and rectum. Two genetic conditions may be involved: familial adenomatous polyposis (FAP), in which many polyps are present that become cancerous (*see* Familial Adenomatous Polyposis [FAP]); and Lynch syndrome, a hereditary form of colorectal cancer in which polyps are not involved.

Henry T. Lynch, a professor of medicine at Creighton University, noted the presence of a cancer that was hereditary but did not involve polyps in the colon; he called it "cancer family syndrome." Later, other authors coined the term

"Lynch Syndrome" in 1984, and Lynch himself used the term HNPCC or hereditary nonpolyposis colorectal cancer in 1985. The terms were used interchangeably,
with Lynch syndrome now being preferred because it relates to the specific genetics of the disorder.

What Is Lynch Syndrome?

Lynch syndrome is a hereditary condition in which cancers that are not related to
polyps or growths in the intestinal region occur. Although there may be an occasional benign growth or polyp, this is not a hallmark of this disorder. However,
those with the condition are at increased risk for other cancers, such as stomach,
small intestine, liver, urinary tract, brain, and skin cancer, and endometrial cancer
in women. Also, people with Lynch syndrome tend to develop the colon cancers at
an earlier age (younger than 50 years). Children, sisters, and brothers of people
with Lynch have a 50% chance of inheriting this condition. Also, it will probably
be seen in other members of the family.

Symptoms of Lynch disease are the same as other intestinal cancers:

- Abdominal pain and tenderness in the lower abdomen
- Blood in the stool
- Obvious change in bowel habits, such as chronic diarrhea or constipation
- Narrow, ribbon-like stools
- Weight loss for no obvious reason

What Are the Genetic Causes of Lynch Syndrome?

Mutations in four different genes cause Lynch syndrome. All the genes relate to
mistakes made when DNA makes copies when the cells divide. The following
genes are involved in Lynch syndrome: *MLH1, MSH2, MSH6*, and *PMS2*.

MLH1

Mutations in the *MLH1* or the "mutL homolog 1, colon cancer, nonpolyposis
type 2 (E.coli)" gene cause Lynch syndrome. Normally, *MLH1* instructs for a protein that repairs DNA. When a mistake is made in a dividing cell, the protein joins
with another protein called the PMS2 protein to coordinate other proteins that
repair errors. The proteins oversee removing the faulty DNA and replace it with
the correct information. Sometimes these genes are known as "mismatch repair"
genes, or MMR. Over half of the cases of Lynch syndrome have one of the over
100 mutations of *MLH1*. When this protein is defective or absent, the mistakes that
are made are left unrepaired during cell division. The result is a growth of abnormal tumorous cells in the colon or other body parts. In addition, the mutations of
MLH1 may cause a condition known as Turcot syndrome, a type of cancer in the
brain called a glioblastoma. *MLH1* is inherited in an autosomal dominant pattern
and located on the short arm (p) of chromosome 3 at position 21.3.

MSH2

The second gene whose mutations cause Lynch syndrome is *MSH2*, or "mutS homolog 2, colon cancer, nonpolyposis type 1 (E.coli)." Normally, this gene makes a protein that is essential for DNA repair. It is also a member of the mismatch repair (MMR) genes. The protein joins with other genes to form a complex that spots where mistakes are made and then takes over to repair the problem. About 40% of Lynch syndrome cases are related to this mutation. The mutation causes the protein to be abnormally short and ineffective. Mistakes in dividing cells are not corrected, and a tumor may form. People with this gene mutation may develop a type of skin cancer known as Muir-Torre syndrome. *MSH2* is inherited in an autosomal dominant pattern and is found on the short arm (p) of chromosome 2 at position 21.

MSH6

A third gene involved in Lynch syndrome is the *MSH6* gene or "mutS homolog 6 (E. coli)." Normally, this gene produces a protein that repairs DNA when it is copied in cell division. It is part of a complex that identifies where mistakes are made and points the way for the MLH1-PMS2 complex to do the actual repair. It is also part of the mismatch repair (MMR) genes. About 10% of families with Lynch syndrome have this variation. People with *MSH6* may also be at risk to develop cancers in other parts of the body. *MSH6* is inherited in an autosomal dominant pattern and is located on the short arm (q) of chromosome 2 at position 16.

PMS2

The fourth gene involved in Lynch syndrome is the *PMS2* gene, or "PMS2post-meiotic segregation increased 2 (S. cerevisiae)." Normally, *PMS2* instructs for a protein that helps fix mistakes in the cell division. This protein is the coordinator of all the other activities of proteins and works to repair mistakes. It is also a member of the mismatch repair or MMR genes. This mutated gene is found in about 2% of families with Lynch syndrome. The mutations allow for costly mistakes to go unrecognized, allowing tumors to form in the colon or other body parts. *PMS2* is inherited in an autosomal dominant pattern and is located on the short arm (q) of chromosome 7 at position 22.2.

What Is the Treatment for Lynch Syndrome?

With proper screening and care, the prognosis for colon cancer is very good. Catching it in very early stages before it has spread to inner layers of the colon is important. Treatment depends on the stage and generally includes surgery to remove the cancer cells, chemotherapy to kill the cancer cells, or radiation therapy to destroy cancerous tissue.

Further Reading

"Colon Cancer." 2011. MedlinePlus. National Library of Medicine (U.S.). http:// www.nlm.nih.gov/medlineplus/ency/article/000262.htm. Accessed 5/22/12.

Lynch, Henry, and Albert de la Chapelle. 2003. "Hereditary Colorectal Cancer." *New England Journal of Medicine*. 348: 919–932. http://www.nejm.org/doi/full/10.1056/ NEJMra012242. Accessed 5/22/12.

"Lynch Syndrome." 2011. Genetics Home Reference. National Library of Medicine (U.S.). http://www.ghr.nlm.nih.gov/condition/lynch-syndrome. Accessed 5/22/12.

M

Macular Degeneration. *See* Age-Related Macular Degeneration (AMD); Stargardt Macular Degeneration

---

Majeed Syndrome

Prevalence   Very rare; reported in three families in the Middle East

Other Names   chronic recurrent multifocal osteomyelitis; congenital dyserythropoietic anemia; neurtrophilic dermatosis

In 2000, Dr. H. J. Majeed reported in the *Journal of Pediatrics* of a family with a condition characterized by consistent bouts of fever and inflamed bones and skin. Majeed, a professor at the University of Jordan, noted the swelling began early in infancy and persisted throughout adulthood. When the gene was located several years later, the researchers referred to the condition as Majeed syndrome.

What Is Majeed Syndrome?

Majeed syndrome is an inflammatory condition of the bones and skin. Children with the disorder have recurring fever and an obvious painful infection in the bones of the joints. The condition is called chronic recurrent multifocal osteomyelitis (CRMO). CRMO affects the bones and slow growth, joint deformities called contractures, and other bone problems.

One symptom of the syndrome is a blood disorder called congenital dyserythropoietic anemia. The term *dys* means "with difficulty," and *erythropoietic* refers to the "red blood cells." Anemia is a condition in which there is a shortage of the red blood cells. When the red blood cells are short in number or dysfunctional, the blood cannot carry the proper amount of oxygen to the body. The person becomes very pale and tired and cannot breathe properly. This congenital condition ranges from mild to severe.

Another symptom experienced by most people with Majeed syndrome is inflammation of the skin, a condition called Sweet syndrome. The syndrome includes fever and painful bumps or blisters on the face, back, and arms.

What Is the Genetic Cause of Majeed Syndrome?

Mutations in the *LPIN2* gene, known also as the "lipin 2" gene, cause Majeed syndrome. Normally, *LPIN2* instructs for a protein called lipin-2. Lipin-2 protein appears to play a role in processing fats and may also be involved in controlling infection and in cell division.

About three mutations in the *LPIN2* cause Majeed syndrome. These mutations disrupt the function of the lipin-2 protein, leading to the inflammation of body tissues and possibly to the shortage of red blood cells and skin disorders. *LPIN2* is inherited in an autosomal recessive pattern and is located on the short arm (p) of chromosome 18 at position 11.31.

What Is the Treatment for Majeed Syndrome?

Treatment of the symptoms may make the person more comfortable. CRMO can be treated with nonsteroidal anti-inflammatory drugs (NSAIDs) and with physical therapy that keeps the muscles from deteriorating. However, because these drugs can cause complications with long-term use, they are usually limited, especially in children. If the anemia is severe, a blood transfusion may be needed. Keeping the person active and avoiding bed rest is important for the well-being of the individual.

Further Reading

"Majeed Syndrome." 2011. Genetics and Rare Diseases Information Center. National Institutes of Health (U.S.). http://rarediseases.info.nih.gov/GARD/Condition/10088/Majeed_syndrome.aspx. Accessed 5/23/12.

"Majeed Syndrome." 2011. Genetics Home Reference. National Library of Medicine (U.S.). http://ghr.nlm.nih.gov/condition/majeed-syndrome. Accessed 5/23/12.

Male Pattern Baldness. *See* Androgenetic Alopecia

Malignant Hyperthermia

Prevalence   About 1 in 5,000 to 50,000 reactions when people are given
                   anesthetic gases; susceptibility probably more frequent, because
                   many people with an increased risk of this condition do not
                   undergo surgery

Other Names   anesthesia related hyperthermia; hyperpyrexia, malignant; hyper-
                   thermia, malignant; malignant hyperpyrexia; malignant hyper-
                   thermia; MHS

It is quite unusual to find a condition such as malignant hyperthermia through research on pigs. Porcine stress syndrome (PSS) occurring in Danish Landrase and other pig breeds affects the muscles, which makes the meat less marketable. The syndrome occurs when animals face great stress. The symptom is called "awake triggering." The pigs develop hyperthermia and die from the stressful situations. To combat the trigger, pig farmers exposed piglets to an anesthesia halothane. Those that were susceptible to the anesthesia halothane died, and the farmer saved the expense of raising stress-susceptible pigs. The breeding stock revealed a defect in the genes. Researchers realized they could use the pig model for a condition called malignant hyperthermia in humans.

What Is Malignant Hyperthermia?

Malignant hyperthermia is a severe reaction of bodily overheating when the person takes general anesthesia. The condition can be life-threatening. Three types of anesthesia appear to trigger the condition:

1. The volatile anesthetic agents, which include the volatile anesthetics sevoflurane, desflurane, isoflurane, halothane, enflurane, and methoxyflurane

2. The neuromuscular blocking drug succinylcholine, used to block sensation of pain

3. A muscle relaxant used to temporarily paralyze a person for a surgical procedure

Generally, other anesthesias are safe.

In malignant hyperthermia, the skeletal muscles experience excess oxygen leading to a condition that overcomes the body's supplying oxygen to the cells, removing carbon dioxide and regulating the body's temperature. The muscle fibers break down, and the increased acid levels in the blood and a rapid heart rate may occur. If the condition is not detected or treated immediately, the individual with malignant hyperthermia will die.

The following symptoms occur:

• Very rapid rise in body temperature to 105 degrees F or higher

• Rigid muscles and joint stiffness

- Muscle ache without exercise
- Bleeding for no apparent reason
- Dark-brown urine

People at risk have susceptibility for this condition. It may occur without other medical problems, but it may be associated with people who have other inherited central core diseases.

What Are the Genetic Causes of Malignant Hyperthermia?

Mutations in the *CACNA1S* and the *RYR1* genes working together cause malignant hyperthermia.

CACNAIS

The first gene is the *CACNAIS* gene, also known as the "calcium channel, voltage-dependent, L type, alpha 1S subunit" gene. Normally, *CACNAIS* instructs for making calcium channels, through which calcium ions move. These positively charged ions help the cell transmit electrical signals. The gene also helps muscles contract and relax, the process than enables them to move.

When the CACNA1S protein is made, it activates another protein called ryanodine receptor 1, which is created by the *RYR1* gene. When signals are given to the muscle cells, this receptor creates a channel, the calcium ions excite the muscle fibers, and the muscles in the body move.

Mutations in the *CACNA1S* gene working with a second gene cause malignant hyperthermia. Two of the mutations are associated with high risk of the condition. In one of the mutations, the amino acid cysteine replaces arginine, and in the other, histidine replaces arginine. These mutations disrupt the proper functioning of the calcium channels when exposed to certain drugs used during surgery. As a result, a large number of calcium ions are released from the muscles, causing them to contract abnormally. Heat and excess acid are also generated and the temperature of the body climbs. *CACNA1S* is inherited in an autosomal recessive pattern and is located on the long arm (q) of chromosome 1 at position 32.

RYR1

The second gene is *RYR1*, known also as the "ryanodine-receptor 1" gene. Normally, *RYR1* instructs for making the protein ryanodine receptor 1. This protein creates the channels that transport calcium ions within the cells and help the muscles move. When muscles are resting, calcium resides within the cell in a special internal structure called the sarcoplasmic reticulum. When the signal excites the RYR1 channels, the calcium ions move into a part of the muscle cell called the T-tubule. When calcium is present in the proper concentrations, the muscles contract to move the body.

The 30 or so mutations in *RYR1* increase the risk of malignant hyperthermia. Most mutations are the result of the exchange of only one amino acid in the ryanodine receptor 1 protein. The mutations disrupt the structure of the channel and make it open and close erratically. As a result the stored calcium surges into the muscles, making them contract more rapidly in response to certain drugs. The increased activity leads to an increase in body heat. *RYR1* is inherited in a dominant pattern and is located on the long arm (q) of chromosome 19 at position 13.1.

What Is the Treatment for Malignant Hyperthermia?

The treatment is in the preparation for surgery. A plan is needed any time a person is exposed to anesthesia. When the plan is in place, if an emergency arises, treatment can be prompt and lifesaving. It is essential that the operating room personnel know the symptoms of malignant hyperthermia and respond at once. To cool the patient, hypothermia blankets are placed under and over the person, and the person is given cold infusions of saline or salt solution. Taking a history or family incidents of MH is essential before the person undergoes surgery. Surgery can be performed, but the MH-triggering anesthetics must be avoided.

Further Reading

"Malignant Hyperthermia." 2011. Genetics Home Reference. National Library of Medicine (U.S.). http://ghr.nlm.nih.gov/condition/malignant-hyperthermia. Accessed 5/23/12.

"Malignant Hyperthermia." 2011. MedlinePlus. National Institutes of Health (U.S.). http://www.nlm.nih.gov/medlineplus/ency/article/001315.htm. Accessed 5/23/12.

Malignant Hyperthermia Association of the United States. 2011. http://www.mhaus.org. Accessed 5/23/12.

"Malignant Hyperthermia—Overview." 2011. University of Maryland Medical Center. http://www.umm.edu/ency/article/001315.htm. Accessed 5/23/12.

Mannosidosis

Prevalence	Alpha-mannosidosis occurs in 1 in 500,000 worldwide; beta-mannosidosis known in only about 20 families
Other Names	LAMAN; lysosomal acid alpha-mannosidase; MA2B1_HUMAN; MANB; mannosidase, alpha B, lysosomal

"Locoweed" is found in many farm areas of the world. When livestock eat it, they begin to act "crazy" or "loco," a term borrowed from Spanish. The weed is actually a flowering plant that produces a harmful toxin known as swainsonine. When cattle or other animals eat the weed, they develop unusual neurological symptoms. The disease of cattle became of interest because the toxic component that paralyzed the nerves was similar to a disorder first described in 1986 in humans. Both the animals and humans had progressive deterioration caused by a lysosomal disorder. However, in humans, the disease was genetic.

What Is Mannosidosis?

Two kinds of mannosidosis exist, alpha and beta. The rare inherited disorders are related to a malfunction in the lysosomes, the garbage collector and recycler of the cells. Both forms are the result of defective enzymes that do not break down complex sugars in the lysosomes. Sugar then builds up in the cell and disrupts bodily functions. Although the manifestations are similar, the conditions are caused by different genes. Both the conditions affect many different body organs and tissues.

Alpha-Mannosidosis

Alpha-mannosidosis is the more severe form of the two. Following are symptoms of alpha-mannosidosis:

- Progressive intellectual disability beginning in the first few months of life
- Distinctive facial features, such as protruding forehead, flat nose bridge, and small face
- Liver and spleen enlargement
- Hearing loss
- Cataracts
- Severe respiratory infections
- Thickening of bones on top of skull
- Spine deformities
- Low bone density, a condition called osteopenia
- Skeletal deformities such as bowed legs and knock knees
- Trouble coordinating movements
- Muscle weakness
- Anxiety or hallucinations

Conditions generally appear in infancy. Children with early severe cases may die in infancy or may be stillborn. Those with less severe forms may live into their 50s.

What Is the Genetic Cause of Alpha-Mannosidosis?

Mutations in the *MAN2B1* gene, officially known as the "mannosidase, alpha, class 2B, member 1" gene, causes alpha-mannosidosis. Normally, *MAN2B1* instructs for the enzyme alpha-mannosidase. This enzyme is active in the lysosomes, working to break up large sugar molecules called oligosaccharides. Specifically, it breaks down a sugar molecule called mannose. About 70 mutations in *MAN2B1* disrupt the process and cause the enzyme to be short or to be put together wrong. Thus, the alpha-mannosidose enzyme cannot break down the sugars, and the sugars accumulate causing the symptoms of the disorder. *MAN2B1* is inherited in an autosomal recessive pattern and is located on the cen-q or long arm end of chromosome 19 at position 13.1.

Beta-Mannosidosis

This condition is the one that was first related to locoweed. The condition is very similar to alpha-mannosidosis with defects in the way the body processes sugar molecules. The symptoms of the condition vary in severity and in the age of onset, which can be from infancy to adolescence. Following are signs of the disorder:

- Extremely introverted personality
- Prone to depression
- Intellectual disability
- Seizures
- Impaired motor development
- Behavioral problems such as hyperactivity and aggression
- Ear infections and hearing loss
- Speech impairment
- Problems swallowing
- Loss of sensation in the legs called neuropathy
- Distinctive facial features
- Small dark red spots on the skin

What Is the Genetic Cause of Beta-Mannosidosis?

Mutations in the *MANBA* gene, officially known as the "Mannosidase, beta A, lysosomal" gene, causes beta-mannosidosis. Normally, *MANBA* instructs for an enzyme beta-mannosidase, which works in the lysosomes. In the lysosomes, the enzyme breaks down the sugar complex called oligosaccharides. This enzyme is the last step in the process that breaks the molecule into a simpler sugar called mannose. About 12 mutations in *MANBA* cause beta-mannosidosis. The mutations disrupt the activity of the enzyme, causing the sugars to not be broken down properly and toxic levels of sugars to build up. *MANBA* is inherited in an autosomal recessive pattern and is located on the long arm (q) of chromosome 4 at position 22-q25.

What Is the Treatment for Mannosidosis?

No cure exists for either alpha-mannosidosis or beta-mannosidosis. Treatment is symptomatic. If the person has seizures, loss of hearing, or muscular weakness or pain, taking care of the symptoms may make life more pleasant. The person may be confined to a wheelchair. Some researchers have found that bone marrow transplants if performed when the disorder is first detected may halt some of the progression of the disorder. The individual may benefit from a variety of strategies and qualifies under the Individuals with Disabilities Education Act (IDEA) to receive special services.

Further Reading

"Alpha-Mannosidosis." 2011. Genetics Home Reference. National Library of Medicine (U.S.). http://ghr.nlm.nih.gov/condition/alpha-mannosidosis. Accessed 5/23/12.

"Mannosidosis." 2011. Lysosomal Learning. http://www.lysosomallearning.com/healthcare/about/lsd_hc_abt_mannosidosis.asp. Accessed 5/23/12.

ISMRD: International Advocate for Glycoprotein Storage Diseases. 2009. http://www.ismrd.org. Accessed 5/23/12.

Maple Syrup Urine Disease (MSUD)

Prevalence   Affects 1 in 185,000 infants; more frequent in Old Order Mennonite population with an incidence of about 1 in 380; also seen often in people of French-Canadian descent; some types in people of Ashkenazi Jewish descent

Other Names   BKCD deficiency; branched-chain alpha-keto acid dehydrogenase deficiency; branched-chain ketoaciduria; ketoacidemia; MSUD

Naming diseases and disorders really follows no pattern. Some are named for the researchers who work diligently on them. Others are named by some chemical or metabolic deficiency. Seldom are they named from a smell. Maple syrup comes from maple trees and is a staple in certain parts of the northeastern United States. People in the area know the distinct aroma. So when the aroma of maple syrup appearing on the baby's diaper began to be associated with a serious disorder, the name maple syrup urine disease (MSUD) was given. Its name is an anomaly among names.

What Is Maple Syrup Urine Disease (MSUD)?

Maple syrup urine disease is a disorder in the body's ability to use certain amino acids in proteins. The building block amino acids are leucine, isoleucine, and valine. These essential amino acids are called the branched-chain amino acids or BCAAs because of their chemical structure. In a normal person, protein from food and muscles are broken down into the branched-chain amino acids and other amino acids. The three amino acids are present in many kinds of food, particularly protein-rich foods such as milk, meat, and eggs. When these foods or are eaten, certain enzymes called BCKAD enzymes use these products for energy and growth.

However, when a child with MSUD eats proteins with the amino acids leucine, isoleucine, and valine, the foods are broken down into branched-chain amino acids and others, but because the BCKAD enzymes are not working properly, the substances are not used for energy and growth. The buildup of the branched-chain amino acids in the blood gets out of control and becomes toxic, causing health problems.

The child appears normal at birth. Various forms of the disease have different symptoms, but the following signs may appear within two to four days as soon as the baby is fed protein:

- Poor appetite
- Poor sucking for nourishment
- Weight loss
- High-pitched cry
- Vomiting
- No energy

Without treatment, the child may go into metabolic crisis and eventual swelling of the brain, which may cause seizures. They may lapse into a coma leading to death.

What Is the Genetic Cause of Maple Syrup Urine Disease (MSUD)?

Mutations in four genes are associated with maple syrup urine disease: *BCKDHA*, *BCKDHB*, *DBT*, and *DLD*.

BCKDHA

The first gene is *BCKDHA*, known officially as "branched chain keto acid dehydrogenase E1, alpha polypeptide." Normally, this gene instructs for make one part of an enzyme group called the branched-chain alpha-keto-acid dehydrogenase or BCKD. To make this complex, two alpha subunits adhere to two beta units produced by another gene. These two enzymes form an important part of the complex called the E1 complex. The BCKD complex is only one step in the breakdown of leucine, isoleucine, and valine. This complex helps in the process that is active in the mitochondria, the powerhouses that produce energy for growth.

More than 40 mutations of *BCKHA* cause the severe, classic cases of MSUD. Most of these mutations result from only one single amino acid change in the alpha subunit of the complex. The one that is most common in the Old Mennonite population involves the replacement of tyrosine with the amino acid asparagine. Any of the mutations stop the normal function of the BCKD enzymes. The amino acids leucine, isoleucine, and valine are not broken down, leading to buildup in the cells and tissues. This toxic buildup can then lead to the serious medical problems associated with MSUD. *BCKDHA* is inherited in an autosomal recessive pattern and is located on the long arm (q) of chromosome 19 at position 13.1-q13.2.

BCKDHB

The second gene associated with MSUD is *BCKDHB*, or "branched chain keto acid dehydrogenase E1, beta polypeptide." Normally, *BCKDHB* instructs for the making of a second part of the BCKD complex called the beta unit. When two beta units combine with two alpha units produced by the first genes, the E1 complex is formed. The two work together to breakdown the three amino acids produced from food.

There are more than 40 mutations in the *BCKDHB* gene, usually associated with most severe forms of MSUD. One of the most common is in the people of Ashkenazi Jewish descent where proline replaces the amino acid arginine. Mutations in this gene disrupt the normal function of the BCKD enzyme complex causing the amino acids from food not to be broken down and toxic materials to accrue. *BCKDHB* is inherited in an autosomal recessive pattern and is located on the long arm (q) of chromosome 6 at position 14.1.

DBT

The third gene related to MSUD is *DBT*, or "dihydrolipoamide branched chain transacylase E2." Normally, this gene instructs for making a critical part of the BCKD complex called the E2 component. Combined with the E1 components from *BCKDHA* and *BCKDHB*, the gene helps in another step in the breakdown of the amino acids leucine, isoleucine, and valine.

Thirty mutations of *DBT* are related to milder forms of MSUD. Usually the mutations are a change in only one amino acid but can result from insertions or deletions of tiny amounts of DNA. These mutations keep the E2 component of the BCKD from working properly. The three amino acids are not broken down, and toxic materials build up in the bloodstream, causing the health problems of MSUD. *DBT* is inherited in an autosomal recessive pattern and is located on the short arm (p) of chromosome 1 at position 31.

DLD

The fourth gene responsible for MSUD is *DLD*, or "dihydrolipoamide dehydrogenase." Normally, this gene instructs for the protein dihydrolipoamide dehydrogenase that is part of a group of enzymes that work together. This enzyme forms a subunit called E3 that is shared by several enzymes. BCKD is only one of the enzyme complexes that share the E3 component.

More than 10 mutations in *DLD* are associated with MSUD. The mutation in the structure disrupts the function of the enzyme complex and keeps the breakdown of the amino acids leucine, isoleucine, and valine from occurring. The toxic buildup can lead to seizures and other medical problems associated with MSUD. *DLD* is inherited in an autosomal recessive pattern and is located on the long arm (q) of chromosome 7 at position 31-q32.

What Is the Treatment for Maple Syrup Urine Disease (MSUD)?

Prompt treatment is essential for treating MSUD. Following are the treatments suggested for infants with the disorder:

- Beginning formula: A special medical formula is given that will be a substitute for milk and have the nutrients necessary for growth.

- Diet: A low-protein diet avoiding any foods that would have the three branch-chain amino acids—leucine, isoleucine, or valine. This means no cow's milk, regular formula, meat, fish, cheese, eggs, flour, dried beans, nuts, and peanut butter. Many fruits and vegetables have BCAAs. These must be carefully measured. So the child will have to be on a diet of special low-protein flours, pastas, and rice made especially for people with MSUD. A special metabolic doctor, along with a dietician, works with patients to determine the diet.

- Supplements: Some children with MSUD may benefit from thiamine, a member of the vitamin B family.

- Tracking BCAA levels: Blood tests are needed regularly for children with MSUD. The diet may have to be adjusted if the BCAAs get too high.

- Other illnesses. Other types of illnesses such as colds can evoke a metabolic crisis. People with MSUD need to inform their physicians at the first sign of other illnesses.

- Liver transplant: The enzymes that are deficient are located in the liver. This extreme measure is recommended for cases that do not respond to diet.

Further Reading

"Maple Syrup Urine Disease." Learn.Genetics. University of Utah. http://learn.genetics.utah.edu/content/disorders/whataregd/msud. Accessed 5/23/12.

"Maple Syrup Urine Disease." 2010. STAR-G. http://www.newbornscreening.info/Parents/aminoaciddisorders/MSUD.html. Accessed 5/23/12.

The MSUD Family Support Group. 2011. http://www.msud-support.org. Accessed 5/23/12.

Marble Bone Disease. *See* Osteopetrosis

Marfan Syndrome

Prevalence   Affects about 1 in 5,000 worldwide; affects both boys and girls
Other Names Marfan's syndrome; MFS

In the latter part of the nineteenth century, France and specially Paris was the hot spot for medical developments. Many French physicians began to specialize in areas such as cardiology, surgery, and obstetrics. Pediatrics also was identified as an area of specialization. One doctor, Antoine Bernard-Jean Marfan, stood out in this field. In 1896, he noted unusual features of in a five-year-old girl. The girl was very tall, had long, thin fingers, and several other physical conditions. He proposed that this was a genetic condition that affected the connective tissue. The condition that he described was given his name, Marfan syndrome. In 1991, Dr. Francesco Ramirez at the Mount Sinai Medical Center in New York City located the gene that causes the disorder.

What Is Marfan Syndrome?

Marfan syndrome is a disorder of the connective tissue, an all-important substance found throughout the body. Connective tissue gives strength to bones, ligaments, heart, and walls of the blood vessels. Because of the presence of connective tissue everywhere in the body, many different symptoms of the disorder may appear.

Following are some of the areas that may show the symptoms of Marfan syndrome:

- Skeletal system: The most obvious signs of Marfan syndrome appear in this system. The person will be tall and have long, slender limbs and long, thin fingers and toes. The arm span may exceed their body height. The joints may be very flexible. Other very thin features may appear in the skeletal system. The face will be long and narrow. The high arched roof of the mouth crowds the teeth. Also, the person may have scoliosis or curvature of the spine. Osteoarthritis may develop early in life.

- Heart and blood vessels: This symptom is the most serious and can be life-threatening. Abnormalities may occur in the heart and aorta, causing tiredness, shortness of breath, racing heartbeats, palpitations, and angina or terrible chest pains. Because of poor circulation, the person's hands and feet may always be cold. Heart valves may leak, and the aorta, because of weakness in the connective tissue, may develop an aneurysm, a condition in which a thin segment of the vessel balloons out. Women with Marfan have serious problems with the heart issue during pregnancy.

- Eyes: Over half of the people with Marfan have problems with the eyes. The most serious is a dislocation of the lens. The person may have nearsightedness and astigmatism. Sometimes, the retina may become detached, causing early-onset glaucoma or early cataracts.

- Lungs: Air may escape from the lungs and be trapped in a space between the chest and the lungs. Pain, shortness of breath, and even death may result.
- Central nervous system: The weak connective tissue sac around the spinal cord may cause headaches, back pain, leg pain, stomach pain, and other neurological symptoms.
- Skin: Individuals may have stretch marks and hernias caused by abnormal growth

What Are the Genetic Causes of Marfan Syndrome?

One gene has been associated with Marfan syndrome, and a second is believed by some researchers to relate in some way. The two genes are *FBN1* and *TGFBR2*.

FBN1

The first gene is the *FBN1* gene, whose official name is the "Fibrillin 1" gene. Normally, *FBN1* instructs for a protein called fibrillin-1. This protein moves out of cells into an area called the extracellular matrix, a complex of proteins that occupies the spaces between cells. In the matrix, fibrillin-1 proteins bind together to form tiny thread-like strands called microfibrils. These microfibrils are quite stretchy and give the skin, ligaments, and blood vessels their elastic properties. Microfibrils can also form more rigid tissues that support the muscles, nerves, and lens of eyes. Certain growth proteins called transforming growth factor-β (TGF-β) also hold the microfibrils. TGF-β factors control the growth and repair of body tissues.

More than 600 mutations of *FBN1* cause Marfan syndrome. Most of the changes are in only one of the building blocks that make fibrillin-1. Other changes are caused by the production of an abnormal protein that simply cannot function. The result of this disruption of fibillin-1 is a decrease in the number of microfibrils, thus weakening the structures that depend on the proper support of the connective tissue. The decrease also allows overactivation of TGF-β factors leading to the symptoms of Marfan syndrome. *FBN1* is inherited in a dominant pattern and is located on the long arm (q) of chromosome 15 at position 21.1.

TGFBR2

The second gene is *TGFBR2*, also known as the "transforming growth factor, beta receptor II" gene. Normally, *TGFBR2* instructs for the protein transforming growth factor TGF-β type II receptor. The structure of the TBF-β protein is such that part of the receptor is inside the cell and part protrudes out to gather substances from outside the cell to control cell division and growth. Sometimes called a tumor suppressor cell, this protein is important in keeping growth under control.

More than 10 mutations in *TGFBR2* cause Marfan syndrome. Almost all are the result of one change of one building block that makes the TGF-β type II receptor. The mutations interrupt the signaling activity of the receptor and keep the tissues from growing normally. The disturbances lead to the abnormal skeletal and heart conditions of Marfan syndrome. *TGFBR2* is inherited in an autosomal dominant pattern and is located on the short arm (p) of chromosome 3 at position 22.

What Is the Treatment for Marfan Syndrome?

Although no cure exists, treatments for the symptoms of Marfan have advanced in the past few years. A team, consisting of a geneticist, cardiologist, ophthalmologist, orthopedist, and cardiopulmonary surgeon, works together to attack specific disorders. The life-threatening heart issues are addressed first. The skeletal conditions can be painful but are generally not life-threatening. A physical therapist may use TENS therapy, ultrasound, or skeletal adjustment. Experiments are being done on animals to find substances that will block the TNFβ over activity.

Further Reading

Dietz, Harry. 2011. "Marfan Syndrome." *GeneReviews*. http://www.ncbi.nlm.nih.gov/books/NBK1335. Accessed 5/24/12.

"Marfan Syndrome." 2011. Genetics Home Reference. National Library of Medicine (U.S.). http://ghr.nlm.nih.gov/condition/marfan-syndrome. Accessed 5/24/12.

National Marfan Foundation. 2011. http://www.marfan.org/marfan. Accessed 5/24/12.

Marie-Struempell Disease. *See* Ankylosing Spondylitis (AS)

MCAD Deficiency

Prevalence   About 1 in 17,000 people; common among people of northern European heritage

Other Names   ACADM deficiency; MCADD; MCADH deficiency; medium chain acyl-CoA dehydrogenase deficiency; medium-chain acyl-coenzyme A dehydrogenase deficiency

The term "sudden infant death syndrome" or SIDS describes a condition that occurs when a child dies suddenly for no apparent reason. An investigation is initiated, and a sample of the blood is taken to measure the presence of certain chemicals called acylcarnitines. Elevated levels of the chemical indicate a serious imbalance in fatty acid metabolism.

Since the 1970s, investigators began to suspect that the genetic errors of metabolism described by Archibald Garrod were probably present in several disorders. In 1973, the first fatty acid condition to be identified was carnitine palmitoyltransferase or CPT. Soon, another condition was found in which patients excreted dicarboxylic acids of chain lengths of C-6-C10 in the urine. Researchers surmised that there was a problem in the beta-oxidation process, a component of fatty oxidation metabolism. In 1983, Gregersen and a team found the deficiency was in a medium-chain acyl-coenzyme A (CoA) dehydrogenase enzyme in a patient. This disorder was referred to as MCAD deficiency. The disorder may be as common in newborns as phenylketonuria.

What Is MCAD Deficiency?

Medium-chain acyl-CoA dehydrogenase (MCAD) deficiency is a disorder that affects the way the body breaks down fats. In the body, fats are stored for reserve energy. When carbohydrates and sugars have been used, an enzyme called MCAD then breaks down fats for use in the body. Babies with MCAD deficiency usually appear healthy at birth but begin to have the following signs in about two months to two years:

- Vomiting
- Extreme tiredness
- Low blood sugar
- Breathing difficulties
- Seizures
- Heart failure
- Coma

The symptoms may appear after prolonged periods without eating, after an illness such as a viral infection, or after vigorous exercise. The condition is sometimes mistaken for Reye syndrome, a disorder that may develop when a child is recovering from chicken pox of from the flu. In rare cases, the symptoms first appear in adulthood.

What Is the Genetic Cause of MCAD Deficiency?

Mutations in the *ACADM* gene, known also as the "acyl-CoA dehydrogenase, C-4 to C-12 straight chain" gene, cause MCAD deficiency. Normally, *ACADM* instructs for an enzyme called medium-chain acyl-CoA dehydrogenase or MCAD. This enzyme works in the mitochondria or powerhouses of the cell to oxidize fatty acids. This first step breaks down fats and converts them to energy. The fats that are involved are called medium-chain fatty acids. During periods of going without food or fasting, fatty acids are important sources of energy for the body's organs, especially the heart, brain, and nervous system.

More than 80 mutations in *ACADM* cause MCAD deficiency. Only a single protein building block can disrupt the process of the proper production of the enzyme. The change reduces or eliminates the activity of oxidation, and the fatty acids are not metabolized. The partially broken down products build up and cause damage to the liver and brain. The toxic buildup causes the signs and symptoms of MCAD deficiency. *ACADM* is inherited in an autosomal recessive pattern and is located on the short arm (p) of chromosome 1 at position 31.

What Is the Treatment for MCAD Deficiency?

Although no cure exists for MCAD deficiency, the unwanted symptoms can be prevented with proper management. An important element is to get a proper diagnosis. The person should never go without food for more than 10 to 12 hours. A low-fat diet is usually prescribed. Doctors may also prescribe daily doses of carnitine to help reduce the toxic buildup of fatty acids. When the patient reaches puberty, the symptoms may decrease, but a long period of fasting can cause a crisis.

Further Reading

"*ACADM*." 2011 Genetics Home Reference. National Library of Medicine (U.S.). http://ghr.nlm.nih.gov/gene/ACADM. Accessed 5/23/12.

Matern, Dietrich, and Piero Renaldo. 2012. "Medium-Chain Acyl-Coenzyme: A Dehydrogenase Deficiency." *GeneReviews*. http://www.ncbi.nlm.nih.gov/books/NBK1424. Accessed 5/23/12.

"Medium-Chain Acyl-CoA Dehydrogenase Deficiency." 2011. Genetics Home Reference. National Library of Medicine (U.S.). http://ghr.nlm.nih.gov/condition/medium-chain -acyl-coa-dehydrogenase-deficiency. Accessed 5/23/12.

McCune-Albright Syndrome

Prevalence	Between 1 in 100,000 and 1 in 1,000,000 people worldwide
Other Names	Albright-McCune-Sternberg syndrome; Albright's disease; Albright's disease of bone; Albright's syndrome; Albright's syndrome with precocious puberty; Albright-Sternberg syndrome; Albright syndrome; fibrous dysplasia, polyostotic; fibrous dysplasia with pigmentary skin changes and precocious puberty; MAS; osteitis fibrosa disseminate; PFD; POFD; polyostotic fibrous dysplasia

The parents of a two-year-old girl find blood on her diapers. She has begun her monthly period—even before she has developed other secondary sex characteristics such as breasts or pubic hair. The rare condition happened as a random mutation in the genes of the young girl in the early stages of embryonic development.

In 1937 Donovan James McCune and Fuller Albright reported in the *New England Journal of Medicine* of five cases with premature puberty in girls, bone disorders, and large areas of coffee-colored blotches on the skin. In 2005, the disease made headlines when doctors at Holtz Children's Hospital in Miami restored a child's face to normal proportions. The condition that has serious bone malformations, premature puberty, and café-au-lait spots was named for the two physicians—McCune-Albright syndrome.

What is McCune-Albright Syndrome?

McCune-Albright syndrome is a condition that affects the bones, skin, and some hormone-producing organs. The disease is suspected if two of the following three conditions are present:

- Premature puberty: Endocrine glands hyperfunction causing precocious puberty, especially in girls.
- Bone disorders: Abnormal scar-like fibrous tissue develops in the bones. This condition is known as polyostotic fibrous dysplasia.
- Mottled skin: Coffee-colored blotches known as café-au-lait spots on the skin, especially on the back.

Individuals with the disease may have several of the following problems:

- Bone fractures, because the bone is displaced by the abnormal fibrous tissue
- Uneven lesions in the skull and jaw, which make the face appear asymmetrical
- Curvature of the spine that may cause scoliosis
- Adrenal abnormalities
- Hyperparathyroidsm and hyperthyroidism
- Liver disease with jaundice
- Ovarian cysts
- Pituitary and thyroid tumors
- Gigantism

The problems with the endocrine glands can develop into some very serious glandular disorders. For example, the thyroid gland may become enlarged, causing a growth known as a goiter. The pituitary gland, located at the base of the brain, may produce too much growth hormone and cause abnormal growth of hands and feet and distinctive facial features. A rare condition called Cushing's syndrome is caused by malfunction of the adrenal glands on top of each kidney. This disease causes weight gain in the upper face and body and serious health problems.

Early menstrual bleeding is believed to be caused by excess estrogen caused by a cyst on one of the ovaries. It is less common for boys to experience early puberty.

What is the Genetic Cause of McCune-Albright Syndrome?

A random mutation in the *GNAS* gene or "GNAS complex locus" gene causes McCune- Albright syndrome. The syndrome is not inherited in the normal genetic pattern but is the result of an unusual occurrence that happens in embryonic development. Normally, *GNAS* instructs for a component called the stimulatory alpha subunit, of a protein complex called a guanine nucleotide-binding protein or the G protein. G proteins are made of three components—alpha, beta, and gamma subunits—and are closely involved in regulating hormones. The *GNAS* gene instructs for an enzyme called adenylate cyclase. This enzyme affects the thyroid, pituitary, ovaries, testes, and adrenal glands. Adenylate cyclase also appears to play a role in regulating bone development.

Mutations in *GNAS* cause McCune-Albright syndrome. It is not inherited in the usual way and operates differently. Some of the cells of the body have a normal version of *GNAS* while others have the mutation. This happening is called mosaicism. The severity depends on the number and location of the cells that have the mutated *GNAS* gene.

The mutations cause the G protein to be abnormal, which in turn causes to the enzyme adenylate cyclase to be turned on constantly in certain cells. This activation leads to the overproduction of many hormones that cause endocrine problems, as well as bone overgrowth and unusual skin pigmentation. *GNAS* is located on the long arm (q) of chromosome 20 at position 13.3.

What is the Treatment for McCune-Albright Syndrome?

No specific treatment for McCune-Albright syndrome exists. Basically, treatment depends on the severity and locations of the disorder. Treating the symptoms is possible. Drugs that block estrogen production may relieve the symptoms of early puberty. Surgery may be used to treat adrenal abnormalities. Certain other glands may be treated with hormones or surgery.

Further Reading

"McCune-Albright Syndrome." 2011. Cleveland Clinic. http://my.clevelandclinic.org/disorders/mccune-albright_syndrome/hic_mccune-albright_syndrome.aspx. Accessed 5/23/12.

"McCune-Albright Syndrome." 2011. MedlinePlus. National Institutes of Health (U.S.). http://www.nlm.nih.gov/medlineplus/ency/article/001217.htm. Accessed 6/15/11.

"McCune-Albright Syndrome—General Information." 2011. Magic Foundation. http://www.magicfoundation.org/www/docs/109.119/mccune_albright_syndrome. Accessed 5/23/12.

McKusick-Kaufman Syndrome (MKS)

Prevalence   Affects about 1 in 10,000 in the Old Order Amish population; incidence not known in general population

Other Names   HMCS; hydrometrocolpos, postaxial polydactyly, and congenital heart malformation; Kaufman-McKusick syndrome; MKS

Finding genes that cause disorders demand dedication and perseverance. The Old Amish population is a group of people who live in many parts of the United States, including Pennsylvania, Michigan, and Ohio. They are known for abiding by their over-200-year-old traditions, including not using modern technology or modern transportation. Because of the intermarriage, the gene pool is restricted. However, the circumstances do aid geneticists in identifying "orphan" genes that are isolated in this population.

In 1990 researchers at the University of Michigan found a gene that affects a molecule called chaperonin or "protein cages." These molecules capture and refold misshapen proteins that would disrupt normal functions. The study of this rare gene has led to greater understanding of both normal and abnormal development.

McKusick-Kaufman syndrome is named for two people who studied the disorder. Victor McKusick, a professor at Johns Hopkins University, has been acclaimed as the father of clinical medical genetics. He first wrote about the disorder in 1964 and 1968. His colleague Robert Kaufman provided more comprehensive information in 1972. The name McKusick-Kaufman syndrome is recognized by the medical profession.

What Is McKusick-Kaufman Syndrome (MKS)?

McKusick-Kaufman syndrome is a disorder of the hands, feet, heart, and reproductive system. Three basic symptoms characterize the syndrome:

- Abnormal fingers and toes: Extra digits are referred to by the term polydactyly; fingers and toes that are grown together are called syndactyly. This symptom is probably the less severe of the two because the extra digits can be removed or fingers separated.

- Heart defects: A congenital heart defect shows up at birth and leads to abnormal heartbeat and several disorders.

- Reproductive disorders: Females are born with a deformity caused by a large accumulation of fluid in the pelvis. This condition, called hydrometrocolpos, arises from the blockage of the vagina before birth. Sometimes the vagina simply fails to develop, or a membrane blocks the opening of the vagina. Because of the blockage, a large fluid-filled mass stretches the vagina and uterus. The compression of the fluid can compress other organs, including the heart and lungs, and cause death. Males may also have reproductive difficulties. The opening of the urethra may be on the underside of the penis. The penis may curve downward, and testicles may not descend normally.

It is estimated that 1% to 3% of Amish people in Lancaster County, Pennsylvania, are carriers of the disease.

The symptoms of MKS may overlap with another genetic condition present in the Old Order Amish called Bardet-Biedl syndrome. However, MKS does not have other disorders seen in Bardet-Biedl, such as vision loss, delayed development, kidney failure, and obesity.

What Is the Genetic Cause of McKusick-Kaufman Syndrome (MKS)?

Mutations in the *MKKS* gene, officially known as the "McKusick-Kaufman syndrome" gene, cause MKS. Normally, the gene instructs for a protein that forms the limbs, heart, and reproductive system. The protein acts as a chaperonin, which helps other proteins fold and create their proper three-dimensional shapes. Other possible functions of the MKKS protein is playing a role in the centrosome of the cell and cell division, as well as transporting other proteins to the cell.

Two mutations in *MKKS* have been found in the Amish population. The mutations are the result of a single exchanged amino acid or building block in the gene. These mutations disrupt the development of the limbs, heart, and reproductive system before birth and lead to the specific conditions of the disorder. *MKKS* is inherited in an autosomal recessive pattern and is located on the short arm (p) of chromosome 12.

What Is the Treatment for McKusick-Kaufman Syndrome (MKS)?

Because this is a genetic disorder, there is no cure. However, the symptoms of the disorder can be managed. The most obvious is surgical repair of the hydrometrocolpos and draining of accumulated fluid so that compression on the diaphragm and breathing can be alleviated. The extra digits or the digits that are grown together can be treated surgically. Also, treatment for congenital heart defects is standard.

Further Reading

"McKusick-Kaufman Syndrome." 2011. Genetics Home Reference. National Library of Medicine (U.S.). http://ghr.nlm.nih.gov/condition/mckusick-kaufman-syndrome. Accessed 5/23/12.

"*MKKS*." 2011. Genetics Home Reference. National Library of Medicine (U.S.). http://ghr.nlm.nih.gov/gene/MKKS. Accessed 5/23/12.

Slavotinek, Anne. 2010. "McKusick-Kaufman Syndrome." *GeneReviews*. http://www.ncbi.nlm.nih.gov/books/NBK1502. Accessed 5/23/12.

Meckel-Gruber Syndrome

Prevalence   Affects 1 in 13,250 to 140,000 live births; high incidence among Finnish population of 1 in 9,000 births; cases have been reported worldwide

Other Names   dysencephalia splanchnocystia; Gruber syndrome; Meckel syndrome

In 1822, Johann Friedrich Meckel wrote about an unusual condition in which the child had a large protrusion of the brain through the soft spot in the skull, growths in the abdomen, and extra digits. Over 100 years later in 1934, G. B. Gruber reported several patients with the same condition and called it dysencephalia splanchnocystia. The words come from the Greek: "Dysencephalia" is made up of *dys*, meaning "with difficulty," and *encephal*, meaning "brain." Protrusions from brain tissue cause bumps and knots on the skull. "Splanchnocystia" is made up of the Greek words *splanchno*, meaning "viscera" or internal organs, and *cyst*, meaning growth. The name Meckel-Gruber syndrome is simpler to remember.

What Is Meckel-Gruber Syndrome?

Meckel-Gruber syndrome is a lethal condition that affects many parts of the body. Because of the unusual symptoms and appearances, infants born with this syndrome can be diagnosed immediately at birth. Many are screened before birth with ultrasonography as early as 10 weeks gestation.

Following are the symptoms of Meckel-Gruber syndrome:

- Serious brain malformation: An encephalocele is the protrusion of the brain through a break in the tissues. Parts of the brain distend through the soft spot or fontanelle at the back of the roof of the skull and in the occipital region (back) of the skull. A sac covers the protruding structures. Fluid usually collects around the area, a condition known as hydrocephalus. Other malformations of the brain include microcephaly or very small head and anencephaly, when part of the brain is missing.

- Kidney malfunction: The kidneys may be polycystic or have many cysts and may be enlarged to 10 to 20 times the normal size. A condition called oligohydramnios may affect the kidneys. In this condition, the lack of amniotic fluid may cause the kidneys to stick to other tissues causing malformation.

- Extra digits: There may be extra fingers and toes, a condition called polydactyly.

- Heart: A defect in the atrium and the aorta may be present.

- Lungs: When the amnion does not have proper fluid, the lung may also malfunction.

- Liver: As one of the internal viscera, the liver is affected and does not work properly.

- Facial features: Cleft lip, cleft palate, small eyes, and a small jaw may be noted.
- Reproductive abnormalities: Incomplete development of sex organs may lead to confusion of sex.

What Is the Genetic Cause of Meckel-Gruber Syndrome?

Mutations in the *MKS* genes or "Merkel syndrome" gene cause Meckel-Gruber syndrome. Normally, these genes instruct for a protein called meckelin that is important in the development of the cilia in ciliated epithelial cells. These cells function to give information about the proper alignment of embryonic cells. When this protein malfunctions, the many symptoms of this disorder develop.

Meckel-Gruber is inherited in an autosomal dominant pattern and has been mapped to six loci of different chromosomes: *MKS1* on the long arm (q) of chromosome 17 at position 21-24; *MKS2* on the long arm (q) of chromosome 11 at position 13; *MKS3* on the long arm (q) of chromosome 8 at position 21.3-22.1; *MKS4* on the long arm of chromosome 12 at position 21.31-21.33; *MKS5* on the long arm (q) of chromosome 16 at position 12.2; and *MKS6* on the short arm (p) of chromosome 4 at position 15.3. All the genes appear to cause the same abnormalities although the types may vary.

What Is the Treatment for Meckel-Gruber Syndrome?

No treatment exists for this many-faceted disorder. Too many physical defects cause the infant to be either stillborn or die at birth. There appears to be little indication of the severity of the problem during pregnancy. Prenatal ultrasound may detect the problems, giving the parent the difficult choice of whether to terminate the pregnancy.

Further Reading

"Meckel-Gruber Syndrome." 2011. About.com Rare Diseases. http://rarediseases.about.com/cs/meckelgrubersynd/a/020804.htm. Accessed 5/23/12.

"Meckel-Gruber Syndrome." 2011. Medscape. http://emedicine.medscape.com/article/946672-overview. Accessed 5/23/12.

Melnick-Needles Syndrome

Prevalence Very rare; fewer than 100 cases worldwide

Other Names Melnick-Needles osteodysplasty; MNS; osteodysplasty of Melnick and Needles

In 1966, J. C. Melnick and C. F. Needles wrote about a study of two families, one through four generations and the other through three generations. The families had a rare disorder with skeletal malformations and many other physical complications, such as hearing loss. The males were most severely affected, but the females also had some of the symptoms. Their report of this previously undiagnosed bone dysplasia was published in the *American Journal of Roentgenology, Radium Therapy, and Nuclear Medicine*. The disorder was named for these two researchers: Melnick-Needles syndrome.

What Is Melnick-Needles Syndrome?

Melnick-Needles syndrome (MNS) is a condition that affects many body parts, especially the bones. MNS is part of a complex group of dysplasias or bone malformations called otopalatodigital spectrum disorders. The word "otopalatodigital" comes from Greek words: *oto*, meaning "ear"; *palato*, meaning "roof of mouth"; and *digit*, meaning "fingers or toes." MNS is the most severe of this spectrum of disorders.

Although the cases appear most severely in males, females can display a weaker version. Of the cases that have been reported, the following symptoms may occur:

- Short stature
- Bowing of legs
- Ribbon-like ribs that can cause breathing problems
- Skull base hardened
- Abnormal curvature of the spine, or scoliosis
- Absent or other abnormal bones
- Distinct facial features with bulging eyes, hair growth on forehead, round cheeks, and misaligned teeth
- Small bones in ear not developed, causing hearing loss
- Unusually long fingers and toes
- Kidney and bladder obstructions
- Heart defects

Males often die soon after birth.

What Is the Genetic Cause of Melnick-Needles Syndrome?

Mutations in the *FLNA* gene, known officially as the "filamin A, alpha" gene, cause MNS. Normally, *FLNA* instructs for a protein called filamin A. This protein, along with another protein called actin, builds the cytoskeleton that provides form and structure to the cells. Changes in the gene structure cause mutations that are referred to as "gain-of-function" because they increase the production of filamin A giving it a new structure and function. Because of this change, the process of skeletal development is disrupted and the many symptoms of MNS occur. *FLNA* is inherited in a dominant X-linked pattern and is found on the long arm (q) of the X chromosome at position 28.

What Is the Treatment for Melnick-Needles Syndrome?

No cure exists for this serious illness. Treatment of the multiple manifestations of hearing aids for deafness, cosmetic surgery for facial deformities, and orthopedic surgery may correct some of the symptoms.

Further Reading

"About Melnick-Needles Syndrome." Melnick-Needles Syndrome Support Group. http://www.melnickneedlesyndrome.com/html/about_mns.html. Accessed 5/23/12.

"Melnick-Needles Syndrome." 2011. Genetics Home Reference. National Library of Medicine (U.S.). http://ghr.nlm.nih.gov/condition/melnick-needles-syndrome. Accessed 5/23/12.

Robertson, Stephen. 2005. "Otopalatodigital Spectrum Disorders." *GeneReviews*. http://www.ncbi.nlm.nih.gov/books/NBK1393. Accessed 5/23/12.

Ménière Disease

Prevalence   About 615,000 people in the United States with Ménière disease, with 45,000 new cases each year; common in people of European descent

Other Names   auditory vertigo; aural vertigo; Meniere disease; Ménière's disease; Meniere's disease; Meniere's syndrome; Ménière's vertigo; otogenic vertigo

Throughout history, physicians have described people with dizziness and balance problems. In 1861, Prosper Ménière, a French physician, reported that inner ear disorders caused vertigo. He noted that the condition varied in intensity and affected individual people in different ways. The condition was named after Dr. Ménière.

What Is Ménière Disease?

Ménière disease is a disturbing disorder of the inner ear that causes dizziness and loss of balance and hearing. Following are the symptoms of the disease:

* Dizziness: Bouts of sudden dizziness or vertigo characterize Ménière disease. The episodes generally last from 20 minutes to 24 hours. Stress, fatigue, emotional upset, illness, and certain foods appear to trigger the condition.

* Roaring sound in the ears, a condition known as tinnitus: The roaring usually begins in one ear but later moves to both ears.

From Greek Mythology to Science

Many science terms have their origin in Greek mythology. The inner structure of the ear is called the labyrinth. When anatomists of the Renaissance began to study the intricacies of the ear, they were reminded of the ancient story of the elaborate structure that Daedalus built for King Minos of Crete at Knossus. The labyrinth was so elaborate that Daedalus himself could barely escape. The labyrinth was the home of the mythical Minotaur, a creature who was half man and half bull. In order to kill this creature, one would have to navigate the maze to the center. Theseus, the Athenian hero, killed the Minotaur and then escaped with the help of the goddess Ariadne, who gave him a skein of thread to leave as a path so he could escape.

In everyday English, labyrinth means the same as maze. A labyrinth has a single, non-branching path to the center and back again. Labyrinths were used on coins as early as 430 BC and on pottery, basketry, and walls.

- Extreme fullness and feeling of pressure in the ears.
- Fluctuation in hearing: The loss is hearing is progressive.
- Nausea and vomiting.
- Periodic attacks with periods of normality.
- Usually appears in adulthood in the 40s and 50s.

The person may have no symptoms of the disorder between episodes, but over time may develop consistent symptoms and permanent hearing loss. When bouts occur, the life of the person is severely disrupted.

Buildup of fluid in the compartments of the inner ear causes the symptoms of Ménière disease. The structure is generally known as the labyrinth, a term borrowed from Greek mythology. The labyrinth contains organs of balance, the semicircular canals and the organs for hearing, called the cochlea. It has two sections: the bony labyrinth and the membranous labyrinth, which is filled with fluid called endolymph. In this structure, balance is created as the body moves and stimulates receptors. These receptors send signals to the brain about the body's position. In the cochlea, fluid responds to sound vibrations, which then sends signals to the brain. In Ménière disease, fluid buildup interferes with normal balance and hearing signals, causing the symptoms of the disorder.

What Is the Genetic Cause of Ménière Disease?

Scientists do not know the exact cause of Ménière disease. In addition to environmental factors, they surmise it is genetic, and although they have researched more than a dozen genes, they have found no specific gene at this time. Most cases occur in people with no family history of the condition. Only a few cases appear to run in families, perhaps as an autosomal dominant pattern. Many risk factors may be responsible for the complex disorder. These include viral infections, ear trauma, noise pollution, allergies, immune responses, and migraines.

What Is the Treatment for Ménière Disease?

Changing lifestyle and diet may reduce the frequency and severity of the symptoms. Several of the following things may be recommended:

- Restricting salt: High-salt diets appear to cause water retention and possibly relate to the buildup in the inner ear.
- Diuretics to facilitate the low-salt diet.
- Avoiding alcohol, caffeine, and tobacco.
- Prescription and over-the-counter medicines for nausea.
- Antivirals for suspected viral infections.
- Surgery to restore balance areas.
- Techniques for managing stress.

The U.S. Food and Drug Administration approved a device that fits into the outer ear and gives air pressure pulses to the middle ear. This device appears to act on the fluid to prevent dizziness.

Further Reading

Hain, Timothy C. 2011. "Ménière Disease." http://www.dizziness-and-balance.com/disorders/menieres/menieres.html. Accessed 11/24/11.

"Ménière Disease." 2010. National Institute on Deafness and Other Communication Disorders (U.S.). http://www.nidcd.nih.gov/health/balance/pages/meniere.aspx. Accessed 11/24/11.

Meniere's Disease Information Center. 2011. http://www.menieresinfo.com. Accessed 11/24/11.

Menkes Syndrome

Prevalence   About one in 100,000 newborns

Other Names copper transport disease; hypocupremia, congenital; kinky hair syndrome; Menkea syndrome; Menkes disease; MK; MNK; steely hair syndrome; X-linked copper deficiency

In the early 1900s, Archibald Garrod first suspected that a disorder characterized by black urine was an "inborn error of metabolism." Since this early discovery, many inborn errors of metabolism have been identified. In 1962, J. H. Menkes and colleagues wrote about an inborn error of the metabolism of the element copper. Writing in the journal *Pediatrics*, the team traced the disorder to a sex-linked recessive gene that led to slow growth, unusual hair, and brain deterioration. The condition was named Menkes syndrome after the lead scientist of the study.

What Is Menkes Syndrome?

Menkes syndrome is an inborn error of metabolism that affects the levels of copper in the body. An essential trace element, copper is required for a number of normal enzymes to work and function. Most people get the necessary copper through a balanced diet. However, modern habits of eating may cause a deficiency in copper. Studies of Menkes syndrome have shed light on the normal use of copper in metabolism.

In Menkes syndrome, copper is not absorbed properly into the body for its use. The disorder is not part of newborn screening and may not appear for several months. Symptoms of the disorder include the following:

- Very slow growth, a condition called failure to thrive
- Sparse, kinky hair
- Nervous system decline
- Weak muscle tone
- Intellectual developmental delay
- Seizures
- Sagging face

If the condition is not diagnosed, the child will probably not live past the age of three. If treatment is begun, the child may live a normal life.

What Is the Genetic Cause of Menkes Syndrome?

Mutations in the *ATP7A* gene, known officially as the "ATPase, CU++ transporting alpha polypeptide" gene, causes Menkes syndrome. Normally, *ATP7A* instructs for a protein that controls the copper levels in the body. Necessary for normal metabolism, copper is dangerous in excessive amounts. In the small intestine, the protein aids in taking copper from food and then sending it to the cells of the body. ATP7A protein normally is located in the cell structure called the Golgi bodies, an apparatus that modifies protein and enzymes. In the Golgi bodies, the protein delivers copper to enzymes that are essential for proper formation of the bones, hair, skin, blood vessels, and nervous system. It is found in all cells of the body, except liver cells.

Over 150 mutations in *ATP7A* causes Menkes syndrome. Most of the mutations have deleted parts of the gene; others have additional material or exchange of one building block for another. The mutations disrupt the function of the absorption of copper from food. Thus, copper accumulates and is not supplied to the essential enzymes. The abnormal protein moves from the Golgi bodies to the cell membrane and gets stuck there. Copper does not get to the cells of the body and gathers up in some tissues, such as the small intestine and kidney. Decreased supply to the brain causes the developmental delay. Disruption of these copper-containing enzymes causes the symptoms of Menkes. *ATP7A* is inherited in a recessive pattern and is located on the long arm (q) of the X chromosome at position 21.1.

What Is the Treatment for Menkes Syndrome?

Treatment for Menkes syndrome depends on early detection. This disorder is not part of newborn screening, and the symptoms at birth are not present or are not detected. If the condition is detected, injections of copper daily can lead to very positive results. If not detected, even with the treatment of symptoms, the child will probably die in the first decade of life.

Researchers have developed a blood test for the disorder. People with a positive history of the disorder should be tested. The test can predict the disorder before symptoms appear and then lead to the daily injections of copper.

Further Reading

"Menkes Syndrome—Overview." 2011. University of Maryland Medical Center. http://www.umm.edu/ency/article/001160.htm. Accessed 5/23/12.

"NINDS Menkes Disease Information Page." 2011. National Institute of Neurological Disorders and Stroke (U.S.). http://www.ninds.nih.gov/disorders/menkes/menkes.htm. Accessed 5/23/12.

Metachromatic Leukodystrophy

Prevalence   Affects about 1 in 40,000 to 160,000 worldwide; high incidence in areas with limited gene pools: 1 in 75 in a group of Jews called Habbanites who migrated to Israel from Arabia; 1 in 2,500 in part of Navajo nation; 1 in 8,000 in Arab groups in Israel

Other Names   ARSA deficiency; arylsulfatase A deficiency disease; cerebral sclerosis, diffuse, metachromatic form; cerebroside sulphatase deficiency disease; Greenfield disease; metachromatic leukoencephalopathy; MLD; sulfatide lipidosis; sulfatidosis

The leukodystrophies are a group of genetic disorders that affect the white matter around nerve fibers. The word "leukodystrophy" comes from Greek roots: *leuko*, meaning "white"; *dys*, meaning "with difficulty"; and *troph*, meaning "nourish."

Thus a leukodystrophy is a disorder of the white matter or fatty myelin sheath that covers the nerve cells. The word "metachromatic" originates in two Greek words: *meta*, meaning "change," and *chrom*, meaning "color." Metachromatic means that the white matter of the brain has changed in color due to certain deposits or granules. This disorder gets its name from the accumulation of sulfatides that appear as colored when stained for microscopic examination.

What Is Metachromatic Leukodystrophy?

Metachromatic leukodystrophy (MLD) is a one of a group of leukodystrophies. Myelin, which appears as white, gives the color to white matter of the brain. Myelin is made up of about 10 different enzymes, and each leukodystrophy is caused by a genetic defect in one of these enzymes. In MLD, fats called sulfatides build up around the nerve fibers and progressively destroy the covering of the myelin sheath. The deficient enzyme arylsulfatase causes this buildup of fats around the myelin.

As the fats build up, the insulating property of myelin does not function and the following symptoms occur:

- Progressive loss of intellectual functions and motor skills
- Loss of feeling of touch, pain, heat, and sound
- Inability to walk
- Loss of feeling in feet and hands called neuropathy
- Incontinence
- Seizures
- Paralysis
- Inability to speak
- Blindness

Eventually the person loses all contact with surroundings and is unresponsive to everything.

Several forms of MLD exist:

- Late infantile form: This most common form affecting about 50% to 60% of individuals appears in the second year. Children first lose muscle tone and then the skills they have acquired. Progression is over 5 to 10 years.

- Juvenile form: Twenty percent to 30% of individuals with MLD develop the condition between the age of four and adolescence. The most telling symptom may be difficulty with school work and behavior problems. Symptoms develop over 10 to 20 years.

- Adult form: About 15% to 20% have first symptoms during teenage years. Alcoholism, drug abuse, and difficulties at school may appear. Individuals may hallucinate. Progression is over 20 to 30 years, with intermittent periods of apparent normal behavior.

What Are the Genetic Causes of Metachromatic Leukodystrophy?

Mutations in two genes cause metachromatic leukodystrophy—*ARSA* and *PSAP*.

ARSA

ARSA is known officially as the "arylsulfatase A" gene. Normally, *ARSA* instructs for the enzyme arylsulfatase A, which is located in the cell in the lysosomes. The lysosomes are the garbage disposal and recycling centers of the cells. The enzyme arylsulfatase A processes the sulfatides, a type of fat that is essential for building cell membranes. Sulfatides are an important part of the myelin sheath.

Mutations in *ARSA* are present in most people with MLD. The mutations disrupt the function of the arylsulfatase A enzyme, keeping it from breaking down the sulfatides. The sulfatides accumulate and cause the breakdown of the myelin sheath and lead to the progressive symptoms of MLD. *ARSA* is inherited in an autosomal dominant pattern and is located at on the long arm (q) of chromosome 22 at position 13.31-qter.

PSAP

The *PSAP* gene is also known as the "prosaposin" gene. Normally, *PSAP* instructs for a protein called prosaposin, a substance involved in many functions in the body. When prosaposin is broken up, it makes four smaller molecules—A, B, C, and D—which are critical parts of the lysosomes that break up fatty acid. Especially involved is Saposin B that activates arylsulfatase A. Only a small number of individuals with MLD have this gene. However, it is a serious component, and its deficiency leads to the accumulation of toxic levels of the sulfatides in the nervous system. *PSAP* is inherited in an autosomal recessive pattern and is located on the long arm (q) of chromosome 10 at position 210q22.

What Is the Treatment for Metachromatic Leukodystrophy?

No cure exists for MLD. Most strategies treat the symptoms and are supportive. Prognosis is poor. Most children with the infantile form die before age five. In the other forms, symptoms may extend for longer periods. Some physicians have recommended bone marrow transplants, which have delayed some of the progression of the disorder. Gene therapy is being tested in animal models. Also, research on the lipid storage diseases is being done in major medical institutions.

Further Reading

"Metachromatic Leukodystrophy." 2011. Mayo Clinic. http://www.mayoclinic.org/ metachromatic-leukodystrophy. Accessed 5/23/12.

Metachromatic Leukodystrophy Foundation. 2011. http://mldfoundation.org. Accessed 5/23/12.

"NINDS Metachromatic Leukodystrophy Information Page." 2011. National Institute of Neurological Disorders and Stroke (U.S.). http://www.ninds.nih.gov/disorders/meta-chromatic_leukodystrophy/metachromatic_leukodystrophy.htm. Accessed 5/23/12.

Methylmalonic Acidemia (MMA)

Prevalence   About 1 in 50,000 to 100,000 people

Other Names methylmalonic aciduria; MMA

Several rare genetic conditions are organic acid disorders (OAs). Basically, organic acids are those that have carbon in their chemical structures. All living things are made of organic compounds, such as proteins and enzymes. Likewise, the building blocks of the body are made of organic compounds called proteins. Enzymes process protein from food to make them work in the body. When the enzymes do not work properly, the proteins are not broken down correctly, causing harmful products to build up. Methylmalonic aciduria (MMA) is one of these organic acid disorders.

What Is Methylmalonic Acidemia (MMA)?

Methylmalonic acidemia is a condition in which the individual cannot process certain amino acids and fatty acids from food. When a person eats food with protein from meat or other sources, the normal digestive system breaks it down into smaller substances called amino acids. Special enzymes then process the amino acids into materials that can be used. In the same way, special enzymes also break down fat in food into fatty acids that the body can use for energy. If one of these special enzymes is missing or not working properly, toxic materials such as glycine and methylmalonic acid build up, causing some serious health problems.

The effects of MMA appear in early infancy and can vary from mild to life-threatening. Babies may appear normal at birth but soon develop the following symptoms once they start eating protein:

- Vomiting
- Failure to gain weight and grow as expected
- Excessive lethargy and tiredness
- Dehydration
- Intellectual decline that worsens with age
- Poor muscle tone
- Skin rashes and yeast infections
- Feeding problems
- Kidney disease
- Pancreatitis
- Osteoporosis
- Vision loss
- Seizures and stroke

Without treatment, death is common.

What Are the Genetic Causes of Methylmalonic Acidemia (MMA)?

Three genes appear to be related to methylmalonic acidemia. Mutations in *MMAA*, *MMAB*, and *MUT* cause the disorder.

MMAA

The first gene, *MMAA*, is known officially as the "methylmalonic aciduria (cobalamin deficiency) cblA type" gene. Normally, *MMAA* instructs for a compound called adenosylcobalamin or AdoCbl. This compound is made from vitamin B12 (cobalamine) and is essential for the making of the methylmalonyl CoA mutase enzyme, which breaks down certain proteins and fats. The MMAA protein is active in transporting vitamin B12 into the mitochondria or powerhouse of the cell

The more than 25 mutations in *MMAA* prevent proteins and fats from being broken down. Some mutations are the result of additions, deletions, or duplications of small amounts of genetic material. The mutations disrupt the creation of the normal protein, making it abnormal and unstable. Thus, unstable compounds build up in the blood and urine, leading to the signs and symptoms of MMA. *MMAA* is inherited in an autosomal recessive pattern and is located on the short arm (q) of chromosome 4 at position 31.21.

MMAB

MMAB, known also as "methylmalonic aciduria (cobalamin deficiency) cb1B," is the second gene related to MMA. Normally, *MMAB* instructs for the same enzyme as *MMAA*—adenosylcobalamin, derived from vitamin B12. There are more than 20 mutations of *MMAB* involved in this disorder. Most involve only a single exchange of an amino acid or missing pieces of the gene. *MMAB* is inherited in an autosomal recessive pattern and is located on the long arm (q) pf chromosome 12 at position 24.

MUT

The third gene that is involved is *MUT*, or the "methylmalonic CoA mutase" gene. Normally, *MUT* instructs for the enzyme call methylmalonic CoA mutase, which is active in the mitochondria. This enzyme targets the amino acids isoleucine, methionine, threonine, and valine, in addition to fats and cholesterol. The breakdown of these materials is what gives the body energy.

About half of the cases caused by the more than 150 mutations are in the *MUT* gene. These mutations affect the production of the enzyme and thus the proteins and fats are not broken down properly. Toxic products buildup in the body tissues and cause the symptoms of methylmalonic acidemia. *MUT* is inherited in an autosomal recessive pattern and is located on the short arm (p) of chromosome 6 at position 12.3.

What Is the Treatment for Methylmalonic Acidemia (MMA)?

Accurate diagnosis and treatment is essential to avoid the serious medical problems and mental retardation. Many states have MMA as one of the screening for newborns. Several strategies are used to treat MMA:

- Low-protein diet. Limiting foods with lucine, valine, methionine, and threonine is essential. Most of the diet will be high in carbohydrates.

- Medication: Treating with injections of vitamin B12 helps some children that are designated as vitamin B12 responsive. Some children are given L-carnitine, a natural substance that creates energy.

- Eating frequently: Children with MMA must avoid going for long periods of time without food. A metabolic crisis may occur after long periods of fasting.

- Blood tests and urine test: The urine is tested for ketones, substances that are formed when body fat uses energy.

- Contact doctor if any of the symptoms occur.

- Organ transplants, if kidney or liver damage occurs.

Children who receive early and prompt treatment may live normal healthy lives.

Further Reading

Manoli, Irini, and Charles P. Venditti. 2010. "Methylmalonic Acidemia." *GeneReviews*. http://www.ncbi.nlm.nih.gov/books/NBK1231. Accessed 6/17/11.

"Methylmalonic Acidemia." 2011. MedlinePlus. National Institutes of Health (U.S.). http://www.nlm.nih.gov/medlineplus/ency/article/001162.htm. Accessed 5/23/12.

"Methylmalonic Acidemia." 2011. STAR-G. http://www.newbornscreening.info/Parents/organicaciddisorders/MMA.html. Accessed 5/23/12.

Milroy Disease

Prevalence   Rare; incidence unknown

Other Names   congenital familial lymphedema; hereditary lymphedema type I; Milroy's disease; Nonne-Milroy lymphedema

Although a condition in which swelling of the legs and extremities in adults has been cited for years, the incidence of the disorder in infants is unknown. Rudolf Virchow, a leading German physician who is known as the founder of social medicine, first described the condition in 1863. Later, an American physician, William Milroy, studied the symptoms of the disorder and wrote about it in 1892. The famous Canadian and American physician Sir William Osler first called the condition Milroy disease, the name recognized today.

What Is Milroy Disease?

Milroy disease is a disorder of the distribution of lymph in the body, causing swelling especially in the legs and feet in newborns. Normally, the lymphatic system, the fluid part of the blood, transports immune cells and other cells throughout the body. However, if the system is not functioning properly, fluid lymph builds up, causing a swelling condition known as lymphedema.

Infants with Milroy disease develop swelling in the lower legs and feet soon after birth. The condition is seen on both sides of the body and does not worsen over time. In addition, other symptoms may occur with Milroy disease:

- A skin infection called cellulitis
- Upslanting toenails with deep creases in the toes
- Prominent leg veins
- Wart-like growths called papillomas
- In males, fluid in the scrotum and abnormalities in tube that caries urine from bladder to outside the body

What Is the Genetic Cause of Milroy Disease?

Mutations in the *FLT4*, or the "fms-related tyrosine kinase 4" gene, cause Milroy disease. Normally, *FLT4* instructs for a protein known as vascular endothelial growth factor receptor 3 (VEGFR-3), an important element in regulating the lymphatic system. Two proteins, VEGF-C and VEGF-D, turn on and bind to VEGFR-3, creating the signals for the lymphatic cells to grow and function.

About 19 mutations of *FLT4* are related to Milroy disease. Most of the changes involve one protein building block in an area known as the tyrosine kinase domain. In Milroy disease, the signaling of VEGFR-3 is interrupted and the tubes that create are small or absent. Lymph is not transported properly and swelling occurs where the fluid collects. Researchers are not clear as to why the other symptoms occur. Milroy disease is inherited in an autosomal dominant patter and is located on the long arm (q) of chromosome 5 at position 35.3.

What Is the Treatment for Milroy Disease?

Because Milroy disease is a genetic disorder, no cure exists. A therapist may recommend fitted stockings and massage. Good hygiene may prevent cellulites, and antibiotics may help prevent reoccurring symptoms.

Further Reading

Brice, Glen, et al. 2009. "Milroy Disease." *GeneReviews*. http://www.ncbi.nlm.nih.gov/books/NBK1239. Accessed 5/23/12.

"Milroy Disease." 2011. Genetics Home Reference. National Library of Medicine (U.S.). http://ghr.nlm.nih.gov/condition/milroy-disease. Accessed 5/23/12.

"Milroy Disease." 2011. Medscape. http://emedicine.medscape.com/article/299840 -overview. Accessed 5/23/12.

Model Organisms: A Special Topic

The knowledge that is required to understand the study of human genetic disorders has developed over many years. From the first studies of plants in the garden of

Mice are often used as model organisms for genetics experiments. (Ralph Morse / TIME & LIFE Pictures / Getty Images)

Gregor Mendel (1822–1884) in the mid-nineteenth century, scientists have used other organisms to investigate genes and chromosomes and find out how these genes are expressed in the signs and symptoms of genetic disorders. The term "basic research" is applied to the investigation of principles. Basic research is not done on humans but with experiments using a variety of model organisms.

What Is a Model Organism?

A model organism is a nonhuman species used to study a biological happening with the expectation that the study will provide information into the workings of another organism. Molecular biologists have used model organisms to study and establish the complex principles of genetics. A well-known adage among scientists is that the fundamental problems are most studied in the simplest and most accessible system that can address the problem. All these model systems have areas in common: They have the important tools making it possible to manipulate them genetically, and they have attracted a critical mass of investigators who can share ideas and facilitate rapid progress. Among the most important model organisms are the following: bacteriophage, bacteria, yeast, the nematode worm, the fruit fly, and the house mouse.

Bacteriophage

Bacteriophage is a virus that enters bacterial cells and disintegrates the cell. It is more commonly called phage. Phages are the simplest organism to offer the examination of life. Their genomes are small and expressed only after entering another cell. During these infections, the genome undergoes recombination. Phages were used especially in the early days of molecular biology and were vital to the development of the field. They still remain a system of choice for studying basic mechanisms of DNA replication, gene expression, and recombination. The important viruses for study include: Phage Lambda; PhiX174, whose genome of a circle of 11 genes and 5,386 base pairs was the first to be sequenced; and the tobacco mosaic virus.

Bacteria

Bacteria are prokaryotes, meaning they have no nucleus and all cell functions are within the cell membrane. Bacteria such as *Escherichia coli* (*E. coli*) and *Bacillus subtilis* (*B. subtilis*) are very simple cells and can be grown and manipulated with relative ease. Because they have no nucleus, these single-cell organisms have all the machinery for DNA, RNA, and protein synthesis in one contained area. Bacteria usually have one chromosome that contains a simple genome. The short generation time of reproducing every 20 minutes is attractive to researchers. Bacteria can inherit traits and undergo mutations. For example, a bacterium can become resistant to infection of a particular phage. Experiments with bacteria and phages led to the discoveries that DNA is the genetic material, that the genetic code is made of triplets, and that genes regulate the production of proteins. Other bacteria used as models are: *Caulobacter crescentus*, *Vibrio fischeri*, and *Pseudomonas fluorescens*. These two simple model organisms—phage and bacteria—have told us the most fundamental things about genetics.

Eukaryotes

Commonly referred to as simple baker's yeast or brewer's yeast, *Saccharomyces cerevisiae* (or *s. cerevisiae*) is a unicellular eukaryote that has an advantage over bacteria in that it has a defined nucleus. Yeasts are examples of one-celled eukaryotes, which multiple by budding. Compared to the other eukaryotes, yeasts have a relatively small genome with a very small number of genes. Within the distinct nucleus are packaged multiple linear chromosomes with chromatin. The cytoplasm of the cell also has a full group of organelles, such as mitochondria.

S. cerevisiae can be grown in the laboratory and will reproduce every 90 minutes. Since the 1930s, the genetics of yeast has been studied and has resulted in the mapping of its genome. In 1996, the genome of *S. cerevisiae* was sequenced, identifying about 6,000 genes. Levels of gene expression have been tested in more than 200 different conditions, such as cell types and growth temperatures. Studies of yeasts have led to the role of genes in many processes such as transcription, genetic expression, DNA replication, translation, and splicing. This model organism has allowed investigators to attack fundamental problems of biology using both genetic and biochemical approaches.

Other simple eukaryotes are used in basic research. Algae, including *Chlamydomonas reinhardtii*, have many known mapped mutants and expresses sequence tags, and *Dictyostelium discoideum* is used in genetics to study cell communication, differentiation, and cell death. Fungi such as *Aspergillus nidulans* and *Neurospora crassa* have added to knowledge of metabolic regulation.

The Nematode Worm: *Caenorhabditis elegans* (*C. elegans*)

The nematode is a member of a group of round worms that can live either on its own or as a parasite. Sydney Brenner (1927–), a biologist and Nobel Prize winner in 2002, was looking for an organism that had a very rapid life cycle, that had differentiated cell types, and that could be used in genetics experiments. He found his model in *C. elegans*, an organism that multiplied rapidly. *C. elegans* is hermaphroditic, meaning that it can fertilize itself and create large numbers of progeny; it also has sexual reproduction capacity so that genetic stocks can be made by mating.

Scientists have created a few, well-studied lineages of *C. elegans*. Several important biological discoveries have been made with this organism. One of the most important is the tracing of the molecular pathway that regulates apoptosis or cell death. Understanding of cell death is extremely important in many genetic conditions that are caused by the proliferation or overgrowth of cells. The model has been used in basic research to develop drugs for cancer and neurodegenerative diseases.

The Fruit Fly: *Drosophila melanogaster*

One of the standard organisms in the study of basic science is the fruit fly. Thomas Hunt Morgan (1866–1945), in his laboratory in Columbia University, placed some rotting fruit on a window ledge and soon had a number of organisms that his team could isolate and study their characteristics. In just two weeks, the adults of this fly produce a large number of progeny. Working with only four chromosomes that the fly possesses, researchers identified 5,000 bands and soon established the correlation between the bands and certain traits. The first genome maps were produced in *Drosophila*. This model enabled scientists to develop a wealth of genetic tools and provided extensive groundwork for decades of investigation. Today, it is still the premier model for study of genetic development and behavior. Morgan won the Nobel Prize in 1933 for his discoveries of the role that the chromosome plays in heredity, stemming from his studies of the fruit fly.

The House Mouse: *Mus musculus*

Mice are still a staple of basic research. Before any pharmaceutical company can ask for classification for an investigational new drug, extensive animal tests must be done; most of these tests are done in mice. Compared to the other model organisms such as *Drosophila* and *C. elegans*, reproduction in the mouse is slow and cumbersome. The embryo develops in a period of 3 weeks, and then it takes 5 to 6 weeks for the mouse to reach adulthood to be able to reproduce. The cycle is roughly 8 to 9 weeks. However, the mouse is a mammal, and the organs and other processes are more closely related to humans. The mouse provides the link between the basic principles learned by studying the lower organisms but adds a new element—that of genetic manipulation. The chromosome compliment of the mouse is similar to humans—19 autosomes, sex chromosome X and Y, and about 30,000 genes. Regions on chromosomes are similar to those of corresponding human chromosomes. The mouse genome has been sequenced.

A great tool for the study of disease is called "knock out" technology. Researchers have the ability to remove ("knock out") specific genes in normal mice and "knock in" genes for study. In this way, scientists have created mice with Alzheimer disease and various dementias, Parkinson disease, and very obese mice. The mouse models are very valuable for basic research.

Many other organisms are used for specific research. Studying model organisms can be valuable, but the scientist must be careful not to generalize from one to another. Other organisms used as models include zebra fish, guinea pigs, and certain primates such as monkeys and chimpanzees.

Further Reading

Fields, S., and M. Johnston. 2005. "Cell Biology: Whither Model Organism Research?" *Science*. 307(5717): 1185–1186.

Watson, James; Tania Baker; et al. 2008. *Molecular Biology of the Gene*. San Francisco: Pearson-Benjamin Cummings.

Mowat-Wilson Syndrome

Prevalence   Unknown; at least 170 people have been reported

Other Names Hirschsprung disease–mental retardation syndrome;
microcephaly, mental retardation, and distinct facial features, with
or without Hirschsprung disease; MWS

As early as 1691, doctors of the time talked about infants who died with swollen
bellies. In 1888, the Danish physician Harald Hirschsprung wrote about two
infants who died with constipation and swollen abdomens. The disorder was later
named for this doctor. Hirschsprung disease is an inherited disease of the colon in
which certain nerve cells are not present, causing chronic constipation. In 1998,
doctors D. R. Mowat and M. J. Wilson described a new genetic disorder in which
children had Hirschsprung disease along with a host of other symptoms. The published their findings in the *Journal of Medical Genetics* and were honored with the
name of the disorder—Mowat-Wilson syndrome.

What Is Mowat-Wilson Syndrome?

Mowat-Wilson syndrome affects many parts of the body. More than 50% of
people have Hirschsprung disease, and those without the disease may still suffer
with chronic constipation. In addition to intestinal complications, children with
Mowat-Wilson may have the following symptoms:

- Distinctive facial features: The face is square with widely spaced, deep-set
  eyes. Their nose is broad. The chin is pointed, and eyebrows point out. As
  adults, they also have facial district features that increase with age. They have
  a pronounced chin and jaw and always look like they are smiling with open-
  mouthed expressions and are typically friendly and happy personalities.
- Delayed growth and development: Small heads with moderate to severe men-
  tal retardation.
- Severe speech impediment.
- Seizures.
- Agenesis or underdevelopment of the corpus callosum of the brain.
- Eye defects.
- Slender build with tapering fingers.
- Short stature.
- Heart defects.
- Abnormalities of the urinary tract or genitalia.

What Is the Genetic Cause of Mowat-Wilson Syndrome?

Mutations in the *ZEB2*, known officially as the "zinc finger E-box binding homeobox 2" gene, cause Mowat-Wilson disease. Normally, *ZEB2* instructs for a protein

that is important in the formation of many organs and tissues before birth, especially the embryonic neural crest. This group of cells is present in the early embryo and forms parts of the nervous system, the endocrine glands, the smooth muscle, and parts of the heart, the pigment, and many parts of the face and skull. The protein is also active in many other body parts including the digestive system, skeleton, muscles, kidneys, and the face and skull.

Over 100 mutations of *ZEB2* have been found. The mutations may delete the entire gene or cause it to be very short. Thus, the protein that is created either does not exist or is dysfunctional. When this shortage of the protein exists, the neural crest and all other structures do not develop properly. Its relationship to the formation of the digestive tract probably explains the Hirschsprung disease. *ZEB2* is inherited in an autosomal recessive pattern and is found on the long arm (q) of chromosome 2 at position 22.3.

What Is the Treatment for Mowat-Wilson Syndrome?

Because this is a serious genetic disorder with so many symptoms and defects, no cure exists. To treat the manifestations of the disorder, proper specialists such as a cardiologist, gastrointestinal specialists, and others should be involved in a care team. Educational intervention should begin early in infancy with speech pathologists working with the child.

Further Reading

Adam, Margaret P.; Lora Bean; and Vanessa Miller. 2008. "Mowat-Wilson Syndrome." *GeneReviews*. http://www.ncbi.nlm.nih.gov/books/NBK1412. Accessed 5/23/12.

Mowat-Wilson Organization. http://www.mowatwilson.org. Accessed 5/23/12.

Moynahan Syndrome. *See* LEOPARD Syndrome

MPD1. *See* Laing Distal Myopathy

MSUD. *See* Maple Syrup Urine Disease (MSUD)

Muckle-Wells Syndrome

Prevalence   Rare with prevalence unknown; reported in many countries
Other Names   MWS; UDA syndrome; urticaria-deafness-amyloidosis syndrome

In 1962, doctors Thomas James Muckle and Michael Vernon Wells described a
condition in which the individuals had periods of chills, fever, and joint pains with
no obvious infection such as the flu. The episodes appeared to be triggered by cold
and other stimuli. Later, these people developed a rash, deafness, and kidney damage. They wrote about the condition of uticaria, deafness, and amyloidosis in several families in the *Quarterly Journal of Medicine*. The syndrome was named after
these two researchers—Muckle-Wells syndrome.

What Is Muckle-Wells Syndrome?

Muckle-Wells syndrome, or MWS, affects many parts of the body. Beginning in
infancy or early childhood, children first have flu-like symptoms of fever and
chills. A non-itchy skin rash and joint pain may also occur. Soon the episode happens again for no apparent reason. The episodes may be eventually connected to
the triggers of cold, heat, tiredness, or other stressors. Over the years, the following
symptoms may then occur:

- Progressive hearing loss during teenage years
- Redness in the whites of eyes, a condition called conjunctivitis
- Amyloidosis, leading to kidney failure

Amyloidosis occurs when abnormal proteins called amyloids build up in the
tissues or organs. These proteins are produced by cells in the bone marrow,
and in MWS affect the kidney and other organs. About one-third of people
with MWS have progressive kidney damage and some damage to other
organs.

What Is the Genetic Cause of Muckle-Wells Syndrome?

Mutations in the *NLRP3* gene, officially called the "pyrin domain containing 3"
gene, cause Muckle-Wells syndrome. Normally, *NLRP3* instructs for a protein
called cryopyrin. This protein is a member of a group of proteins known as
"nucleotide-binding domain and leucine-rich repeat containing" proteins, or
NLR proteins. Cells of animals are made up of the nucleus or center, the cytoplasm
or main body of the cell, and the cell membrane. The NLR proteins are located in
the cytoplasm of the cells. Specifically, cryopyrin is located in the cytoplasm of
white blood cells and in the cells that make cartilage.

  NLR proteins and cryopyrins are important to the immune system. They
respond to injury, toxins, and any invaders from the outside, such as microorgan-
isms. Cryopyrin is active in locating bacteria, and chemicals such as asbestos,

silica, and uric acid, as well as proteins released by injured cells. When an injury occurs or when outside invaders attack the body, groups of cyropyrins rush to help other proteins form inflammasomes. This strong line of defense begins the process of fighting off outside insults.

About 10 mutations in *NLRP3* cause MWS. The mutations cause the cyropyrins to become overactive, resulting in fever and inflammation. The overactive cryopyrins also result in hearing loss and kidney problems. *NLRP3* is inherited in an autosomal dominant pattern and is located on the long arm (q) of chromosome 1 at position 44.

What Is the Treatment for Muckle-Wells Syndrome?

A correct diagnosis of MWS is essential and is done through genetic testing. Because there is no cure, treating the symptoms can alleviate some pain and problems. A hearing aid may compensate for hearing loss. Anti-inflammatories may be used to treat joint pain although side effects may occur after long-term use. No medications are currently indicated for MWS.

Further Reading

"Muckle-Wells Syndrome." 2011. Genetics and Rare Diseases Information Center (GARD). National Institutes of Health (U.S.). http://rarediseases.info.nih.gov/GARD/Condition/8472/MuckleWells_syndrome.aspx. Accessed 5/23/12.

"Muckle-Wells Syndrome." 2011. Orphanet (UK). http://www.orpha.net/data/patho/GB/uk-MWS.pdf. Accessed 5/23/12.

"Muckle-Wells Syndrome." 2011. RightDiagnosis.com. http://www.rightdiagnosis.com/m/muckle_wells_syndrome. Accessed 5/23/12.

Mucopolysaccharidosis Type I

Prevalence	About 1 in 100,000 newborns; attenuated MPS I about 1 in 500,000 newborns
Other Names	alpha-L-iduronidase deficiency; Hurler-Scheie syndrome; Hurler syndrome; IDUA deficiency; MPS I; MPS I H; MPS I H-S; MPS I S; mucopolysaccharidosis I; Scheie syndrome

Around the turn of the twentieth century, Archibald Garrod was the first to note genetic disorders that were caused by problems with the way the body used certain elements. He called this condition an "inborn error of metabolism." Since that time, many inborn or genetic errors have been described. Mucopolysaccharidosis is one of those errors.

What Is Mucopolysaccharidosis Type I?

Mucopolysaccharidosis (MPS) is an inborn error of metabolism that affects many body parts. Children appear normal at birth but then begin to develop symptoms during the first year. The child is missing an enzyme that breaks down huge sugar molecules called glycoaminoglycans or GAGs. At one time the GAGs were called mucopolysaccharides; The term GAG is used now. When sugars are not broken down, toxic materials begin to build up especially in the lysosomes, the cell structures responsible for recycling and disposing of unwanted materials. MPS is a member of a large group of conditions called lysosome storage disorders.

Mucopolysaccharidosis was once considered to be three syndromes named after the people who first described them. The conditions were called:

- Hurler syndrome, or MPS I-H
- Hurler-Scheie syndrome, or MPS I-H/S
- Scheie syndrome, or MPS I-S

No real distinction exists between the three disorders, and currently physicians divide MPS I into severe and attenuated types.

MPS I affects many body parts and has many symptoms. Most children appear normal at birth, although some may have an umbilical hernia, which is an out-pouching around the navel area or an inguinal hernia located in the lower abdomen. The following symptoms begin to appear before the first birthday:

- Large head or macrocephaly caused by a buildup of fluid in the brain
- Unusual appearing face, which some describe as coarse
- Enlarged tongue and vocal cords that cause voice difficulties
- Clouding of the cornea of the eye that may lead to vision loss
- Recurrent ear infections
- Abnormal heart valve
- Enlarged liver and spleen
- Narrowed airways that lead to frequent respiratory infections and sleep apnea
- Short stature and skeletal abnormalities
- Carpal tunnel syndrome, leading to weakness in hand and fingers
- Spinal stenosis in neck, causing damage to spinal cord

The child begins with intellectual and developmental decline, losing skills he has acquired. Children with severe cases seldom live past the age of 10. Children with the attenuated types may or may not display some learning disabilities and also may or may not have a shortened life span. The major causes of death in both the severe or attenuated types are heart disease and airway obstruction.

What Is the Genetic Cause of Mucopolysaccharidosis Type I?

Mutations in the *IDUA* gene, or "iduronidase, Alpha-L" gene, cause MPS. Normally, *IDUA* instructs for an enzyme alpha-L-iduronidase that breaks down large sugar molecules called glycoaminoglycans or GAGs. In the large sugar molecules are two smaller GAGs that have sulfate. The two GAGs, called heparin sulfate and dermatan sulfate, are located in a huge molecule known as sulfated alpha-L-iduronic acid. The process of the removal of the sulfate is done in the lysosomes, which then make the proper substances available for use in the body.

The more than 100 mutations in in *IDUA* cause MPS I. Only one small chain in a DNA building block can disrupt the function of the gene and cause no or little alpha-L-iduronidase to be produced. When the enzyme is not present, heparin sulfate and dermatan sulfate build up in the lysosomes, which become large. For this reason, in MPS I the cells and tissues of the body are enlarged. The accumulated GAGs disrupt the functions of other proteins in the lysosomes, causing the many symptoms of the disorder. *IDUA* is inherited in an autosomal recessive pattern and is located on the short arm (p) of chromosome 4 at position 16.3.

What Is the Treatment for Mucopolysaccharidosis Type I?

No cure exists for MPS I, although some researchers deem it is a prime candidate for gene therapy. Treatment of the symptoms is essential. These children qualify under the Individuals with Disabilities Education Act for special education and infant learning programs. For the many manifestations of the disorder, surgery may be needed for heart valve disorders and bone/joint disorders.

Stem cell transplantation may increase survival in children with severe MPS I. For people with attenuated MPS I, some physicians have suggested enzyme replacement therapy with Aldurazyme, licensed for the treatment of central nervous system disorder, liver, and growth disorders. Children with the disorder should be monitored continuously for head growth, nerve conditions, orthopedic disorders, educational function, and other potential disorders.

Further Reading

Clarke, Lorne A., and Jonathan Heppner. 2011. "Mucopolysaccharidosis." *GeneReviews*. http://www.ncbi.nlm.nih.gov/books/NBK1162. Accessed 5/23/12.

"Mucopolysaccharidosis Type I." 2011. Genetics Home Reference. National Library of Medicine (U.S.). http://ghr.nlm.nih.gov/condition/mucopolysaccharidosis-type-i. Accessed 5/23/12.

National MPS Society. 2011. http://www.mpssociety.org. Accessed 5/23/12.

Multiple Endocrine Neoplasia

Prevalence   Two major types affect about 1 in 30,000; subtype 2A most common

Other Names   adenomatosis, familial endocrine; endocrine neoplasia, multiple; familial endocrine adenomatosis; MEA; MEN; multiple endocrine adenomatosis; multiple endocrine neoplasms

Physicians have known for many years that glands of the endocrine system are important to body functions. Disorders of the endocrine glands were treated in many ways with folk medicines. In 1903, Erdheim first traced a case of acromegaly, a condition in which the person has a large face and hands, to a tumor in the pituitary glands. Other tumors were traced to the adrenal glands and the thyroid glands. In 1954, Wermer noted the syndrome was transmitted as a dominant trait. In 1968, Steiner and colleagues introduced the term "multiple endocrine neoplasia" or MEN to describe conditions when endocrine tumors occurred. They also proposed that MEN1 be named "Wermer syndrome" and MEN2 named "Sipple syndrome."

What Is Multiple Endocrine Neoplasia?

Multiple endocrine neoplasia (MEN) is a group of disorders that affect the endocrine glands. Endocrine glands produce hormones, the chemical messengers that travel through the bloodstream telling other tissues and cells what to do. The endocrine glands are most important in regulating mood, metabolism, growth and development, and the reproductive process. Growths in the glands can be cancerous or malignant and life-threatening, or noncancerous and benign.

Two forms of MEN are called type 1 and 2. The forms are caused by different genes and affect different glands. They also have different symptoms.

MEN1

Type 1 was first known as MEN1 or Wermer syndrome. More frequently involved are tumors of the parathyroid, pituitary, and pancreas. The tumors that are present disrupt the normal function of the gland and lead to overactivity. The overproduction of hormones causes serious body disorders. For example, an overactive parathyroid causes excess calcium to be released into the bloodstream, leading to kidney stones, thin bones, high blood pressure, weakness, and general malaise.

Symptoms vary from person to person and depend on which gland is involved. They may include:

- Abdominal pain and bloating after meals
- Fatigue
- Loss of appetite
- Headache

- Lack of menstrual periods and infertility
- Loss of body hair in men
- Muscle pain and weakness
- Vision problems

MEN2

Type 2 was first known as MENII or Sipple syndrome. A type of thyroid cancer called medullary thyroid carcinoma is the most common MEN type 2. This disorder can also lead to pheochromocytoma, a tumor located in the adrenal glands that can cause very high blood pressure. Type 2 is divided into three subtypes: Type 2A, Type 2B, and familial medullary thyroid carcinoma (FMTC). The features of the disorders may vary but are relatively consistent within a family.

What Are the Genetic Causes of Multiple Endocrine Neoplasia?

Two different genes cause Multiple Endocrine Neoplasia—*MEN1* and *RET*.

MEN1

Mutations in the *MEN1* gene or the "multiple endocrine neoplasia I" gene cause MEN type 1. Normally, *MEN1* instructs for making the protein menin. Menin is a tumor suppressor, keeping cells from dividing wildly and causing tumors. Menin is located in the nucleus of certain cells and may play a role in copying and repairing DNA.

Over 400 mutations of MEN1 cause multiple endocrine neoplasia type 1. Changes in the gene lead to the production of a very short menin protein. If one copy of the altered gene is inherited, the protein does not work properly; if two copies are present, the protein does not exist. Cells begin to divide wildly, forming tumors and the many symptoms of multiple endocrine neoplasia 1. *MEN1* is inherited in an autosomal dominant pattern and is located on the long arm (q) of chromosome 11 at position 13.

RET

The second gene involved in MEN2 is the *RET* gene or "ret proti-oncogene." Normally, *RET* instructs for the RET protein that sends signals for the normal development of several kinds of nerve cells. Many of these nerve cells control the involuntary functions. For example, these nerves lead to the normal function of the intestine or enteric neurons. RET protein is also necessary for heart rate, normal kidney development, and production of sperm. RET is a signaling protein conducting its activity by having one end projecting out of the cell and the other end located within the cell. The protein receives signals from elements in the bloodstream and then transfers it to specific areas within the cell. Growth factors attach to the RET protein and causes a cascade of reaction that cause the cell to divide and mature.

The more than 25 mutations in *RET* cause MEN2. Most of the changes are caused by one single building block. For example, in MEN type 2A, arginine replaces cysteine. In type 2B, threonine replaces methionine. The mutations cause the proteins to become overactive. Cells grow rapidly, causing tumors to form. Mutations in *RET* cause Hirschsprung disease, the disorder that causes severe constipation, and is often seen in people with MEN type 2. *RET* is inherited in an autosomal dominant pattern and is located on the long arm (q) of chromosome 10 at position 11.2.

What Is the Treatment for Multiple Endocrine Neoplasia?

Depending on the gland, the option of surgery to remove the malfunctioning gland is used in some cases. A drug called bromocriptine may help reduce tumors of the pituitary gland. Removal of the parathyroid gland is always possible, but because it is so essential to regulate calcium, it is done as a last resort. Hormone replacement therapy is used when entire glands are removed or do not produce enough hormones. Recurrent tumors may develop.

Further Reading

"Multiple Endocrine Neoplasia." 2011. M. D. Anderson Cancer Center. University of Texas. http://www.mdanderson.org/patient-and-cancer-information/cancer-information/cancer-types/multiple-endocrine-neoplasia/index.html. Accessed 5/23/12.

"Multiple Endocrine Neoplasia (MEN) I." 2011. MedlinePlus. National Institutes of Health (U.S.). http://www.nlm.nih.gov/medlineplus/ency/article/000398.htm. Accessed 5/23/12.

"Multiple Endocrine Neoplasia Type 1." 2009. National Endocrine and Metabolic Diseases Information Services. U.S. Department of Health and Human Services. http://endocrine.niddk.nih.gov/pubs/men1/men1.htm. Accessed 5/23/12.

Multiple Epiphyseal Dysplasia

Prevalence   About 1 in 10,000; incidence of recessive type unknown; may be more common because people with mild symptoms are never diagnosed

Other Names EDM1; EDM2; EDM3; EDM4; EDM5; epiphyseal dysplasia, Fairbank type; epiphyseal dysplasia, multiple, 1; epiphyseal dysplasia, multiple, 2; epiphyseal dysplasia, multiple, 3; epiphyseal dysplasia, multiple, 4; epiphyseal dysplasia, multiple, 5; epiphyseal dysplasia, ribbing type; MED; multiple epiphyseal dysplasia, autosomal dominant; multiple epiphyseal dysplasia, autosomal recessive; rMED

Attention to bone irregularities and the field of orthopedics began to develop late in the nineteenth century. Doctors became aware that bone and cartilage made up the skeletal system, and also that as a child ages, cartilage in certain areas become bone. That process is called ossification. The process happens at the end of certain bones called the epiphysis. The epiphyses are the areas where cartilage turns to bone. The cartilage that is left protects the ends of the bones. The word "epiphysis" comes from Greek words: *epi*, meaning "upon," and *physis*, meaning "to grow." The word dysplasia comes from two Greek words: *dys*, meaning "with difficulty," and *plas*, meaning "to form."

Radiology and X-rays were just coming into prominence in 1935 when Thomas Fairbank described a patient with an irregular pattern of bone forming at the end of many areas of the bones. In 1947 he coined the term *dysplasia epiphysealis multiplex* and wrote about the clinical and radiological features in a journal. Physicians today drop the Latin name in favor of multiple epiphyseal dysplasia.

What Is Multiple Epiphyseal Dysplasia?

Multiple epiphyseal dysplasia (MED) is a condition in which the cartilage at the end of the bones does not form properly, and this malformation occurs in many of the epiphyses.

Children with MED have some or all of the following symptoms:

• First noted in late childhood
• Pain in joint areas, especially elbows, knees and hips
• Problems walking, perhaps with an awkward gait
• Tiredness after exercise
• Shoulder pain

Two types of MED are noted and are distinguished by their pattern of inheritance. There are both dominant and recessive types, and both can have relatively mild symptoms. Most people are of normal height, although a few have relatively short stature. Some cases are not diagnosed until adulthood.

The recessive type is characterized by dysplasia in the hands, feet, and knees, and usually the presence of scoliosis or curvature of the spine. About half of the children are born with at least one kind of the following malformations:

• Clubfoot in which the foot turns inward
• Cleft palate or a hole in the roof of the mouth
• Curved fingers or toes, a condition called clinodactyly
• Swollen ears
• A double-layered kneecap
• Height generally normal after puberty

The autosomal dominant MED type is probably the most common. Although it has some symptoms in common with the recessive types, the following differences may exist:

- Irregular bone formation, mostly in hips and knees
- Spine normal with only a few irregularities
- Short stature in adults or in low normal range
- Short limbs relative to trunks
- Limited movement at large joints, especially the elbows
- Early-onset osteoarthritis

What Are the Genetic Causes of Multiple Epiphyseal Dysplasia?

Mutations in six different genes cause multiple epiphyseal dysplasia. The genes are *COMP*, *MATN3*, *COL9A1*, *COL9A2*, *COL9A3*, and *SLC26A2*. All the genes instruct for proteins that are found in the epiphyseal area and interact in special ways with other proteins to convert cartilage to bone.

COMP

The most common causes of MED are mutations in the *COMP* gene, or the "cartilage oligomeric matrix protein" gene. Normally, *COMP* instructs for the COMP protein, found in the extracellular matrix between cells. COMP protein is located in the spaces around the cells of the ligaments, tendons, and the cartilage cells where bone is formed. COMP protein plays a major part in the process of bone formation called osteogenesis. COMP protein may also play a role in cell division and telling cells when to self-destruct—a process known as apoptosis. COMP protein binds to calcium, an element essential for cartilage formation.

   About 20 mutations in *COMP* cause MEP. Most mutations result from changes in only one protein building block or from small additions or deletions in the gene. The mutations may cause an improper folding of the COMP protein in the endo-plasmic reticulum, which is a series of mazes in the cells that are involved in pro-tein processing and transport. If the COMP protein is abnormal, it cannot move through the endoplasmic reticulum, causing it to accumulate and enlarge. This important part of the cartilage- building cell—called a chondrocyte—becomes so large that it cannot function, so it dies. The death of the cartilage-building cells keeps the long bones from growing, causing the short stature and possibly early-onset osteoarthritis. *COMP* is inherited in an autosomal dominant pattern and is located on the short arm (p) of chromosome 19 at position 13.1.

MATN3

About 10% of the cases of MEP are caused by mutations in the *MATN3* gene, or "matrilin 3" gene. Normally, *MATN3* instructs for a protein called matrilin-3.This protein is also found in the extracellular matrix, a mixture of proteins surrounding the cells that make the ligaments, tendons, and cartilage-forming cells. *MATN3* may also instruct for the organization of collagen, which provides strength and support to the cartilage. Matrilin-3 interacts with several other substances including COMP protein, Type II collagen, and Type I collagen, which are essential for the formation of cartilage and bone.

About 14 mutations in *MATN3* cause a mild form of MED. The change in one amino acid with tryptophan replacing arginine accounts for about 40% of the mutations in *MATN3*. The way a protein folds is critical in its formation. Mutations keep the protein matrilin-3 from folding properly. Instead of moving through the corridors of endoplasmic reticulum, it stops and accumulates. The cell enlarges and then dies. It is believed that the death of the chondrocytes keep the long bones of the person from growing, accounting for the short height. *MATN3* is inherited in an autosomal dominant pattern and is located on the short arm (p) of chromosome 2 at position 24-p23.

COL9A1

A third gene that is responsible for MED is *COL9A1*, or the "collagen, type I, alpha 1" gene. *COL9A1* instructs for making part of a molecule called type IX collagen. Type IX is an important component of cartilage and also adds strength and support to skin, bone, tendons, and ligaments. Type IX collagen consists of the following three important proteins made from three separate genes:

- *COL9A1* makes the α1(IX) protein chain. Only one mutation in *COL9A1* or "collagen, type IX, alpha 1" gene has been found. This mutation is called a splice-site, which means the addition of a single amino acid causes a deletion in several of amino acids in the protein. The single change disrupts the formation of the protein in several building blocks in the α1(IX) chain, causing the symptoms of MED. *COL9A1* is inherited in an autosomal dominant pattern and is located on the long arm (q) of chromosome 6 at position 12-q14.

- *COL9A2* makes the α2(IX) protein chain. *COL9A2*, or the "collagen, type IX, alpha2" gene, has about five mutations that disrupt the α2(IX) chain. This gene is inherited in an autosomal dominant pattern and is located on the short arm (p) of chromosome 1 at position 33-p32.

- *COL9A3*, or the "collagen, type IX, alpha 3" gene, makes the α3(IX) protein chain. About three mutations disrupt the α3(IX) chain. The gene is inherited in an autosomal dominant pattern and is located on the long arm (q) of chromosome 20 at position 13.3.

These twisted chains form a flexible molecule that is then closely associated with the more rigid type II collagen. Type IX collagen forms a bridge to other collagen components. This collagen also interacts with proteins from *COMP* and *MATN3*.

SLC26A2

The last gene, *SLC26A2*, or "solute carrier family 26 (sulfate transporter), member 2," is different from the other dominant genes. It is inherited in a recessive pattern.

Normally, *SLC26A2* instructs for a protein that carries sulfate ions across the cell membranes. Active in many body tissues, the protein is especially seen in developing cartilage. Cartilage must have sulfate molecules to build up proteoglycans, which have several sugars that give cartilage its rubbery, gel-like appearance. Sulfate ions are critical to the construction of cartilage.

Mutations in *SLC26A2* cause MED. Usually only one exchange disrupts the making of the protein. The mutations appear to have less serious effects on cartilage and bone formation than the other genes. This form of MED is relatively mild. This is inherited in an autosomal recessive pattern, which means that both parents must carry the gene and pass on to the offspring. *SLC26A2* is located on the long arm (q) of chromosome 5 at position 31-q34.

What Is the Treatment for Multiple Epiphyseal Dysplasia?

An evaluation by an orthopedic sturgeon for potential treatment of limb deformities is critical. Treatment of the symptoms includes pain management, and possible realignment includes an osteotomy (bone surgery) to limit pain and osteoarthritis. The person should have psychosocial support for short stature and training and education under the Individuals with Disabilities Education Act.

Further Reading

Briggs, Michael D.; Michael J. Wright; and Geert R. Mortier. 2011. "Multiple Epiphyseal Dysplasia." *GeneReviews*. http://www.ncbi.nlm.nih.gov/books/NBK1123. Accessed 5/23/12.

"Multiple Epiphyseal Dysplasia." 2011. Medscape. http://emedicine.medscape.com/article/1259038-overview#showall. Accessed 5/23/12.

"Multiple Epiphyseal Dysplasia." 2011. OrthoPreferred Duke Orthopaedics. http://www.wheelessonline.com/ortho/multiple_epiphyseal_dysplasia. Accessed 5/23/12.

Muscular Dystrophy. *See* Duchenne/Becker Muscular Dystrophy; Emery-Dreifuss Muscular Dystrophy (EDMD); Fascioscapulohumeral Muscular Dystrophy (FHSMD); Ullrich Congenital Muscular Dystrophy (UCMD)

Myopathy, Centronuclear

Prevalence	Centronuclear myopathy rare, with fewer than 50 families affected; autosomal dominant type more common than recessive

type; X-linked type or myotubular myopathy, about 1 in 50,000 newborn males worldwide:

Other Names   CNM, autosomal dominant myotubular myopathy autosomal recessive centronuclear myopathy; myotubular myopathy; myotubular myopathy 1; myopathy, myotubular; myotubular myopathy, X-linked; X-linked recessive centronuclear myopathy

The word myopathy comes from two Greek words: *myo*, meaning "muscle," and *path*, meaning "disease." The three forms of this disorder relate to the way they are inherited and the different genes involved. There is a type inherited in a dominant pattern, one in a recessive pattern, and one in an X-linked pattern.

What Is Centronuclear Myopathy?

Myopathies affect the large voluntary skeletal muscles, which are used for movement. The centronuclear myopathies are a group of congenital disorders that are distinguished from others because of the way the cells appear under the microscope. In this disorder, the nuclei of the cells are located in the center of the muscle cell instead of the normal location at the edge.

The following three different types exist:

- *Autosomal dominant centronuclear myopathy*: People with this type usually develop muscle weakness during exercise during their teen years or early adulthood. The condition is slowly progressive. By mid-to-late adulthood these people may need a wheelchair or other assistive devices. Some other symptoms may occur such as eye weakness, loss of nerve function, or neuropathy. Some people may be more severely affected, experiencing muscle weakness from infancy.

- *Autosomal recessive centronuclear myopathy*: People with this type have progressive muscle weakness, which begins in infancy or early childhood. They also may have a number of abnormalities: foot disorders, high arch in the roof of the mouth, scoliosis, breathing problems, weak heart muscles, and eye disorders.

- *X-linked myotubular myopathy*: This form of the disorder is the most severe form, occurring at birth or in early infancy. Motor skills are impaired. One of the most serious symptoms is that breathing may be affected, and the child cannot breath on his own.

What Are the Genetic Causes of Centronuclear Myopathy?

The different types of centronuclear myopathy are caused by three different genes, inherited in different ways.

Dominant Form

Mutations in the *DNM2* gene or the "dynamin 2" gene cause the dominant form. Normally, *DNM2* instructs for making the protein dynamin 2. Dynamin 2 makes tube-like structures called microtubules, which make up the cell's framework or cytoskeleton. The cytoskeleton determines cell shape and organizes the cell. Muscle cells are long and rod-shaped, with the nucleus located near the edge of the cell. This protein interacts with the BIN1 protein to form the cell membrane structures.

About 14 mutations of *DNM2* cause the dominant form of centronuclear myopathy. Most of the changes are in a single building block. These mutations lead to a change in dynamin 2 and disrupt its interaction with other cell structures, especially the microtubules. *DMN2* is located on the short arm (p) of chromosome 19 at position 13.2.

Recessive Form

Mutations in the *BIN1* gene, or the "bridging integrator 1" gene, cause the recessive form of the disorder. Normally, *BIN1* instructs for a protein found in body tissues that interacts with other proteins. The BIN1 protein is an important molecule for transporting materials from the cell surface into the center of the cell for cell processes. The protein also may act as a tumor suppressor protein. The BIN1 protein is expressed in muscle cells and is involved in forming structures in the membrane of the cells called transverse tubules or T tubules. The structures are found only in the muscle cells and are responsible for contracting and relaxing the muscles.

The three mutations in *BIN1* cause the recessive form of centronuclear myopathy. The mutations are generally a result of a change in one building block. The change does not allow the BIN1 protein to interact with other proteins, thereby disrupting the formation of the T tubules. The abnormalities in the T tubules disrupt muscle function and cause the signs and symptoms of this disorder. *BIN1* is located on the long arm (q) of chromosome 2 at position 14.

X-Linked Myotubular Myopathy

The *MTM1* gene, or "myotubularin 1," causes linked myotubular myopathy. Normally, *MTM1* instructs for an enzyme called myotubularin. This enzyme develops and maintains the muscle by acting as a phosphatase. A phosphatase removes groups of oxygen and phosphorus atoms from other molecules. This process allows molecules to move in transporting other substances within the cells.

Over 200 mutations in *MTM1* are related to the X-linked condition. Most of the gene mutations involve a change in only one amino acid building block, resulting in a dysfunctional or shortened enzyme. Individuals with the mild form make some of the needed enzymes; the more severe cases are without any of the enzymes. *MTM1* is located on the long arm (q) of the X chromosome at position 28.

What Is the Treatment for Myopathy?

Treatment of all three types of disorders is symptomatic. For the more severe X-linked type, the person requires a team of specialists including a pulmonologist, physical therapist, and rehabilitation specialist. The other two types are treated for symptoms requiring assistive devices such as wheelchairs.

Further Reading

Das, Soma, James Dowling, and Christopher Pierson. 2011."X-Linked Centronuclear Myopathy." *GeneReviews*. http://www.ncbi.nlm.nih.gov/books/NBK1432. Accessed 5/23/12.

"Myotubular Myopathy." 2009. Family Village. Wisconsin Education. http://www.familyvillage.wisc.edu/lib_myot.htm. Accessed 5/23/12.

Myotubular Myopathy Resource Group. 2001. http://www.mtmrg.org. Accessed 5/23/12.

Myotonic Dystrophy

Prevalence   Affects about 1 in 8,000 people worldwide; type 1 most common; type 2 prevalent among people in Germany and Finland

Other Names   dystrophia myotonica; myotonia atrophica; myotonia dystrophica

The muscular dystrophies are a group of disorder of the muscle. The word "dystrophy" comes from two Greek words: *dys*, meaning "with difficulty," and *troph*, meaning "nourish." The muscles have difficulty getting their supply of materials for sustenance and thus waste away. The word "myotonic" comes from two Greek roots: *myo*, meaning "muscle," and *tonic*, meaning "tension" or "spasm." The condition was first recognized by Steinert, who called it by a Greek name—*dystrophia myotonica*.

What Is Myotonic Dystrophy?

Myotonic dystrophy (DM) is one of the muscular dystrophies. It affects the muscles with both weakness and spasms, as well as other parts of the body. DM differs from other muscular dystrophies in the myotonic symptom. Other forms of muscular dystrophy do not have the muscular spasms. DM also most frequently begins in adulthood and usually the person with milder forms learns to cope. However, an infant type usually born to parents with DM can be very serious. The severity of the disorder varies greatly even among affected families.

The two main characteristics of DM are progressive muscle wasting and weakness and muscle contractions. When the person uses the certain muscles, the muscles appear to lock in place and the individual cannot relax them. Simple activities become an issue. For example, a person may put his hands on a doorknob to open it. The muscles freeze and he cannot release the grip. The following other parts of the body may also be affected:

- Slurred speech caused by locking of the jaw muscles
- Muscle pain

N

NAGA Deficiency. *See* Schindler Disease

Nail-Patella Syndrome

Prevalence   About 1 in 50,000

Other Names   Fong disease; hereditary onycho-osteodysplasia; hereditary osteo-onychodysplasias, Osterreicher syndrome; pelvic horn syndrome; Turner-Kieser syndrome

In 1920, the French physician Chatelain described a condition in which seemingly unrelated symptoms occurred. The condition involved an abnormality in the nails as well as bone disorders. Many years later, Little documented that this disorder appears to be inherited. They called the condition hereditary osteo-onychodysplasia (HOOD), but later it became known by a simpler term, nail-patella syndrome.

What Is Nail-Patella Syndrome?

Nail-patella syndrome (NPS) is a disorder of the nails, patella (kneecap), elbows, and pelvis. These seemingly unrelated abnormalities may vary in severity even among those of the same family. Following are the symptoms of nail-patella syndrome:

- Nail disorders: Almost all people with this syndrome have abnormal nails. There may be no nails, or they may appear as split, pitted, ridged, or discolored. The area at the base of the nail may have a triangular shape. Also, the fingernails are usually more affected than the toenails.

- Kneecap or patella disorders: The kneecap may be very small with an irregular shape. A dislocation of the patella is common.

- Elbows: The elbows may lock, and the individual cannot extend the arms or turn the palms up with their elbows straight. Elbows may angle outward or be webbed.

- Hips: The ilium is the top part of the pelvic girdle. Many people with this syndrome have horn-like outgrowths at the top of the iliac bones. These horns appear as projections on X-rays but cause no symptoms.

- Kidneys: Some individuals develop renal disease, which can lead to end-stage renal disease.

- Eyes: A tendency toward pressure building in the fluid of the eyes leading to glaucoma is common.

What Is the Genetic Cause of Nail-Patella Syndrome?

Mutations in the *LMX1B* gene, or the "LIM homeobox transcription factor 1, beta" gene, causes nail-patella syndrome. Normally, *LMX1B* instructs for a protein that binds to certain areas of DNA and controls the activity of several genes. This protein is called a transcription factor. The *LMX1B* protein is essential for the normal development of the limbs, kidneys, and eyes during the embryo's development.

About 145 mutations in *LMX1B* cause nail-patella syndrome. Most mutations produce short, nonfunctioning proteins. Some mutations involve only one substitution of one amino acid for another. When the mutations occur, they disrupt the protein's ability to bind to DNA and to do its work as a transcription factor. The genes that they regulate are no longer functioning, which lead to the signs and symptoms of nail-patella syndrome. *LMX1B* is inherited in a dominant pattern and is located on the long arm (q) of chromosome 9 at position 34.

What Is the Treatment for Nail-Patella Syndrome?

Obviously this syndrome requires annual monitoring for development of serious kidney disease and glaucoma. Treating many of the symptoms is possible. Orthopedic problems can be helped with painkillers, splinting, bracing, or surgery. The person must avoid the chronic use of NSAIDs for pain because of the possible damage to kidneys.

Further Reading

"Dermatologic Manifestations of Nail-Patella Syndrome." 2011. Medscape. http://emedicine.medscape.com/article/1106294-overview. Accessed 7/4/11.

"*LMX1B*." 2011, Genetics Home Reference. National Library of Medicine (U.S.). http://ghr.nlm.nih.gov/gene/LMX1B. Accessed 5/23/12.

"Nail-Patella Syndrome." 2011, Genetics Home Reference. National Library of Medicine (U.S.). http://ghr.nlm.nih.gov/condition/nail-patella-syndrome. Accessed 5/23/12.

Nance-Insley Syndrome

Prevalence Rare; only a few families worldwide

Other Names chondrodystrophy with sensorineural deafness; Insley-Astley syndrome; mega-epiphyseal dwarfism; Nance-Sweeney chondrodysplasia; oto-spondylo-megaepiphyseal dysplasia

Although disorders in which individuals were extremely short have been written about throughout history, in 1970, Nance and Sweeney observed a condition that added many other symptoms. The height of the person was 51 inches, but the patient also had a sunken nasal bridge, fusion of the carpal bones, and progressive deafness. The same symptoms were present in siblings and female cousins. In 1974, Insley and Astley described two affected sisters. Eventually, the condition was called Nance-Insley syndrome, although many health care workers refer to it as otospondylomegaepiphyseal dysplasia (OSMED).

What Is Nance-Insley Syndrome?

Nance-Insley syndrome, or otospondylomegaepiphyseal dysplasia (OSMED), is a disorder of the skeletal system combined with distinct facial appearance and progressive hearing loss. Otospondylomegaepiphyseal dysplasia is best described with the acronym OSMED:

- O for *oto*, the Greek root meaning ear. Severe progressive hearing loss, especially with high tones, is part of this disorder.
- S for *spondyl*, the Greek root for spine. People with the condition experience joint and back pain.
- MED for megaepiphyseal dysplasia. The epiphyses are the ends of the long bones in the arms and legs. These areas are where cartilage present at birth becomes bone as one grows. Some cartilage remains to protect the ends of the bones. In this condition, the ends of the bones are enlarged and do not grow normally. People are shorter than normal because their leg bones do not grow.

The distinct facial appearance includes a flat nose, protruding eyes, a turned-up nose, and a small lower jaw. All affected infants are born with a hole in the roof of the mouth or cleft palate. Pain, limited joint movement, and arthritis also may be symptoms of the disorder and persist throughout the life of the individual. Another milder skeletal disorder, Weissenbacher-Zweymüller syndrome, is similar to this syndrome.

What Is the Genetic Cause of Nance-Insley Syndrome?

Mutations in the *COL11A2* gene, known also as the "collagen, type XI, alpha 2" gene causes Nance-Insley syndrome. Normally, *COL11A2* instructs for making a protein that is part of type XI collagen called the pro-alpha(XI) chain. This collagen chain adds strength and structure to cartilage, the tough, rubbery material that covers the ends of bones and is present in the ears and nose. Cartilage makes up most of the skeletal system at birth but then is converted to bone as one grows. Type XI collagen is also found in the clear gel that fills the eyeball, the inner ear, and area between the discs in the vertebrae.

To create the strong formation of XI collagen, the pro-alpha2(XI) chain joins with two other chains to form a rope-like chain that is processed by enzymes to become long fibrils that crosslink in spaces around the cells. Proper formation and arrangement of these fibrils is critical for the normal structure of cartilage.

About 10 mutations in *COL11A2* cause Nance-Insley syndrome. Most mutations disrupt the production and assembly of the pro-alpha2(XI) chain. The loss of the chain function leads to the characteristic signs of OSMED. *COL11A2* is inherited in a recessive pattern and is found on the short arm (p) of chromosome 6 at position 21.3.

What Is the Treatment for Nance-Insley Syndrome?

Because this syndrome is a genetic disorder, there is no cure. Treating the symptoms with medication can help the back pain and arthritis. People with hearing loss may have a cochlear implant. Surgery can correct cleft palate.

Further Reading

"Nance-Insley Syndrome." 2011. RightDiagnosis.com. http://www.rightdiagnosis.com/medical/nance_insley_syndrome.htm. Accessed 5/23/12.

"Otospondylomegaepiphyseal Dysplasia." 2011. Medic8. http://www.medic8.com/genetics/otospondylomegaepiphyseal-dysplasia.htm. Accessed 5/23/12.

"Otospondylomegaepiphyseal Dysplasia." 2011. OMIM Online Mendelian Inheritance in Man. http://www.omim.org/215150. Accessed 5/23/12.

Narcolepsy

Prevalence   Affects 1 in 2,000 people in United States and Western Europe; most common in Japan 1 in 600; mild cases may be undiagnosed

Other Names   Gelineau syndrome; narcoleptic syndrome

Jean-Baptiste Edouard Gelineau, a French physician, was fascinated by neurological conditions like epilepsy. While studying epilepsy, he encountered an unusual disorder in which a person could just be talking with another and go to sleep. He began to study the condition and in 1880 named it "narcolepsy." He used two

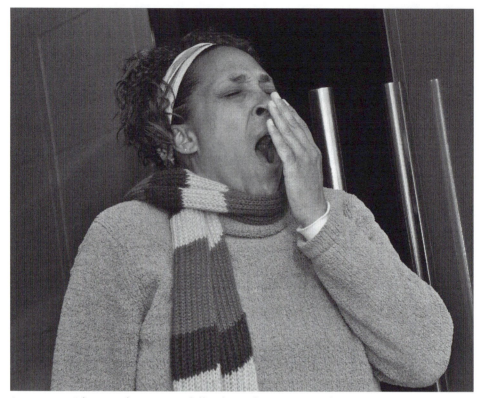

A person with narcolepsy may fall asleep during normal conversation.
(rSnapshotPhotos/Shutterstock)

Greek root words: *nar*, meaning "numbness" or "stupor," and *lepsis*, meaning "seizure" or "attack." The name "narcolepsy" has been in use since the late 1800s for this condition.

What Is Narcolepsy?

Narcolepsy is a neurological sleep disorder. The brain has a problem regulating normal sleep-wake cycles. At times during the day, the person with narcolepsy will just fall asleep. The "sleep attacks" may last from several seconds to minutes and can occur at unusual times such as during a meal or a conversation. When the person wakes, he or she may feel refreshed for a short period of time. This excessive daytime sleepiness is referred to as EDS.

The following additional symptoms of narcolepsy may occur:

- Extreme urge to sleep: The person feels always tired and slow even if they do not completely nod off.
- REM stage almost immediately: The REM is the deepest stage of sleep, where one dreams. People with narcolepsy experience with REM stage within 10 minutes, in contrast to most people who experience REM in about 90 minutes.

- Cataplexy: This condition occurs with most people with narcolepsy. When persons with narcolepsy experience strong emotion such as laughing, anger, or surprise, they may lose muscle tone and become like a rag doll. They may slump over and fall down. Although the attacks may last only momentarily, they are very dangerous. There are a few people who do not exhibit cataplexy, which have led researchers to think there are possibly two types of narcolepsy: one with cataplexy, and another without.

- Problems with nighttime sleep: Most of the individuals do not sleep for more than a few hours overnight. They often have hallucinations when waking up or falling asleep. Other sleep anomalies include vivid dreams, which may be very disturbing, and sleep paralysis. With sleep paralysis, the person may not be able to move during sleep and may not be able to speak or walk for a short period after waking.

- Problems throughout life: Most people with narcolepsy battle the condition through life. Some may have only one or two of the symptoms

What Is the Genetic Cause of Narcolepsy?

Mutations in the *HLA-DQB1* gene, or the "histocompatibility complex, class II, DQ beta 1" gene, causes narcolepsy. Normally, *HLA-DQB1* instructs for a protein important in the immune system. *HLA-DQB1* is part of a family of genes called the HLA or leukocyte antigen complex. Like other complexes of the immune system, HLA aids the body in determining its own cells from invaders like bacteria and viruses. The HLA gene family is a member of a large groups of complexes called major histocompatibility complex, or MHC. The complex occurs in many species and is divided into three classes: I, II, and III. The *HLA-DQB1* gene is a member of MHC class II.

What does a gene that controls the immune system have to do with narcolepsy? It appears that narcolepsy is related to a malfunction of the immune system. One version of the *HLA-DQB1* gene is found in people with narcolepsy, especially those who also have catalepsy. How this exactly fits into the symptoms is unknown. Sleep abnormalities probably result from the loss of neurons in the hypothalamus, part of the brain. This part produces hypocretins or orexins, which may control the sleep-wake cycle. An abnormality in the hypothalamus may trigger a problem within the neurons producing hypocretins. Certain environmental causes, such as bacterial or viral infections, may be responsible for developing the disorder. *HLA-DQB1* is located on the short arm (p) of chromosome 6 at position 21.3.

What Is the Treatment for Narcolepsy?

Treatment of this mysterious condition is based on symptoms and how the person responds. Stimulant medications are the mainstay of treatment, and these must be adjusted according to the way the person responds.

Further Reading

"About Narcolepsy." 2010. Narcolepsy Network. http://www.narcolepsynetwork.org/about-narcolepsy. Accessed 5/23/12.

"NINDS Narcolepsy Information Page." 2011. National Institute of Neurological Disorders and Stroke (U.S.). http://www.ninds.nih.gov/disorders/narcolepsy/narcolepsy.htm. Accessed 5/23/12.

Narcoleptic Syndrome. *See* Narcolepsy

Nephrogenic Diabetes Insipidus

Prevalence   Rare; about 45 families affected in Dutch population; acquired form more common than hereditary form; studies from Quebec, Canada, show 8.8 in 1,000,000 males

Other Names   ADH-resistant diabetes insipidus; congenital nephrogenic diabetes; diabetes insipidus, nephrogenic; diabetes insipidus renalis; insipidus; NDI; vasopressin-resistant diabetes insipidus

Diabetes is a common disease, whose name is mentioned in advertisements and in public and private discussions. The diabetes that most people refer to is diabetes mellitus or "sugar diabetes." This disease is characterized by the presence of glucose, a simple sugar, in the blood and by major effects on almost every body system. Diabetes insipidus is a completely different disease with different symptoms and should not be confused with the more common disease.

Williams and Henry coined the term "nephrogenic diabetes insipidus" in 1947.

The name "nephrogenic diabetes insipidus" is derived from Greek and Latin root words. The word "nephrogenic" comes from the root words *nephr*, meaning "kidney," and *gen*, meaning "origin." Diabetes literally means "passing through a siphon," inferring excessive excretion of urine. When people pass lots of urine, it is known as polyuria. In both diabetes mellitus and diabetes insipidus, the individuals pass lots of urine; this is the common element of their names. The word "insipidus" comes from Latin words: *in*, meaning "not," and *sapere*, meaning "to taste." In contrast to diabetes mellitus, the urine of the person with diabetes insipidus does not have sugar or elevated glucose content.

What Is Nephrogenic Diabetes Insipidus?

Nephrogenic diabetes insipidus (NDI) is a disorder of water balance. The person with this disorder must go to the bathroom often to eliminate urine. The problem originates with production of a hormone in the brain called the vasopressin or anti-diuretic hormone (ADH), which directs the kidneys to concentrate urine by putting some of the water in the bloodstream and releasing it in the urine. In NDI, because of genetically acquired malformations, the kidneys do not respond properly to the hormone. People with NDI are constantly passing urine and are constantly thirsty. If they do not drink enough water, especially in hot weather, they can become dehydrated, creating a host of difficulties.

NDI can be hereditary of acquired. Mutations in the genes cause the hereditary form, and the following signs usually appear in a few months after birth:

- Poor eating with a failure to gain weight
- Irritability
- Fever
- Diarrhea
- Vomiting
- Symptoms of dehydration leading to delayed development
- Damage to bladder and kidney, leading to kidney failure
- Pain

With proper treatment, individuals usually have few complications and a normal lifespan.

The acquired form is more common of the two. The condition is brought on by use of certain drugs such as lithium. Chronic diseases or an obstruction in the urinary tract can cause the disorder, as well of low levels of potassium and high levels of calcium in the blood. The acquired form can occur at any time of life.

What Are the Genetic Causes of Nephrogenic Diabetes Insipidus?

Mutations in two genes—*AQP2* and *AVPR2*—cause the hereditary form of NDI. Both of the genes instruct for the proteins that determine how water is eliminated in the urine.

AQP2

Mutations in the *AQP2* gene, officially known as the "aquaporin 2 (collecting duct)" gene, cause NDI. Normally, *AQP2* instructs for making the protein aquaporin 2. This protein creates a canal to transport water molecules across cell membranes. Found in the collecting tubes in the kidneys, the protein enables the reabsorption of water from the kidneys into the bloodstream, thereby maintaining water balance. Control of these canals is enabled by a chemical messenger, a hormone called vasopressin or the antidiuretic hormone (ADH). If the person takes too little fluid or if fluid is lost through sweating of fever, the body calls for more ADH, which then creates the necessary water channels into the membranes of the collecting ducts in the kidneys. Less ADH is produced when the amount of fluid is adequate, the channels are removed, and less water is reabsorbed into the bloodstream.

About 40 mutations in *AQP2* are responsible for 90% of the NDI cases. Most of the mutations keep the aquaporin 2 protein from being folded properly, trapping it within the cells of the kidney. As a result, the cell membrane cannot transport water molecules. Kidneys are not able to respond to the ADH signals, disrupting the water reabsorption process. The problems with the balance of water cause the symptoms of NDI. The *APQ2* gene can be inherited in an autosomal recessive pattern or less commonly a dominant pattern and is located on the long arm (q) of chromosome 12 at position 12-q13.

AVPR2

Mutations in *AVPR2*, officially known as the "arginine vasopressin receptor 2" gene, cause NDI. Normally, *AVPR2* instructs for the protein vasopressin V2 receptor. This receptor works with the hormone ADH in the kidneys and is found in the collecting ducts of the kidneys. The two together trigger the reactions that control the water balance in the body. When the fluid is low, more ADH is produced. The hormone binds to the vasopressin V2 receptor, telling the kidneys to concentrate urine and reabsorb some of the water back into the bloodstream. When fluid intake is normal, less ADH is needed to put water into the bloodstream.

More than 200 mutations in *AVPR2* cause NDI. The mutations are the result of improper folding of the vasopressin V2 receptor protein into the proper three-dimensional shape. When the misshapen protein is trapped in the cell, it cannot get to the surface to interact with ADH. Without this process, the kidneys cannot respond, and the collecting ducts make an excessive amount of urine. *AVPR2* is inherited in a recessive pattern and is located on the long arm (q) of the X chromosome at position 28.

What Is the Treatment of Nephrogenic Diabetes Insipidus?

Treatment of the symptoms by a team is essential. This should include a nutritionist, pediatric nephrologist, and clinical geneticist. The person should have free access to drinking water and toilet facilities. Certain thiazide diuretics, such as hydrochlorothiazide, could be prescribed. Diets should restrict sodium and nonsteroid anti-inflammatory drugs. Individuals must have enough water.

Further Reading

Knoers, Nine. 2010. "Nephrogenic Diabetes Insipidus." *GeneReviews*. http://www.ncbi.nlm.nih.gov/books/NBK1177. Accessed 5/23/12.

NDI Foundation. http://www.ndif.org/public/pages/36-Welcome_to_NDI_Foundation. Accessed 5/23/12.

"Nephrogenic Diabetes Insipidus." 2011. Genetics Home Reference. National Library of Medicine (U.S.). http://ghr.nlm.nih.gov/condition/nephrogenic-diabetes-insipidus. Accessed 5/23/12.

Neurofibromatosis Type 1 (NF1)

Prevalence   Affects 1 in 3,000 to 4,000 people worldwide; the most common single gene disorder

Other Names   Neurofibromatosis 1; NF1; peripheral neurofibromatosis; Rechlinghausen disease, nerve; von Recklinghausen disease

In 1872, Friedrich Daniel von Recklinghausen, a German pathologist, became fascinated by a condition in which nerve cells and fibrous tissue were mingled together. He published a paper describing the condition as neurofibromatosis type 1. The term neurofibromatosis comes from Greek root words: *neuro*, meaning "nerve"; *fibro*, meaning "tissues that are string-like"; *oma*, meaning "growth" or "tumor"; and the suffix *osis*, meaning "condition of." At first the condition was named for the German scientist, but later became known as the term he invented—neurofibromatosis.

"Fibromatosis" is also used to describe a separate distinct condition known as neurofibromatosis type 2 or NF2, a disorder of the nerves in the ears.

What Is Neurofibromatosis Type 1 (NF1)?

Neurofibromatosis type 1 is a disorder of the nerves in the skin, brain, and other body parts. It is the most common disorder caused by a single gene. Although most of the signs and symptoms vary among people, almost all cases have multiple splotches called café-au-lait spots, which are present from infancy. These flat spots are darker than the rest of the skin and look like the name—the color of coffee with milk. The splotches grow in size and number as the child ages. Eventually, they will appear as freckles in the underarms and groin area.

As the child gets older, other symptoms appear. Following are the signs of neurofibromatosis 1:

- Neurofibromas: These noncancerous growths are located on top of the skin and under the skin. The growths occur in nerves anywhere in the body, especially near the spinal cord. Some people do get cancerous growths called malignant nerve sheath tumors (MPNST). Also, individuals with NF1 have a tendency to get other cancers, especially brain tumors and leukemia.

- Lisch nodules: Small nodules may occur in the colored part of iris of the eye. Generally, vision is not impaired unless the nodules begin to grow on the optic nerve leading from the eye to the brain. These growths are called optic gliomas and may or may not affect vision.

- High blood pressure.

- Short stature.

- Large head, a condition known as macrocephaly.

- Abnormal curvature of the spine or scoliosis.

- Seizures.
- Speech disorders.
- Normal intelligence: However, children with NF1 may have attention deficit disorders (ADHD) or learning disabilities.

Some other complications may arise that are the result of the effect on the blood vessels, which are related to the central and peripheral nervous system.

What Is the Genetic Cause of Neurofibromatosis Type 1 (NF1)?

Mutations in the *NF1* gene, or the "neurofibromin 1" gene, cause neurofibromatosis type 1 gene. Normally, *NF1* instructs for making the protein neurofibromin. Neurofibromin is found in many types of cells, especially nerve cells and cells that form the myelin sheath. These cells are called oligodendrocytes and Schwann cells. Neurofibromin is also a tumor suppressor, which keeps cells from wildly dividing. It also appears to turn off another protein called *ras* that causes cells to grow and divide.

More than 1,000 *NF1* mutations cause NF1. Most of the mutations are related to a specific family. Most mutations result in an extremely short version of the protein. The short version disrupts the normal functions of inhibiting growth. If the mutations are in the Schwann cells, the loss of the protein causes the noncancerous tumors to form. *NF1* is inherited in an autosomal dominant pattern and is located on the long arm (q) of chromosome 17 at position 11.2.

What Is the Treatment for Neurofibromatosis Type 1 (NF1)?

At this time NF1 has no cure, but the symptoms can be monitored and managed. A person with NF1 should have an eye examination annually. Neurofibromas can be surgically removed. Children with behavioral and learning disorders have responded to lovastatin. Support from psychologists and counselors may be useful. People with the malignant cancers have responded to a therapy that targets MPNST, called pirfenidone.

Further Reading

"Neurofibromatosis Type 1." 2011. Cancer.Net. American Society of Clinical Oncologists. http://www.cancer.net/patient/Cancer+Types/Neurofibromatosis+Type+1. Accessed 5/23/12.

"Neurofibromatosis Type 1." 2011. Genetics Home Reference. National Library of Medicine (U.S.). http://ghr.nlm.nih.gov/condition/neurofibromatosis-type-1. Accessed 5/23/12.

"Type 1 Neurofibromatosis." 2011. Medscape. http://emedicine.medscape.com/article/1177266-overview. Accessed 5/23/12.

Neurofibromatosis Type 1–Like Syndrome. *See* Legius Syndrome

---

Neurofibromatosis Type 2 (NF2)

Prevalence   Affects 1 in 25,000

Other Names BANF; bilateral acoustic neurofibromatosis; central neurofibromatosis; familial acoustic neuromas; NF2; schwannoma, acoustic,
bilateral

Neurofibromatosis type 2 (NF2) is often confused with neurofibromatosis 1 (NF1).
NF2 is caused by a different gene on a different chromosome and has very
different clinical manifestations. People with NF2 do not have learning difficulties
or behavioral issues that may be seen in NF1. Neither do people with NF2 have the
brown splotches or café-au-lait spots. The tumors that appear in the ears seldom
become cancerous. One problem that sometimes leads to confusion is that the
origin of both kinds of tumors is located at the spinal roots.

What Is Neurofibromatosis Type 2 (NF2)?

Neurofibromatosis type 2 is a disorder of tumors that may grow in certain areas of
the nervous system. The most common tumors are called vestibular schwannomas
or acoustic neuromas. The Schwann cells cover the sheaths of the nerves and
provide protection and insulation. A schwannoma is a tumor that grows on the
Schwann cells. The vestibular schwannoma develops on the auditory nerve that
carries impulses from the sensory receptors in the ear to the brain. Other nerve
cells in the brain and spinal cord may also be affected.

Following are some of the symptoms of neurofibromatosis 2, which usually
occur before the age of 20:

- Hearing loss
- Ringing in the ears
- Problems with balance
- Clouding of the lenses of both eyes, which often begins in childhood
- Tumors in other areas of brain and spinal cord
- Numbness in arms and legs
- Fluid buildup in the brain

What Is the Genetic Cause of Neurofibromatosis Type 2 (NF2)?

Mutations in the *NF2* gene, or "neurofibromin 2 (merlin)" gene, cause neurofibromatosis 2. Normally, *NF2* instructs for a protein called merlin, sometime called schwannomin. The protein is abundant in the specialized cells called Schwann cells that are located in the myelin sheath that wrap around nerves. Merlin appears to work with the cytoskeleton or framework of the cell to control the shape of the cells. Merlin is also involved in movement and communication between the cells and may function as a tumor suppressor to keep cells from growing wildly.

The *NF2* gene has more than 200 mutations that cause neurofibromatosis 2. About 90% of the mutations result in a short version of the merlin. The short version disrupts the function of the protein, making it unable to suppress growths on the Schwann cells and thus causing them to multiply. *NF2* is inherited in an autosomal dominant pattern and is located on the long arm (q) of chromosome 22 at position 12.2.

What Is the Treatment for Neurofibromatosis Type 2 (NF2)?

Early diagnosis is important. Therapy for the vestibular schwannomas usually requires surgery. Strategies using lip-reading skills, sign language, and hearing aids may be used. Also, certain people may benefit from a cochlear implant.

Further Reading

Kugler, Mary. 2008. "Neurofibromatosis Type 2." About.com Rare Diseases. http://rarediseases.about.com/od/neurofibromatosis/a/neurofibroma2.htm. Accessed 5/23/12.

"Neurofibromatosis Type 2." 2011. Mayo Clinic. http://www.mayoclinic.org/neurofibromatosis-type-2. Accessed 5/23/12.

"Neurofibromatosis Type 2." 2011. Medscape. http://emedicine.medscape.com/article/1178283-overview. Accessed 5/23/12.

Nevoid Basal Cell Carcinoma Syndrome

Prevalence	About 1 in 56,000–64,000; more than a million new cases of basal cell carcinoma diagnosed each year, but fewer than 1% related to nevoid basal cell carcinoma
Other Names	Gorlin syndrome; basal cell nevus syndrome; BCNS; Gorlin-Goltz syndrome; NBCCS

In 1960, researcher Robert Gorlin described a condition in which the person developed not only basal cell carcinomas, the most common kind of skin cancer, but also had distinct facial appearance because of jaw cysts, and deformed ribs. He wrote about his findings in an article in the *New England Journal of Medicine* called "Multiple Nevoid Basal-Cell Epithelioma, Jaw Cyst, and Bifib Rib: A Syndrome." Sometimes the disorder is referred to as Gorlin syndrome.

What Is Nevoid Basal Cell Carcinoma Syndrome?

Nevoid basal cell carcinoma syndrome (NBCC) is a type of skin cancer that affects many parts of the body. The term comes from the description of the appearance of the skin. The word nevoid means "like a nevus, birthmark, mole, or spot." A basal cell carcinoma is a slow-growing neoplasm or growth derived from the basal cells or outer layer of the epidermis of the skin. Individuals with NBCC start developing basal cell carcinomas on the face, chest, and back during adolescence or early adulthood. Although a few people will not develop the cancers, some will develop thousands. Like most skin cancers, they are more common among those with very light skin.

Other symptoms are associated with this syndrome:

- Noncancerous or benign tumors of the jaw, called keratocystic odontogenic tumors. These tumors begin in adolescence and recur until about age 30.
- Swelling of the face because of the tumors.
- Tooth displacement because of the jaw tumors.
- High risk of developing a brain tumor, called a medulloblastoma, in childhood.
- Benign tumors or fibromas in the heart. These growths do not cause symptoms but may block blood flow, causing irregular heartbeat.
- For women, increased risk for fibromas on the ovaries.
- Skin pits in soles of feet and palms of hands.
- Large size head.
- Skeletal abnormalities, especially of the ribs.

What Is the Genetic Cause of Nevoid Basal Cell Carcinoma Syndrome?

Mutations in *PTCH1*, officially called the "patched 1" gene, cause nevoid basal cell carcinoma syndrome. Normally, *PTCH1* instructs for the Patched-1 protein, a receptor protein. These proteins have special receptor sites that lock into other proteins called ligands. They bind together like a key fits into a lock. Together with the ligand protein called Sonic Hedgehog, the ligands and receptors call signals that affect cell development and function, especially in the early years. The two create a pathway for cell growth, cell differentiation, and body formation. By itself, Patched-1 keeps cells from growing and dividing. When Sonic Hedgehog is present, Patch-1 stops suppressing cell growth. The gene controlling the process

PTCH1 is considered a tumor suppressor gene because of its role in keeping cells from growing wildly.

About 150 mutations in *PTCH1* cause NBCC. The mutation creates an abnormal version of Patched-1 or causes it to be missing completely. Because of the abnormal or missing protein, the receptor cannot bind to the ligand to suppress cell growth and regulation. The cells grow rapidly to form the tumors and other symptoms of NBCC. *PTCH1* is inherited in an autosomal dominant pattern and is located on the long arm (q) of chromosome 9 at position 22.3.

What Is the Treatment for Nevoid Basal Cell Carcinoma Syndrome?

The manifestations of the condition can be treated, but it is critical to identify NBCC early in life. Special surgical excision for keratocysts and for the basal cell carcinomas can completely eradicate and preserve normal tissue to prevent disfiguring. Fibromas of the ovaries require surgery. The head of the child should be measured throughout childhood to detect the presence of a brain tumor. Prevention includes avoiding sun exposure and use of radiation therapy because of developing multiple carcinomas in the treated area.

Further Reading

Basal Cell Carcinoma Nevus Life Support Network. 2011. http://www.bccns.org. Accessed 5/23/12.

Evans, D. Gareth, and Peter A. Farndon. 2011. "Nevoid Basal Cell Carcinoma Syndrome." *GeneReviews*. http://www.ncbi.nlm.nih.gov/books/NBK1151. Accessed 5/23/12.

"Gorlin Syndrome." 2011. Genetics Home Reference. National Library of Medicine (U.S.). http://ghr.nlm.nih.gov/condition/gorlin-syndrome. Accessed 5/23/12.

Newborn Screening: A Special Topic

In the early 1960s, Dr. Robert Guthrie developed a blood test that could determine if a newborn had a metabolic condition called phenylketonuria or PKU. This disorder if left unchecked can lead to serious brain damage and a lifetime disability. The person with PKU cannot process an amino acid called phenylalanine, a substance necessary for normal growth and well-being. Dr. Guthrie's simple test involved pricking the heel of the newborn on the second day after birth and collecting blood samples on a piece of filter paper. The test can determine if the child has PKU, and the doctor can put the child on a special diet immediately. Adhering to this diet will help the person avoid a serious disability that could be life-threatening.

Since the development of the PKU test, additional blood tests can screen newborns for several disorders that if left untreated can cause physical problems, developmental delay, and death. In the 1970s, the second disorder to be tested was congenital hypothyroidisim, a condition in which the thyroid gland does not produce hormone essential for growth and well-being. In the early 1990s, Edwin Naylor and a team of scientists developed a sophisticated test called tandem mass spectrometry, which enables several conditions to be tested.

If a disorder is detected, treatment can be started before the baby is harmed by the condition. Some diseases are so dangerous that infants may die the first time they experience the illness. For example, the condition called medium-chain acyl-coenzyme dehydrogenase deficiency or MCAD causes death in 25% of babies with the first symptoms. This disorder can be avoided by simply giving the child frequent meals and avoiding fasting. Although the individual cases are rare, if all the diseases were added together, they would affect about 1 in 1,500 children born in the United States.

In 1968, J. M. G. Wilson and F. Juenger published proposed screening criteria in their book *Principles and Practice of Screening for Disease*. The following four principles are suggested:

- Having an acceptable treatment for the disorder: Many metabolic conditions can be treated immediately if the condition is identified. Some conditions cannot.

- Understanding of the disorder: Scientists have researched the basic natural history of the disorder and know about the best evidence-based treatment.

- Understanding the individual patient.

- Having a test that will be reliable for both affected and unaffected patients that is acceptable to the public.

Within the past two decades, many diseases have been identified, and some disorders may not meet all the criteria listed above. For example, Duchenne muscular dystrophy has been added to many screening programs around the world. However, there is no real evidence that early detection improves the outcome.

Why Screen?

The baby with one of these disorders may appear healthy at birth, but by the time symptoms appear, permanent damage could have occurred. The number and types of screens is controlled and determined in the United States by the state where the baby is born. All states now have mandatory screening but may vary in the kinds of tests. Most states now screen for more than 30 disorders. A technique called tandem mass spectrometry or MS/MS can use only one drop of blood to screen for 20 disorders.

Here is how the test works: The mother signs a consent form for testing. Blood from the newborn is taken by pricking the child's heel when the baby is between 24 and 48 hours old. Some of the conditions, such as PKU, do not show up immediately, and if the mother is dismissed within 24 hours, the doctor may ask the test to be repeated after 48 hours. Many of the tests are done in the hospital. If the test comes back as "normal," there is an indication the baby does not have the disease. If it is abnormal, further tests may be ordered. The test could be a false positive. Also, the child will be referred to a specialist who can assess and begin treatment.

According to the National Newborn Screening and Genetics Resource Center, the only tests mandated in every state are for congenital hypothyroidism, benign hyperphenylalaninemia, PKU, hearing, and galactosemia. Different states have different testing procedures, and these demands change with legislation that might be submitted each year. In 2005 a "core panel' of the American College of Medical Genetics (ACMG) published a recommended list. The list includes the incidences from the 2005 report, although there may be some variation among different populations. Following are the recommended groups of disorders and their specific diseases that are recommended for testing (those in boldface type have entries in this book):

Blood Cell Disorders

- Sickle-cell anemia: 1 in 5,000; among African Americans, 1 in 400
- **Sickle-cell disease**: 1 in 25,000
- **Beta-thalassemia**: 1 in 50,000

Inborn Errors of Amino Acid Metabolism

- Tyrosinemia: 1 in 100,000
- Arginosuccinic aciduria: 1 in 100,000
- Citrullinemia: 1 in 100,000
- **Phenylketonuria**: 1 in 25,000
- **Maple syrup urine disease (MSUD)**: 1 in 100,000
- Homocystinuria: 1 in 100,000

Inborn Errors of Organic Acid Metabolism

- Glutaric acidemia type 1: 1 in 75,000
- Hydromethylglutrayl lyase deficiency: 1 in 100,000
- **Isovaleric acidemia**: 1 in 100,000
- 3-methylcrotonyl-CoA carboxylase deficiency: 1 in 75,000
- **Methylmalonic acidemia (aciduria)**: 1 in100,000
- Beta-ketothiolase deficiency: 1 in 100,000
- **Propionic acidemia**: 1 in 75,000
- Multiple-CoA carboxylase deficiency (**Holocarboxylase Synthetase Deficiency**): 1 in 100,000

Inborn Errors of Fatty Acid Metabolism

- Long-chain hydroxacyl-CoA dehydrogenase deficiency: 1 in 75,000
- Medium-chain acyl-CoA dehydrogenase deficiency: 1 in 25,000
- Very-long-chain acyl-CoA dehydrogenase deficiency: 1 in 75,000
- Trifunctional protein deficiency: 1 in 100,000
- Carnitine uptake defect: 1 in 100,000

Miscellaneous Multisystem Diseases

- **Cystic fibrosis**: 1 in 5,000
- Congenital hypothyroidism: 1 in 5,000
- Biotinidase deficiency: 1 in 25,000
- Congenital adrenal hyperplasia: 1 in 25,000
- Classic **galactosemia**: 1 in 50,000

Congenital Infections

- TORCH complex
- HIV

Newborn Screening by Methods Other than Blood Testing

- Congenital deafness (**Hereditary Hearing Disorders and Deafness**): 1 in 5,000

Some Controversies

With the development of tandem mass spectrometry in the early 1990s, the number of diseases that can be identified grew. Although the American College of Medical Genetics has recommendations for a large panel of testing, many states do not mandate various tests. It is up to the legislatures of the states to determine whether their health departments will require certain tests. So the number of tests does vary by state.

Issues have arisen that have made screening the subject of controversy. One of the most poignant stories was presented at the 2011 Newborn Screening Symposium in San Diego. Two California babies, Zachary Wyvill and Zachary Black, were born with glutaric acidemia, type 1. When the two were presented at the symposium, Black appeared as a normal healthy child; Wyvill appeared in a wheelchair and has no motor skills. Wyvill has been in an out of the hospital since birth. The tale of the two boys illustrates what a difference newborn screening can make.

At the time of the birth of the two babies, California mandated that only four diseases be screened; however, some hospitals asked to be part of a pilot program for screening. Wyvill's hospital tested only for the four diseases mandated by state law; Black was at a hospital that was in the expanded testing program. He was immediately treated with diet and vitamins; Wyvill went undetected for over six months and had irreversible damage. The news media picked up the story, and the tale of the two Zachs in California went viral. The California legislature responded and now, according to the website of the California Department of Health, the state mandates that about 30 required screenings. Other states also responded. In addition in 2010, California joined Wisconsin and Massachusetts in testing their over 520,000 babies born each year for SCID, an immune system disorder.

Although the costs are coming down, the cost for MS/MS screening is somewhat expensive for an initial output. Many health care providers also do not want screening for disorders for which effective health care programs are not available for follow-up and treatment.

Working through the issues will take time. Some advocates are pushing for national standards that will see all states testing the same conditions with similar procedures. That will be far in the future. Most countries in Europe and throughout the world have some screening requirements, although the lists vary greatly.

Further Reading

Levy, Harvey, MD. 2007. "The History of Newborn Screening." A-Flash Cast, a 40-minute talk. http://www.newenglandconsortium.org/flashcast/nbs_home.html. Accessed 5/23/12.

"Newborn Screening." 2011. MedlinePlus. National Library of Medicine (U.S.). http://www.nlm.nih.gov/medlineplus/newbornscreening.html. Accessed 1/26/12.

"Newborn Screening Tests." 2011. Kids' Health from Nemours. http://kidshealth.org/parent/system/medical/newborn_screening_tests.html. Accessed 1/26/12.

NF1. *See* Neurofibromatosis Type 1 (NF1)

NF2. *See* Neurofibromatosis Type 2

NFLS. *See* Legius Syndrome

Niemann-Pick Disease

Prevalence Type A, about 1 in 40,000 among Ashkenazi Jews; types A and B in general population, about 1 in 250,000; Type C, about 1 in 150,000 among people of French-Acadian descent

Other Names classical Niemann-Pick disease; DAF syndrome; lipoid histiocytosis (classical phosphatide); neuronal cholesterol lipidosis; NPD; ophthalmoplegia, supraoptic vertical; sphingomyelinase deficiency; sphingomyelin/cholesterol lipidosis; sphingomyelin lipidosis

Around the turn of the twentieth century, knowledge of metabolic disorders was expanding. Ever since Archibald Garrod coined the term "inborn error of metabolism," investigators using newfound chemical knowledge discovered other disorders. Many of the rare genetic disorders were found in the population of Ashkenazi Jewish descent from eastern Europe. In 1914, Albert Niemann, a German chemist, first described a condition in which the person could not metabolize fats and cholesterol. In the 1930s, Ludwig Pick described the same disorder in a series of papers. The scientific community dubbed the disorder Niemann-Pick disease.

What Is Niemann-Pick Disease?

Niemann-Pick disease is a disorder in which the body does not properly use lipids or fats. The individual does not break down the fats eaten in food for body use. The materials that are not broken down then accumulate in the lungs, spleen, liver, bone marrow, and brain, and cause a life-threatening condition.

The following four basic types of Niemann-Pick disease exist:

- *Type A, or the neurological type*: The child fails to grow at the expected rate and begins to show signs of progressive disorder of the nervous system. During early infancy, the spleen and liver become enlarged. Children with this type seldom live past early childhood.

- *Type B, or the non-neurological type*: Symptoms of this type include an enlarged liver, growth retardation, and lung infections. High levels of cholesterol and other fats, along with a decreased number of blood platelets, show up in blood tests. People with this type usually live into adulthood.

- *Type C*: People with this type have a host of problems: severe liver disease, developmental delay, seizures, poor muscle tone, lack of coordination, breathing difficulties, and problems with feeding. They may also be unable to move their eyes up and down. The condition usually appears in childhood but may show up during any stage of life. Even with all the disorders, individuals can survive into adulthood.

- *Type C1 and 2*: At one time researchers called this Type D, but now use the subtypes because the subtypes are caused by a different gene mutation rather than by a different gene.

What Are the Genetic Causes of Niemann-Pick Disease?

Mutations in three genes cause Niemann-Pick disease. They are *SMPD1*, *NPC1*, and *NPC2*.

SMPD1

Mutations in *SMPD1*, officially known as the "sphingomyelin phosphodiesterase 1, acid lysosomal" gene, cause types A and B of Niemann-Pick disease. Normally, *SMPD1* instructs for an enzyme called acid sphingomyelinase. This protein is found in the lysosomes, the cell structures that are responsible for disposing of and recycling cell parts. The enzyme also must convert a certain lipid called sphingomyelin into another fat called a ceramide. This process is essential for the normal development of cells and tissues.

A unique pattern is involved in the inheritance of this gene. Although a copy of the gene is inherited from both parents, only the one from the mother is active. This process is known as genomic imprinting. If the mother's gene has a mutated copy, the child will have the disorder because the normal gene from the father is inactive.

Over 100 mutations in *SMPD1* cause types A and B. The mutations cause the reduction or even complete absence of the sphingomyelinase enzyme and leads then to the accumulation of sphingomyelin, cholesterol, and other fats within the cells. The more severe type A is caused when the enzyme is inactive. The milder form, type B, is caused by a defective enzyme. When the enzyme activity fails, lethal products pile up, causing the cell to die and the symptoms of Niemann-Pick disease. *SMPD1* is inherited in an autosomal recessive pattern and is found on the short arm (p) of chromosome 11 at position 15.4-p15.1.

NPC1

Mutations in the *NPC1* gene, officially called the "Niemann-Pick disease, type C1" gene, cause type C1. Normally, *NPC1* instructs for a protein that is located in the membranes of the lysosomes and endosomes. The protein plays a role in the movement of cholesterol and other types of fats across cell membranes.

More than 250 mutations of *NPC1* cause type C1 of Niemann-Pick disease. Most of the mutations involve a change in only one amino acid. The mutations cause a shortage of the protein, keeping the cholesterol and fats from moving across the cell membranes. The blockage of the fats prevents the cell from functioning, which lead to cell death and to the symptoms of type C1. *NPC1* is inherited in an autosomal dominant pattern and is located on the long arm (q) of chromosome 18 at position 11-q12.

NPC2

Mutations in the *NPC2*, officially called the "Niemann-Pick disease, type C2" gene, cause type C2. Normally, *NPC2* instructs for a protein that is located inside the lysosomes that binds to cholesterol. The protein has a big role in moving fats out of the lysosomes to other parts of the cell.

More than 15 mutations in *NPC2* are responsible for type C2. The mutations are usually a change in only one amino acid building block. If the change is only one amino acid, the person usually retains some of the protein and has a less severe form that progresses slowly. If there is no NPC2 protein, the result is a severe disease occurring in the infant or in early childhood. *NPC2* is inherited in an autosomal recessive pattern and is located on the long arm (q) of chromosome 14 at position 24.3.

What Is the Treatment for Niemann-Pick Disease?

Because this is a serious genetic disease, treatments are limited. Most of the interventions include supportive care through nutrition and medical and physical therapy to improve the quality of life. Type A has an extremely poor prognosis because most cases are fatal before the age of 18 months. Types B and C have a better prognosis for many of the individuals, who may live into the teens or adulthood. This condition is the subject of lots of investigations paid for by private foundations. Organ transplants, enzyme replacement, gene therapy, and bone marrow transplants have been used with limited success.

Further Reading

McGovern, Margaret M., and Edward H. Schuchman. 2009. "Acid Sphingomyelinase Deficiency." *GeneReviews*. http://www.ncbi.nlm.nih.gov/books/NBK1370. Accessed 5/23/12.

National Niemann-Pick Disease Foundation. 2010. http://www.nnpdf.org/npdisease_01.html. Accessed 5/23/12.

"Niemann-Pick Disease." 2011. Genetics Home Reference. National Library of Medicine (U.S.). http://ghr.nlm.nih.gov/condition/niemann-pick-disease. Accessed 5/23/12.

Nijmegen Breakage Syndrome

Prevalence	Exact number unknown; estimated 1 in 100,000 newborns; common in Slavic populations of Eastern Europe
Other Names	ataxia-telangiectasia variant 1; Berlin breakage syndrome; Seemanova syndrome

In 1981, Weemas described two sons of second-cousin parents who had small heads, stunted growth, mental retardation, brown spots on the body, and immune disorders. Later in 1987, Seemanova, a professor of medical genetics, found nine patients in six families that had similar traits and described it as a "new" syndrome. Most of the patients were from countries of West Slavic origins with a large number of them living in Poland. Although the condition is sometimes named after Dr. Seemanova, the name most recognized is Nijmegen breakage syndrome, derived from the Dutch city where Weemas first described the condition.

What Is Nijmegen Breakage Syndrome?

Nijmegen breakage syndrome is a disorder caused by damage or breakage to the DNA that manifests in many health problems. It is most prevalent among people who live in the part of eastern Europe, especially the Czech Republic, Slovakia, and Poland, where people of Slavic origin live.

The condition has many faces. Following are the symptoms that characterize Nijmegen breakage syndrome:

- Short stature: Children grow slowly during early years. After a slow start, they grow normally, but are still shorter than their peers.

- Very small head, a condition known as microcephaly: The small head is noted at birth and grows slower than rest of the body. As children get older, it appears that the head is getting smaller as the body grows.

- Distinctive facial features: The facial features, which appear about age three, include a sloping forehead, large nose, small jaw, prominent ears, with the corners of the eyes slanting upward.

- Immune system disorders: People with NBS have very low levels of immunoglobulin G (IgG), immunoglobulin A (IgA), and T cells, the necessary proteins to fight off infections. Therefore, children have many bouts of upper respiratory infections, including bronchitis, pneumonia, and sinusitis.

- Risk of developing cancer: The common cancer of the immune system called non-Hodgkin lymphoma occurs in over half of the people with NBS before the age of 15. Other cancers include brain tumors, glioma, and rhabdomyosarcoma, a cancer of the muscle tissue. Chances of people with NBS developing cancer are 50 times more likely than those without the condition.

- Intellectual decline: For the first year or two of life, children have normal intellectual development, but as skills decline, they have mild to moderate disability.

- Premature failure of ovaries: Women with the disorder have later menstruation that begins around the age 16. Most women with NBS are infertile.

What Is the Genetic Cause of Nijmegen Breakage Syndrome?

Mutations in the *NBN* gene, officially called the "nibrin" gene, cause Nijmegen breakage syndrome. Normally, *NBN* instructs for a protein nibrin that is involved in involved in many cellular actions and is vital for repair of damaged DNA. Two other genes, *MRE11A* and *RAD50*, produce proteins that interact with nibrin to form a complex. Nibrin regulates this complex carrying the proteins of the two genes into the nucleus of the cells and guiding them to sites where DNA damage has occurred. DNA can be broken by a variety of things such as radiation, toxic chemicals, or simply at random breakage when chromosomes divide. When DNA is repaired, the cell processes proceed normally, keeping the cell from dying or from dividing wildly.

Another protein is also at play. The *ATM* gene produces a protein that finds the broken DNA and directs the repair. This protein tells the MRE11A/RAD50/NBN complex where to go and what to repair. This role for *NBN* is preventing numerous cell divisions that may become cancerous; this activity makes nibrin a tumor suppressor.

About 10 mutations in the *NBN* gene cause Nijmegen breakage syndrome. The mutations usually result from a very short version of nibrin protein. The most common mutation found among the Slavic people of eastern Europe deletes five DNA building blocks from the gene. The shortened version of the protein disrupts the normal function of nibrin and consequently affects the work of the complex in repairing damage. Materials accumulate, causing the cells to grow uncontrollably. Also, this defective protein leads to the disorders of the immune system by decreasing the number of immune cells and to other symptoms of the syndrome. *NBN* is inherited in an autosomal recessive pattern and is located on the long arm (q) of chromosome 8 at position 21.

What Is the Treatment for Nijmegan Breakage Syndrome?

Treatment of the condition is primarily symptomatic. The most serious initial complication is that of the immune system. Use of IVIg (intravenous immunoglobulin) may treat severe and frequent infections. Treatment is mostly supportive with use of education and training for lost intellectual skills.

Further Reading

Concannon, Patrick, and Richard Gatti. 2011."Nijmegen Breakage Syndrome." *GeneReviews*. http://www.ncbi.nlm.nih.gov/books/NBK1176. Accessed 5/23/12.

"Nijmegen Breakage Syndrome." 2009. International Birth Defects Information Systems. http://www.ibis-birthdefects.org/start/nijmeg.htm. Accessed 5/23/12.

"Nijmegen Breakage Syndrome." 2011. Genetics and Rare Disease Information (GARD). National Institutes of Health (U.S.). http://rarediseases.info.nih.gov/GARD/Condition/3904/Nijmegen_breakage_syndrome.aspx. Accessed 5/23/12.

Noack Syndrome. *See* Pfeiffer Syndrome

Noonan Syndrome

Prevalence   Affects about 1 in 1,000 to 2,500 people

Other Names   familial Turner syndrome; female pseudo-Turner Syndrome; male
Turner Syndrome; Noonan-Ehmke syndrome; pseudo-Ullrich-
Turner syndrome; Turner-like syndrome; Turner's
phenotype, karyotype normal; Turner syndrome in female with
X chromosome; Ullrich-Noonan syndrome

In 1962, Jacqueline Noonan, a pediatric heart specialist at the University of Iowa,
noticed that both boys and girls with a certain rare heart defect also had a distinct
facial appearance, short stature, and webbed neck. Suspecting this condition was a
variation of Turner syndrome, which occurs mostly in females, she studied the
genetic patterns of 833 patients in her clinic and found a new syndrome. Her
student, Dr. John Opitz, named the condition after her in a symposium in 1971.
Noonan syndrome is now the recognized name of the disorder.

What Is Noonan Syndrome?

Noonan syndrome is a disorder that affects many parts of the body, including
certain malformations of the heart. The following symptoms are characteristics
of Noonan syndrome:

- Heart defects: The cardiologist Dr. Noonan first found the disorder in children with heart defects. The condition called pulmonary valve stenosis
involves the narrowing of the valve that leads from the heart to the lungs.
The defect may lead to an enlargement of the heart, a condition known as
hypertrophic cardiomyopathy, which makes the heart muscle work harder
to pump blood to the body.

- Short stature: At birth the children are normal, but as growth hormones do not
function, growth slows. Between 50% and 70% of people with Noonan
syndrome have short stature.

- Distinct facial characteristics: Children develop a deep groove in the area
called the philtrum between the mouth and nose. They also may have widely
spaced eyes with a blue or blue-green color, low-set ears that are often back-
wards, high-arched palate, poor teeth alignment, and very small lower jaw.

- Short neck: Excess neck skin called webbing and a low hairline at the back of
the neck may be present in both children and adults.

- Skeletal disorders: The child may have scoliosis and either a sunken or
protruding chest.

- Bleeding disorders: People may have excessive bleeding with surgery or
injury, bruising, or nosebleeds. Affected females may have excessive

bleeding during monthly period or at childbirth. Fertility does not appear to be affected.

- Reproductive disorders for males: Starting at ages of about 13 or 14, males grow more slowly compared to their peers. Most males have undescended testicles and infertility.

- Intelligence: Most people with Noonan syndrome have normal intelligence; however, they may have special educational needs.

- Hearing or vision problems.

Infants may be born with lymphedema, a condition characterized by puffy hands with fluid buildup. This symptom disappears as they reach the first year. Infants may also develop feeding problems, which also improve after the first year.

What Are the Genetic Causes of Noonan Disease?

Mutations in six different genes cause Noonan syndrome. The genes are *PTPN11*, *SOS1*, *RAF1*, *KRAS*, *NRAS*, and *BRAF*. Most cases involve one of the first three genes. However, all of these genes have a common method of action. The genes all instruct for proteins that are essential to cell development. Mutations in the genes cause the proteins to be active all the time rather than switching off when signals in the cells tell them to. The active dysfunctional protein disrupts cell functions and leads to the symptoms of Noonan syndrome.

PTPN11

Mutations in *PTPN11*, also called the "protein tyrosine phosphatase, non-receptor type 11" gene, cause about 50% of all cases of Noonan syndrome. Normally, *PTPN11* instructs for a protein SHP-2, which controls the Ras/MAPK signaling pathway. This pathway regulates cell growth and division, cell differentiation, cell movement, and cell self-destruction, a process called apoptosis. The SHP2 proteins are critical in embryonic development of heart, bones, blood cells, and several other tissues. *PTPN11* is an oncogene, which means that mutations in their proteins can cause cells to grow out of control and cause cancer.

About 50 mutations cause Noonan syndrome. Most mutations cause one amino acid in the protein to replace another. The pathway that is regulated by the mutated *PTPN11* protein continues to be active and does not switch off, thereby disrupting the regulation of Ras/MAPK signaling pathway. This disruption results in the heart defects, skeletal defects, and growth problems of Noonan syndrome. *PTPN11* is inherited in an autosomal dominant pattern and is located on the long arm (q) of chromosome 12 at position 24.

SOS1

Mutations in the *SOS1* gene, also known as the "son of sevenless homolog 1 (Drosophila)," cause about 15% of the cases of Noonan syndrome. Normally, *SOS1* instructs for making a protein that regulates the Ras/MAPK signaling pathway. It regulates the growth, division, differentiation, movement, and cell death. *SOS1* controls another protein called Ras, which is related to the growth and division of cells and is essential in early embryonic development. Mutations cause the protein to be continuously active and thereby disrupt the signaling pathways that controls cell growth and division, resulting in the features of Noonan syndrome. *SOS1* is inherited in a dominant pattern and is located on the short arm (p) of chromosome 2 at position 21.

RAF1

Mutations in the *RAF1* gene, known also as the "v-raf-1 murine leukemia viral oncogene homolog 1," cause about 10% to 15% of cases of Noonan disease. Normally, *RAF1* instructs for a protein that is also part of the Ras/MAPK signaling pathway. Ras/MAPK controls growth, division, differentiation, and self-destruction of cells. *RAF1* is an oncogene, which when it is mutated can cause cells to grow wildly.

More than 10 mutations in *RAF1* cause about 5% to 10% of cases of Noonan syndrome. Mutations cause a change in just one amino acid in the RAF1 protein and disrupt the processes of cell function. This gene appears to play a role in heart abnormalities. People with this mutation have a greater incidence of cardiac defects, especially the enlarged heart. *RAF1* is inherited in an autosomal dominant pattern and is located on the short arm (p) of chromosome 3 at position 25.

KRAS

Mutations in the *KRAS* gene, officially known as "v-Ki-ras2 Kirsten rat sarcoma viral oncogene homolog" gene, cause about 1% to 2% of the cases of Noonan syndrome. Normally, *KRAS* instructs for a protein K-Ras, whose primary function is controlling cell division. The process of signal transduction occurs when a protein gets a signal from outside the cell and then sends it to the nucleus. These signals tell the cells when to perform the vital functions of growth. K-Ras protein acts like a switch, which is controlled by other molecules. Because of its role in regulating cell division, *KRAS* belongs to a group of gene known as oncogenes.

Only a small number of mutations have been located in the *KRAS* gene. Each *KRAS* gene mutation changes only one amino acid in the protein. The altered protein leads to prolonged activation when the protein switch does not work. Intellectual disability is present more when people have this gene. *KRAS* is inherited in an autosomal dominant pattern and is found on the short arm (q) of chromosome 12 at position 12.1.

BRAF

Mutations in the *BRAF* gene, officially known as the "v-raf murine sarcoma viral oncogene homolog B1" gene, cause only a small number of cases of Noonan syndrome. Normally, *BRAF* instructs for a protein that is part of the Ras/MAPK pathway. Like the other gene proteins that cause Noonan syndrome, the protein regulates cell functions. *BRAF* is also an oncogene.

About four mutations in *BRAF* cause Noonan syndrome. The pattern is similar to the five other genes and occurs when changes in the gene cause a change in a single amino acid in the protein. The mutations disrupt cell function and cause the abnormalities of the disorder. *BRAF* is inherited in an autosomal dominant pattern and is located on the long arm (q) of chromosome 7 at position 34.

NRAS

The *NRAS* gene, officially known as the "neuroblastoma RAS viral (v-ras) oncogene homolog" gene, is the last of the six genes that are involved in Noonan syndrome. Normally, *NRAS* instructs for a protein N-Ras that regulates cell division. The activity and pattern of this gene and its protein is similar to the others in this group. The two mutations replace single building blocks in the protein, causing it not to work properly and leading to the symptoms of Noonan syndrome. *NRAS* is inherited in an autosomal dominant pattern and is located on the short arm (p) of chromosome 1 at position 13.2.

What Is the Treatment for Noonan Disease?

The treatments for the serious cardiovascular disorder are the same as those in the general population. Sometimes growth hormone is used to increase growth. The presence of any of the symptoms must be treated according to the individual case. For example, if there are developmental disabilities, early education programs are recommended.

Further Reading

Allanson, Judith E., and Amy E. Roberts. 2011."Noonan Syndrome." *GeneReviews*. http://www.ncbi.nlm.nih.gov/books/NBK1124. Accessed 5/23/12.

"Noonan Syndrome." 2011. Genetics and Rare Diseases Information Center (GARD). National Institutes of Health (U.S.). http://rarediseases.info.nih.gov/GARD/Condition/7223/Noonan_syndrome_1.aspx. Accessed 5/23/12.

"Noonan Syndrome." 2011. Medscape. http://emedicine.medscape.com/article/947504-overview. Accessed 5/23/12.

Norrie Disease

Prevalence	Rare; exact incidence unknown; not associated with any ethnic group
Other Names	Anderson-Warburg syndrome; atrophia bulborum hereditaria; congenital progressive oculo-acoustico-cerebral degeneration;

Episkopi blindness; fetal iritis syndrome; Norrie's disease; Norrie syndrome; Norrie-Warburg syndrome; oligophrenia microphthalmus; pseudoglioma congenital; Whitnall-Norman syndrome

In 1961, Mette Warburg, a Danish eye specialist, noted that a family in her care had a common disorder through seven generations. She examined a three-month-old child whose eye lenses were opaque, and the irises of the eye were deteriorating. Later when she noted the lens was filled with a growing yellow mass, she removed the eye, fearing cancer. However, she found the mass was not a tumor but a group of undifferentiated cells. Cells are undifferentiated when they are in the stem cell stage but then become specific tissue as they develop. She concluded that this suspected tumor was a developmental condition due to the malformation of the eye.

Warburg found that five to seven males in this Danish family became deaf in later life and also had intellectual disabilities. Researching the literature, she found 48 similar cases of this "new" disorder. She named it after Gordon Norrie, a famous Danish ophthalmologist who had dedicated his life to working with the blind.

What Is Norrie Disease?

Norrie disease affects the eyes and later other body systems. The condition is usually noted in boys soon after birth when the doctor sees a problem with the retina of the eye. The retina is the area at the back of the eye that has a layer of sensory cells called rods and cones for detecting light and color. Accumulating on the retina are masses of immature, undifferentiated cells, which make the pupils of the eye appear white when light is flashed upon it. Other things that lead to complete blindness happen to the eye. The iris, the colored part of the eye, or the complete eyeball shrinks, and cataracts develop on the lens.

The following other symptoms related to other body systems may develop:

- Progressive hearing loss: About one-third of the individuals with Norrie disease lose their hearing as they get older.
- Developmental delays: In over half of the children, motor skills such as sitting up, crawling, and walking is delayed.
- Intellectual disability: Cognitive disability is mild to moderate.
- Psychosis.
- Other body functions: Circulation, breathing, digestion, and excretion may be affected.

What Is the Genetic Cause of Norrie Disease?

Mutations in the *NPD* gene, known also as the "Norrie disease (pseudoglioma)" gene, cause Norrie disease. Normally, *NDP* instructs for a protein, norrin, that is essential in developing signaling pathways for the proper development of cells.

Norrin plays a role in Wnt signaling, the process that tells cells when to divide, when to attach to one another, when to move, and other cellular activities. Norrin, a ligand, binds to other proteins called frizzled receptors, which are found in the outer cell membrane. The protein grabs onto the frizzled receptor, called frizzled-4, which is produced by the *FZD4* gene. The two fit like a lock and key and together regulate many other genes. Working together, norrin and frizzled-4 affect the developmental process of the normal development of the eye and other body systems. Norrin appears to play a valuable role in the specialization of cells to form the retina and the blood supply to both eye and inner ear.

About 75 mutations in *NDP* affect the ability of norrin to bind with frizzled-4. These abnormal proteins then interfere with the development of the rods and cones in the back of the eye. Large clumps of immature retinal cells collect in the back of the eyes, and tissue begins to break down. Norrin may be found in other body systems affecting the intellectual ability and the other symptoms of the disorder that do not appear to be related to the eye. Some mutations that are the result of large deletions in the *NDP* gene affect the production of norrin and cause the widespread problems not only to the eyes but to other systems. Some mutations that are the result of only one exchange or deletion do not have the widespread effects on other systems. *NDP* is inherited in an X-linked recessive pattern and is located on the short arm (p) of the X chromosome at position 11.4.

What Is the Treatment for Norrie Disease?

Norrie disease is a difficult disease to treat because of its many manifestations. If the retina is not detached, surgery or laser therapy may save the eye. Treatment for hearing loss may include aids or cochlear implants. The intellectual and behavioral issues demand supportive therapy and education.

Further Reading

"Medical Information on Norrie's Disease." Scottish Sensory Center (UK). http://www.ssc.education.ed.ac.uk/resources/vi&multi/eyeconds/Norr.html. Accessed 5/23/12.

"Norrie Disease." 2011. Genetics and Rare Diseases Information Center. National Institutes of Health (U.S.). http://www.rarediseases.org/rare-disease-information/rare-diseases/byID/568/viewAbstract. Accessed 7/17/11.

"Norrie Disease." 2011. Genetics Home Reference. National Library of Medicine (U.S.). http://ghr.nlm.nih.gov/condition/norrie-disease. Accessed 5/23/12.

NPD. *See* Niemann-Pick Disease

O

OA. *See* Ocular Albinism

Obesity: A Special Topic

According to *Webster's New World Medical Dictionary* (New York: Wiley, 2008), obesity is defined as the state of being more than 20% over a person's ideal weight. That ideal takes into account the person's height, age, sex, and build. More precisely, however, the National Institutes of Health calls the body mass index, or BMI, a useful way to define overweight and obesity. Body mass index is calculated using this formula:

BMI = weight (pounds × 703) divided by height (in inches) squared. A person is obese if he or she has a BMI of 30 or more. It should be noted however, that the body mass index does not necessarily measure fat, and some people who have higher levels of muscle mass may have higher BMI levels.

To calculate your BMI, using pounds and inches, do the following:

1. Take your weight in pounds and multiply by 703.
2. Take your height in inches and multiply it by itself.
3. Divide the weight by the height.

Example: A person is 5'4" and weighs 140 pounds.

1. 140 × 703 = 98,420
2. 5'4" is 64 inches, so 64 × 64 = 4,096
3. 98,420 divided by 4,096 = 24

A serious obesity problem is shown with excess abdominal fat. (iStockPhoto)

Here is what this means:

BMI of < 18.5: Underweight

BMI of 18.5–24.9: Normal

BMI of 25.0–29.9: Overweight

BMI of 30.0–34.9: Class I obesity

BMI of 35.0–39.9: Class II Obesity

BMI of > 40.0: Class III extreme or morbid obesity.

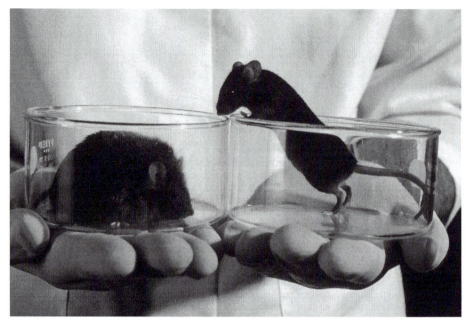

In a laboratory, a fat mouse and a skinny mouse exhibit results of research on the genetic causes of obesity. (Photo by Remi BENALI/Gamma-Rapho via Getty Images) (Getty Images)

The International Obesity Task Force, an international group studying obesity, has created the Class I, II, and III obesity levels.

At first glance, obesity seems simple: people eat too much, and they get fat; if they cut back on food and exercise enough, they will not get fat. But observe two different people who eat and exercise the same amount: one is overweight and the other is not. A simple disease is one that can be traced to a single cause. A few genetic disorders, such as Berardinelli syndrome and Prader-Willi syndrome, might each be called a simple disorder because they can be traced to a single gene.

On the other hand, obesity has many factors and is quite complex. Lifestyle choices, such as food intake, adequate activity, environment, and family traditions, are part of the picture, but many genes are also involved. Genes may even control responses to environment choices, such as appetite and overeating. Although news stories break about the elusive "fat gene" that holds the secret to weight loss, in reality it is not one gene but several genes that affect the weight of a person. (Some say the search should be on for a variety of genes with interesting names, such as "couch potato gene," "stop eating gene," "can't resist gene," and even a "party platter gene," exhibited by people who follow platters at parties!) Obesity is not simple, but is a complex condition controlled not only by environmental, psychological, and social factors, but also by a host of interacting genes.

O.B.'s Story

The fat little fur ball violently clawed at the bar of the food dispenser. After hours of eating, the mouse huddled in the corner to sleep. It did not play mouse games and only ate and slept. Caretakers though the mouse was pregnant, but when she turned out to be a "he," they realized he was a mutant. The mouse was originally named "obese" but was later changed to "O.B." O.B. became one of the first in line for serious study of the genetics of obesity. He passed the obesity trait to his offspring. In the 1970s, scientists at Jackson Harbor in Maine found a blood-borne factor, which they called the satiety factor, which helps control fat storage. That substance later became known as leptin.

Gene Potential

Scientists have used three techniques to try to locate genes: transgenic animals, which uses animals that have been genetically altered to exhibit certain traits; twin studies; and quantitative trait loci, which use maps of generations to trace and locate certain genes.

Following is a table of selected genes and their effects on obesity:

Obesity

Gene	Mechanism	Gene Effect
Leptin	Appetite; energy expenditure	Major
Leptin receptor	Appetite; energy expenditure	Minor
Beta-2 adrenergic receptor	Energy expenditure	Probably minor
Uncoupling protein-1 or UCP-1	Energy	Minor
Uncoupling protein-2 or UCP-2	Raised body temperature requiring increased need for calories	Minor
Proopiomelanocortin (POMC)	Appetite	Major
Melanocortin-4 receptor (MC4R)	Appetite	Major
Peroxisome proliferator-activated receptor	Fat cells; insulin	Major
Hormone-sensitive lipase	Lipid traffic/metabolism	Minor
Low-density lipoprotein receptor	Lipid traffic/metabolism	Minor

Source: Evelyn Kelly, *Obesity* (Santa Barbara CA: Greenwood Press, 2006).

Obesity Syndromes

Several syndromes have fat accumulation as part of the disorder. Many of the disorders include mental retardation. Following are a few of the syndromes (those in boldface type have entries in this book):

Achondroplasia

Age-related macular degeneration

Bardet-Biedl syndrome: There are several different types of this syndrome and all have obesity characteristics

Berardinelli syndrome

Carpenter syndrome

Cohen syndrome: This syndrome is an autosomal recessive pattern and characterized by facial, mouth, eye, and spinal abnormalities

Down syndrome

Factor V Leiden thrombophilia

Hypercholesterolemia

McKusick-Kaufman syndrome

Prader-Willi syndrome: Children eat excessively; found on chromosome 15

Rubinstein-Taybi syndrome

Spina bifida

Genetics and Obesity

1. For those who are genetically predisposed to obesity, prevention is the best strategy. An individualized plan and greater support are required to maintain a healthy weight.

2. Obesity is a chronic, lifelong condition that is the result of an environment of abundance of calories and low physical activity, combined with genetic proclivities. Recognizing the predisposition in a family is important in developing a strategy for prevention.

3. Genes are not destiny. Obesity can be managed with a combination of diet, exercise, and medication.

4. Drugs that will aid in losing weight are being developed and are expected to be available in a few years. However, these will be only for morbidly obese cases and not be a substitute for healthy diet and activity.

Further Reading

"Aim for a Healthy Weight." National Heart, Lung, and Blood Institute. National Institutes of Health (U.S.). http://www.nhlbi.nih.gov/health/public/heart/obesity/lose_wt/index.htm. Accessed 3/7/12.

Kelly, Evelyn B. *Obesity*. Westport, CT: Greenwood Press, 2006.

"Obesity." 2012. MedlinePlus. National Library of Medicine (U.S.). http://www.nlm.nih.gov/medlineplus/obesity.html. Accessed 5/28/12.

"Obesity: Search Results" [lists genetic conditions related to obesity]. Genetics Home Reference. National Library of Medicine (U.S.). http://ghr.nlm.nih.gov/search?query=obesity. Accessed 2/10/12.

Ocular Albinism

Prevalence   Type 1 affects 1 in 60,000 males
Other Names   albinism, ocular; OA; XLOA

In 1908, Archibald Garrod, the doctor who coined the term "inborn error of metabolism," first described a condition in which the skin and hair are very light and the eyes are weak. He referred to the condition as albinism, coming from the Latin word *albus* meaning "white." Later, albinism was divided into two forms: one relates only to the eyes, called ocular albinism; and another form involves the skin and eyes, called oculocutaneous albinism. This article deals only with ocular albinism (*see also* Albinism).

What Is Ocular Albinism?

Ocular albinism is a condition that primarily affects the eyes. The eyes lack melanin, a pigment that gives color to the iris of the eyes. Skin and hair have nearly normal coloration. Pigment in the eye is essential for normal vision, and the lack of it causes the following vision problems:

- Reduced vision: Normal acuity is 20/20. People with ocular albinism may have visual acuity reduced to 20/60 to 20/400. The reduced vision will cause problems with reading, sports, and driving.
- Depth perception: Both eyes do not work together to perceive depth, a condition called stereoscopic vision.
- Rapid eye movements: The eyes may move back and forth, a condition called nystygmus. This movement is involuntary.
- Crossed or lazy eyes: Eyes do not look in the same direction, a condition called strabismus.
- Sensitivity to light: The person is very sensitive to bright light and glare, a condition called photophobia.
- Optic nerve disorder: The optic nerve is the nerve that carries the visual information from the eye to the brain, where the image is perceived. In OA, nerves from the back of the eye do not follow the usual pattern of going to both sides of the brain. The basic pattern is that nerves from the left eye go to the left side of the brain and from the right eye to the right side of the brain. In the eyes with ocular albinism, the nerve fibers may cross over to the opposite side of the brain.

What Is the Genetic Cause of Ocular Albinism?

Mutations in the *GPR143* gene, or "G protein-coupled receptor 143" gene, cause ocular albinism. Normally, *GPR143* instructs for a protein that makes the colored part of the eyes and skin. Made in the light-sensitive retina of the eye and in the skin, the *GPR143* protein makes up the signaling pathway that controls the melanosomes, structures that produce and store melanin. Melanin gives color to the eyes, hair, and skin, and plays a critical role in normal vision.

About 60 mutations of *GPR143* have been identified in type 1 OA. The mutations change the size and shape of the GPR143 protein and in most cases prevent the abnormal protein from reaching the melanosomes. In other cases, the abnormal proteins reach the melanosomes but disrupt the function. In either case, the protein does not act with normal pathways and can cause the retina to grow abnormally, causing the problems of ocular albinism. *GPR143* is inherited in an X-linked recessive pattern and is located on the short arm (p) of the X chromosome at position 22.3.

What Is the Treatment for Ocular Albinism?

Early detection and correction of eye problems are essential. Person must wear sunglasses or special filters for bright light. Strabismus or crossed eyes may be corrected with surgery. Special visual aids and other considerations should be addressed in educational settings.

Further Reading

Lewis, Richard Allan. 2011."Ocular Albinism, X-Linked." *GeneReviews*. http://www.ncbi.nlm.nih.gov/books/NBK1343. Accessed 5/24/12.

National Organization for Albinism and Hypopigmentation. 2002. "Ocular Albinism." http://www.albinism.org/publications/ocular.html. Accessed 5/24/12.

"Ocular Manifestations of Albinism." 2011. Medscape. http://emedicine.medscape.com/article/1216066-overview. Accessed 5/24/12.

Oculodentodigital Dysplasia

Prevalence   Fewer that 1,000 people worldwide; Exact incidence unknown because many cases are probably undiagnosed

Other Names   oculo-dento-digital dysplasia; oculodentodigital syndrome; oculodentoosseous dysplasia; oculo-dento-osseous dysplasia; ODDD; ODD syndrome; ODOD; osseous-oculo-dental dysplasia

The long name "oculodentodigital dysplasia" is made up of several terms. The first word is made of Latin terms: *oculo*, meaning "eye"; *dento*, meaning "tooth"; and *digit*, relating to the fingers and toes. Dysplasia comes from two Greek roots: *dys*, meaning "with difficulty," and *plas*, meaning "form."

What Is Oculodentodigital Dysplasia?

Oculodentodigital dysplasia affects many parts of the body, especially the eyes, teeth, fingers, and toes. Following are the symptoms of the condition:

- Eyes: Eyes are small, a condition known as microphthalmia; individual may have several eye abnormalities that cause poor vision.
- Teeth: Person may have very small or mission teeth, many cavities, and early tooth loss.
- Fingers and toes: A condition known as syndactyly in which the skin of the fourth and fifth fingers are webbed or grown together. Toes may also grow together. Fingers may have an unusual curve ending with brittle nails.
- Small head with very thin nose.
- Sparse hair growth.
- Lack of bladder control.
- Cleft palate.
- Muscle disorders: Person may have difficulty coordinating activities, muscle stiffness.
- Impaired speech and hearing.
- Skin disorders: Skin on the palms and soles become thick, scaly, and calloused.

Although some of the symptoms may be present at birth, others may appear as the person ages.

What Is the Genetic Cause of Oculodentodigital Dysplasia?

Mutations in the *GJA1* gene or "gap junction protein, alpha 1, 43kDa" cause oculodentodigital dysplasia. Normally, *GJA1* instructs for a protein called connexin43, a member of a family of proteins that form communication channels between cells. These channels or gaps permit the transport of nutrients and other important molecules between cells. Connexin43 is found in many tissues, especially those of the eyes, skin, bone, heart, and brain.

About 45 mutations in *GJA1* cause oculodentodigital dysplasia. Most mutations cause a change in only one amino acid in connexin43. Different mutations may cause specific conditions such as the skin condition on the palms and feet. The mutations usually cause a deletion that makes for a short, abnormal protein. The channels are closed or they never develop, causing no molecules to get through. This disruption impairs the communication between cells that cause the abnormalities of oculodentodigital dysplasia. *GJA1* is inherited in an autosomal dominant pattern and is located on the long arm (q) of chromosome 6 at position 21-q23.2.

What Is the Treatment for Oculodentodigital Dysplasia?

Oculodentodigital dysplasia has so many symptoms and affects so many different systems that it is not treated as a single disease. A medical team will treat the manifestations according to the symptoms that can be treated.

Further Reading

"Oculodentodigital Dysplasia." 2011. Genetics Home Reference. National Library of Medicine (U.S.). http://ghr.nlm.nih.gov/condition/oculodentodigital-dysplasia. Accessed 5/24/12.

"Oculo-Dento-Digital Dysplasia." 2011. Health.com. http://www.health.com/health/library/mdp/0,nord912,00.html. Accessed 5/24/12.

OI. *See* Osteogenesis Imperfecta (OI)

Open Spine. *See* Spina Bifida

Opitz G/BBB Syndrome

Prevalence	X-linked type, about 1 in 50,000 to 100,000 males; autosomal dominant form about 1 in 4,000
Other Names	autosomal dominant Opitz syndrome (ADOS); hypertelorism-hypospadias syndrome; hypertelorism with esophageal abnormalities and hypospadias; Opitz BBB/G syndrome; Opitz BBB syndrome; Opitz-Frias syndrome; Opitz G syndrome; 22q11.2 deletion syndrome; X-linked Opitz syndrome (XLOS)

In 1969, J. M. Opitz and a team of doctors noted a disorder with several symptoms that appeared along the midline of the body. They called it the G syndrome of multiple congenital anomalies. Because the names of the complicated symptoms were rather cumbersome, and after several attempts at naming the syndrome, Neri and Capp in 1988 suggested the name just be Opitz syndrome. It is now proposed that the X-linked form for the condition be named Opitz G/BBB or type one, and the autosomal form, type two.

What Is Opitz G/BBB Syndrome?

Opitz G/BBB syndrome is unique in that it appears to affect several body areas along the midline of the body. Two forms of Opitz G/BBB are noted for their different genetic origins. However, the signs and symptoms are comparable. Following are the symptoms of this disorder that are noted along the midline:

- Wide-spaced eyes, a condition known as hypertelorism
- Defects of the larynx and esophagus, causing difficulty in swallowing

- Defects in the trachea, causing difficulty breathing
- Defects in the heart
- Defects in the midline area of the brain with absence of the corpus callosum, the area connecting the right and left side of the brain; fewer than 50% of individuals are affected with this symptom
- In males, opening of the urethra on the underside of the penis
- Cleft palate in autosomal type
- Cleft lip in X-linked type
- Facial abnormalities—person may have a flat nasal bridge, thin upper lip, and low-set ears
- Mild intellectual disability in fewer than 50% of the cases
- Females only mildly affected

What Is the Genetic Cause of Opitz G/BBB Syndrome?

Two different genetic patterns cause Opitz G/BBB syndrome: the X-linked form caused by a mutation in the gene *MID1*, and the autosomal dominant form caused by an unidentified gene on chromosome 22.

MID1

Mutations in the *MID1* gene or "midline 1 (Opitz/BBB syndrome)" cause Opitz G/BBB syndrome. Normally, *MID1* instructs for a protein called midlin or midline-1 that regulates the work of the microtubules, small structures that make up the framework of the cells. These microtubules help cells maintain their shape and play a role in cellular function and transport. Midlin also plays a role in cellular breakdown. Midlin breaks down an enzyme called protein phosphatase 2A (PP2A), which activates microtubule proteins. The proteins mark unwanted material with a molecule called ubiquitin. *MID1* is a member of a family of genes called TRIM, which controls cell functions, especially the destruction of unwanted material. When the unwanted cells are tagged, they are then moved into cells called proteasomes where the proteins are broken down.

About 40 mutations in *MID1* are related with the X-linked Opitz G/BBB syndrome. Most of the mutations cause changes in a single building block of the midlin protein. Other mutations include additions or deletions that cause the protein to work incorrectly. When the protein responsible for breaking down PP2A does not work, the substance builds up in cells, disrupting normal function and causing the birth defects of Opitz G/BBB. *MID1* is inherited in an X-linked dominant pattern and is located on the short arm (p) of the X chromosome at position 22.

Chromosome 22

A deletion in chromosome 22 at the short arm (q) position 22 causes the autosomal dominant form of the disorder. The gene itself has not yet been identified.

What Is the Treatment for Opitz G/BBB Syndrome?

Treating the serious symptoms of this disorder is first priority. Surgical correction of the larynx and trachea may help breathing problems. Other corrections should be made on an individual basis. The individual will also need speech therapy and psychological and educational support.

Further Reading

Meroni, Germana. 2011. "X-Linked Opitz C/BBB Syndrome." *GeneReviews*. http://www.ncbi.nlm.nih.gov/books/NBK1327. Accessed 5/24/12.

"Opitz C/BBB." 2011. Genetics Home Reference. National Library of Medicine (U.S.). http://ghr.nlm.nih.gov/condition/opitz-g-bbb-syndrome. Accessed 5/24/12.

"Opitz Syndrome." 2011. University of Kansas Medical Center. http://www.kumc.edu/gec/support/opitz.html. Accessed 5/24/12.

Oral-Facial-Digital Syndrome

Prevalence   About 1 in 50,000 to 250,000 newborns; type 1 most common; other forms identified in only a few families

Other Names   dysplasia linguofacialis; OFDS; orodigitofacial dysostosis; orodigitofacial syndrome; orofaciodigital syndrome

This condition is one of the inborn errors of metabolism that appeared in the literature only in the mid-1990s. The diagnosis is made at birth but then confirmed in later childhood or adulthood when the person develops polycystic kidney disorder. There are about 13 types of this disorder, with type 1 being the most common. Researchers are finding out more about this disorder as they discover cases that fit the syndrome and more carefully define the disorder.

What Is Oral-Facial-Digital Syndrome?

Oral-facial-digital syndrome affects the mouth, digits, and facial features. The 13 or more forms may have distinct patterns, but the signs and symptoms overlap. Following are some of the general signs of the syndrome:

- Tongue abnormalities: The child is born with a split or cleft tongue, nodules or noncancerous tumors, or an unusual lobed shape tongue.
- Teeth: Person may have missing teeth, extra teeth, malformed enamel, and malocclusion.
- Mouth disorders: The child may have a cleft palate or bands of extra tissue that attach the lips to the gums. A split lip may be present.

- Distinctive facial features: The person has a wide nose, a flat nasal bridge, and widely spaced eyes.
- Abnormal fingers and toes: Finger and toes may be fused together, a condition called syndactyly; very short fingers; and extra fingers, a condition called polydactyly.
- Some degree of intellectual disability.
- Kidney disease: Type 1, the most common type, has a condition in which the kidneys have fluid-filled cysts that keep the kidneys from filtering blood. This condition is called polycystic kidney disease.
- Other forms may have brain, bone, vision, and heart abnormalities.

What Is the Genetic Cause of Oral-Facial Digital Syndrome?

Mutations in the *OFD1* gene, or the "oral-facial-digital syndrome 1" gene, causes oral-facial-digital syndrome. Normally, *OFD1* instructs for a protein that appears to be important in the early development of many parts of the body. Sticking out from the surface of cells are tiny hairlike projections called cilia. These cilia function in cell movement and in many chemical signaling pathways. The OFD1 protein stays at the base of the cilia and is probably essential for the normal formation of cilia. *OFD1* may have other roles in the early stages of embryonic development in the line that separates the left and right sides of the body.

Around 100 mutations in *OFD1* have been found in people with type 1. The mutations result from a change in a single DNA building block or in deletions from the gene. When the mutations occur, the protein is abnormally short and nonfunctional. Most researchers speculate that the disruption of the normal function of the cilia leads to the many abnormal symptoms of oral-facial-digital syndrome. *OPD1* is inherited in an X-linked dominant pattern and is located on the short arm (p) of the X chromosome at position 22.

What Is the Treatment for Oral-Facial Digital Syndrome?

Treatment of oral-facial-digital syndrome includes surgery for the cleft lip and palate, removal of extra teeth, and correction for malocclusion. Routine treatment for renal disease and seizures is essential. Speech therapy and special education may be necessary.

Further Reading

"Oral-Facial-Digital Syndrome." 2007. National Organization for Rare Diseases (NORD). http://www.rarediseases.org/rare-disease-information/rare-diseases/byID/531/view Abstract. Accessed 5/24/12.

"Oral-Facial-Digital Syndrome." 2011. Genetics Home Reference. National Library of Medicine (U.S.). http://ghr.nlm.nih.gov/condition/oral-facial-digital-syndrome. Accessed 5/24/12.

Ornithine Transcarbamylase Deficiency

Prevalence   About 1 in every 80,000 people

Other Names   ornithine carbamoyltransferase deficiency disease

Ornithine transcarbamylase deficiency, or OTC, is a relatively rare disease that has become well known because of a serious mistake. In 1999, 18-year-old Jesse Gelsinger of Tucson, Arizona, was recruited to take part in a momentous gene therapy experiment, which was conducted by the University of Pennsylvania. Gelsinger had a liver disease called ornithine transcarbamylase deficiency. Children with the disorder usually die at birth, but Jesse had a random mutation that produced less severe mutations because some of his cells were normal. He controlled his disorder with diet and medication.

The purpose of the trial was to treat infants with the severe inherited form of the disorder, and it never would have benefited Jesse. On September 13, 1999, Jesse received a corrected OTC gene injected by means of a vector called an adenovirus. He died four days later when his immune system shut down. The Food and Drug Administration found egregious errors in this trial, which breached ethics and protocol and set gene therapy research back for many years.

What Is Ornithine Transcarbamylase Deficiency?

Ornithine transcarbamylase deficiency is a serious metabolic disease in which the urea cycle in the body does not work properly, causing ammonia to build up in the blood and damage tissues. Ammonia is formed when certain components of the urea cycle are not broken down, and the levels of the chemical build up to toxic levels. The nervous system is especially susceptible to ammonia.

The condition usually becomes evident in the first few days or life, and the infant may exhibit the following symptoms:

- Unwilling to eat
- Erratic breathing rate
- Lethargic with no energy
- Unusual body movements
- Developmental delay
- Seizures or coma
- Progressive liver damage, skin lesions, brittle hair as the child ages

Some people may have less severe signs that appear later in life.

What is the Genetic Cause of Ornithine Transcarbamylase Deficiency?

Mutations in the *OTC* gene, officially called the "ornithine carbamoyltransferase" gene, causes ornithine transcarbamylase deficiency. Normally, the *OTC* gene instructs for the enzyme ornithine transcarbamylase. This enzyme is active in the liver in the urea cycle. The role of this cycle is to process excess nitrogen that is made when proteins are used in the body. The cycle in the liver cells makes a compound called urea that the kidneys eliminate. When the excess ammonia is excreted, it does not build up or cause damage to body organs. The role of ornithine transcarbamylase in the urea cycle is to control the compounds carbamoyl phosphate and ornithine and make a new compound called citrulline.

Over 200 mutations in the *OTC* gene cause ornithine transcarbamylase deficiency. The mutations program for a misshapen or shorter-than-normal gene. The enzyme cannot play its role in the urea cycle, and excess nitrogen is not converted to urea for excretion. Ammonia, which is NH_2, builds up in the cells, causing neurological damage. *OTC* is inherited in an X-linked autosomal dominant pattern and is located on the short arm (p) of the X chromosome at position 21.1.

What Is the Treatment for Ornithine Transcarbamylase Deficiency?

Because nitrogen compounds may be a problem, the person will have a life-long struggle to maintain proper nutrients. Following are the suggestions for treatment:

- Very low-protein diet: Proteins are made of nitrogen compounds.
- Prevention of acute illnesses: If the person has an illness, he must take care to prevent dehydration and malnutrition.
- Medications: Some medications, such as sodium benzoate and sodium phenylbutyrate, reduce the amount of nitrogen in the system.
- Supplemental amino acids, such as arginine, citrulline, valine, leucine, and isoleucine.
- Biotin: This compound may stimulate the OTC enzyme.

If diet and medication does not control OTC deficiency, the person may require a liver transplant.

See also Gene Therapy: A Special Topic

Further Reading

"Ornithine Transcarbamylase Deficiency." 2011. Genetics Home Reference. National Library of Medicine (U.S.). http://ghr.nlm.nih.gov/condition/ornithine-transcarbamylase-deficiency. Accessed 11/26/11.

Ornithine Transcarbamylase Deficiency Website. 2007. Children's National Medical Center. http://ureacycle.cnmcresearch.org/otc. Accessed 11/26/11.

Ornithine Translocase Deficiency

Prevalence   Very rare; only 100 cases reported worldwide

Other Names   HHH syndrome; hyperornithinemia-hyperammonemia-homocitrullinemia syndrome; hyperornithinemia-hyperammonemia-homocitrullinuria syndrome; triple H syndrome

In the early 1900s, Archibald Garrod coined a term, "inborn error of metabolism," that now refers to a group of genetic disorders. This disorder, ornithine translocase deficiency, is one of these disorders that usually result from the inability to process certain elements in foods. Sometimes the condition is called triple H syndrome because of the development of hyperornithinemia-hyperammonemia-homocitrullinemia. The root word "hyper" means above and beyond, indicating the presence of a toxic amount of the elements in the blood. This disorder is one that is routinely tested in newborn screening in many states.

What Is Ornithine Translocase Deficiency?

Ornithine translocase deficiency is a disorder of metabolism. The baby appears normal at birth but begins to display symptoms when food with protein is added. At this time ammonia, formed when proteins are broken down in the body, builds up to toxic levels. There is a milder late-onset form of the disorder that occurs in adults.

Ornithine translocase deficiency is most serious in infants. The symptoms usually begin when solid food, especially when high-protein formula or pureed meat is introduced into the diet. Nitrogen, a component of protein, is processed in the kidney. If excess nitrogen is not disposed of, ammonia builds up. Ammonia is made up of nitrogen and hydrogen (NH_2). This built-up ammonia is toxic to the nervous system, causing the following symptoms:

- Lack of energy
- Breathing disorders
- Problems with maintaining proper body temperature
- Seizures
- Random body movements
- Coma

People with late-onset ornithine translocase deficiency have a somewhat milder form. The symptoms appear after eating very high-protein meals or after going without food for a long period of time. Going without food causes ammonia to build up more rapidly and may evoke the following symptoms:

- Vomiting
- Tiredness and lack of energy

- Problems with coordination
- Confusion
- Blurred vision
- Stiffness of muscles
- Intellectual disorder

What Is the Genetic Cause of Ornithine Translocase Deficiency?

Mutations in the *SLC25A15* gene, known officially as the "solute carrier family 25 (mitochondrial carrier; ornithine transporter) member 15," cause ornithine translocase deficiency. Normally, *SLC25A15* instructs for making the protein called the ornithine transporter, which is located in the mitochondria. The mitochondria are bean-shaped structures that are called the powerhouses of the cell. The transporter is essential for making urea, a process that occurs in the liver. In processing protein, nitrogen is one of the products of food breakdown. The urea cycle takes the excess nitrogen and converts it into urea, which ultimately is eliminated by the kidneys. The ornithine transporter molecule located in the mitochondria of the kidney cells then becomes an integral part of the urea cycle.

About 17 mutations in *SLC25A15* cause ornithine translocase deficiency. Several things can go wrong in the process. The transporter can have the wrong shape, be unstable, or malfunction in many ways. The changed protein disrupts the urea cycle. The kidneys do not get rid of the excess nitrogen, and ammonia, which is toxic to the nervous system, builds up, causing the symptoms of ornithine translocase deficiency. *SLC25A15* is inherited in an autosomal recessive pattern and is located on the long arm (q) of chromosome 13 at position 14.

What Is the Treatment for Ornithine Translocase Deficiency?

Treating any of the urea cycle disorders usually includes dialysis to reduce the amount of ammonia in the blood. Certain drugs called nitrogen scavengers may combat the amount of ammonia also. Restriction of the amount of protein in the diet is essential. Calories are usually provided as carbohydrates or fats.

Further Reading

Lanpher, Brenden C.; Andrea Gropman; Kimberly A. Chapman; Uta Lichter-Konecki; and Marshall L. Summar. 2011. "Urea Cycle Disorders Overview." *GeneReviews*. http://www.ncbi.nlm.nih.gov/books/NBK1217. Accessed 5/24/12.

"Ornithine Translocase Deficiency." 2011. Genetics Home Reference. National Library of Medicine (U.S.). http://ghr.nlm.nih.gov/condition/ornithine-translocase-deficiency. Accessed 5/24/12.

"*SLC25A15*." 2011. Genetics Home Reference. National Library of Medicine (U.S.). http://ghr.nlm.nih.gov/gene/SLC25A15. Accessed 5/24/12.

Osler-Rendu-Weber Disease (OWRD)

Prevalence   About 1 in 5,000 to 10,000 people worldwide

Other Names   hereditary hemorrhagic telangiectasia; HHT; Osler-Rendu disease; Osler's disease; Rendu-Osler-Weber; Weber-Osler

In 1864, Sutton noted a condition of disorders in the blood vessels in several members of the same family. The next year in 1865, Benjamin Guy Babington wrote about "Hereditary Epitaxis," a condition with the same symptoms. Henri Rendu described the important features of the disorder and recognized it a separate disease from hemophilia. Later, Sir William Osler, known as the father of modern medicine, along with Frederick Parks Weber, published detailed descriptions of the disorder that now bears the name Osler-Weber-Rendu disease. Sometimes it is called hereditary hemorrhagic telangiectasia (HHT).

What Is Osler-Weber-Rendu Disease (OWRD)?

Osler-Weber-Rendu is a disorder that causes many abnormalities in the blood vessels. The more descriptive "hemorrhagic telangiectasia" is also used. The word

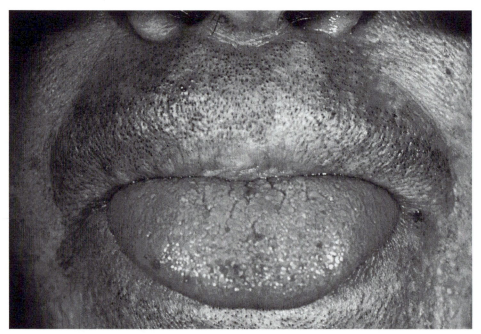

Osler-Rendu-Weber Disease, also known as hereditary hemorrhagic telangiectasia. Note the spots on this person'ls tongue. (CDC / Robert E. Sumpter)

"telangiectasia" is made up of three Greek root words: *telos*, meaning "end"; *angion*, meaning "blood vessel"; and *ektasis*, meaning "dilation." Hemorrhagic is derived from the Greek roots *hemo*, meaning "blood," and *rhag*, meaning "break." In OWRD, certain groups of blood vessels are very thin, and the expansion or dilation of the vessels allow lesions to form. The condition may appear anywhere but is especially common in lining of the nose, digestive tract, and skin.

In a normal circulatory system, blood is pumped into the arteries, which then moves to smaller and smaller arterioles to reach the capillaries where the oxygen–carbon dioxide exchange takes place. To reach the destination in the capillaries, the pressure in the arteries is strong. From the capillaries, the blood carrying carbon dioxide moves through the veins back to the heart. In Osler-Weber-Rendu disease, because of malformation, some vessels bypass the capillaries and go directly from artery to vein. This condition is called an arteriovenous malformation. When these abnormalities are near the skin, they appear as red marks known as telangiectases.

Pressure appears to be the problem. When the buffer of the capillaries is bypassed, the high pressure in the arteries then hits the thin walls of the veins. The pressure causes them to dilate and possibly break through, causing bleeding. The person may have nosebleeds and internal bleeds in the brain, liver, lungs, or other organs where the abnormality occurs.

Four forms of Osler-Weber-Rendu disease are known. They are the following:

- Type 1: Symptoms develop early. Women with the condition appear to develop the abnormalities in the lungs and are at risk for liver involvement.

- Type 2: Development is not as early, but the people with the condition are also at risk for liver involvement.

- Type 3: This type occurs later, and people with the condition are at risk for any of the hereditary problems.

- Juvenile polyposis/hereditary hemorrhagic telangiectasia syndrome: With this form, the person may have the artery-vein malformations and also develop growths or polyps in the gastrointestinal tract.

Many of the symptoms of Osler-Weber-Rendu disease are common in the population. A lot of people have nosebleeds. Researchers think the condition may be undiagnosed in many cases.

What Are the Genetic Causes of Osler-Weber-Rendu Disease (OWRD)?

Mutations in three genes cause Osler-Weber-Rendu disease. The mutations relate to the different types of the disease. All these genes are connected with the formation of the lining of the blood vessels.

ENG

Mutations in the *ENG* gene, or "endoglin" gene, cause type 1 Osler-Weber-Rendu disease. Normally, *ENG* instructs for a protein called endoglin that is found especially in the lining of arteries as they develop. The protein forms an important complex with other proteins to help differentiate blood vessels into arteries or veins. The changes occur as the result of only one substitution of amino acids, causing endoglin to have an abnormal structure that impairs the function of the protein. The dysfunctional protein disrupts the formation of the boundaries between the arteries and veins, resulting in the hemorrhaging and bleeding. *ENG* is inherited in an autosomal dominant pattern and is located on the long arm (q) of chromosome 9 at position 33-q34.1.

ACVRL1

Many mutations in the *ACVRL1* gene, or the "activin A receptor type II-like 1" gene, cause OWRD type 2. Normally, this gene instructs for making the protein called activin receptor-like kinase 1, which is found on the surface of cells and in the lining of arteries. This protein is a signaling molecule that sits as a lock on the surface of cells waiting for the key transforming growth factor beta to come along. The interaction then aids in the development of blood vessels and the differentiation into arteries.

As in the other genes, only one change in the protein can cause a dysfunctional protein that interferes with the development of the boundaries between arteries and veins, causing the symptoms of the disorder. *ACVRL1* is inherited in an autosomal dominant pattern and is found on the long arm (q) of chromosome 12 at position 11-q14.

SMAD4

Mutations in *SMAD4*, or the "SMAD family member 4" gene, cause juvenile polyposis/hereditary hemorrhagic telangiectasia syndrome. Normally, *SMAD4* instructs for a protein that transmits signals from the surface of the cell to the nucleus. This protein binds with the transforming growth factor beta to form a protein complex that moves to the cell nucleus. The SMAD4 complex then binds to certain areas of DNA that control growth and division. Thus, *SMAD4* could be considered a tumor suppressor.

About five mutations in *SMAD4* are related to juvenile polyposis/hereditary hemorrhagic telangiectasia syndrome. People with this disorder have the blood vessel problems but also are at an increased risk for developing intestinal polyps at an early age. The disruption of the pathway appears to interfere with the tumor suppressor function and enable growth to occur. *SMAD4* is inherited in an autosomal dominant pattern and is located on the long arm (q) of chromosome 18 at position 21.1.

Type 3

The gene for type 3 has not been determined but is suspected to be located on chromosome 5.

What Is the Treatment for Osler-Weber-Rendu Disease (OWRD)?

Treating the symptoms is important. Nosebleeds are treated by adding a humidifier, nasal lubricants, or topical hormones, and possibly laser surgery. Other types of bleeding may be treated with surgery. Sometimes for individuals with severe liver damage, a liver transplant is necessary.

Further Reading

"Hereditary Hemorrhagic Telangiectasia." 2011. Genetics Home Reference. National Library of Medicine (U.S.). http://ghr.nlm.nih.gov/condition/hereditary-hemorrhagic-telangiectasia. Accessed 5/24/12.

McDonald, Jamie, and Reed E. Pyeritz. 2012. "Hereditary Hemorrhagic Telangiectasia." *GeneReviews*. http://www.ncbi.nlm.nih.gov/books/NBK1351. Accessed 5/24/12.

"Osler-Weber-Rendu Disease." 2011. Medscape. http://emedicine.medscape.com/article/461689-overview. Accessed 5/24/12.

Osteogenesis Imperfecta (OI)

Prevalence	Affects 6 to 7 per 100,000 people worldwide; types I and IV most common, affecting 4 to 5 per 100,000
Other Names	brittle bone disease; Ekman-Lobstein; Fragilitas ossium; glass bone disease; Lobstein syndrome; Vrolik disease

A mysterious condition in which bones were brittle like glass has been known for many centuries. Archeologists have found an Egyptian mummy from 1000 BC with very fragile bones. Norse history recounts a king named Ivar the Boneless, who may have had the condition. The Swedish physician Olof Jakob Ekman began studies of the disease and mentioned cases of the condition going back to 1678. In 1831, Edmund Axmann believed that he and two other brothers had glass bone disease.

During the latter half of the nineteenth century, several investigators added their names to the list researching this unusual bone condition: Jean Lobstein, William Vrolik, and Martin Benno Schmidt. However, in 1895 the term "osteogenesis imperfecta" was given and has been the accepted medical term throughout the twentieth century to the present.

What Is Osteogenesis Imperfecta (OI)?

Osteogenesis imperfecta (OI) affects the bones. The term "osteogenesis imperfecta" combines Greek and Latin forms: *osteo*, meaning "bones"; *genesis*, meaning "giving rise to"; and *imperfecta*, meaning "not perfect or right." OI is a condition in which the bones are not formed properly and are not strong.

The problem arises with the production of a type of collagen. Collagens are a family of proteins that are essential for proper function of bones, cartilage, tendons, skin, and the white part of the eye or sclera. When collagen does not form the bones properly, bones are weak and break easily. Just the slightest trauma may cause a fracture. Some fractures may even occur before birth or during the birth process.

Researchers have identified eight types of OI. They are designated Types I through VIII. Type I is the mildest kind with Type II being the most severe. The other types range in between Types I and II; some of the symptoms overlap.

Handle With Care

Brittle bone disease or glass bone disease has often piqued the imaginations of writers and authors. Although it is relatively rare, it is known enough to get attention from the public. Several writers and film producers have addressed this disease in various ways, but none as compelling as Jodi Picoult in her book *Handle with Care*. In this book she addresses the reality of living with a child with a serious disability such as osteogenesis inperfecta.

When Willow is born with severe osteogenesis imperfecta, Charlotte and Sean O'Keefe struggle to make ends meet as they cover her medical expenses. Charlotte devises an answer. She will file a wrongful birth lawsuit against her ob-gyn for not revealing that her child would be severely disabled. The money would then pay for the medical expenses for care of the child. Of course, this could mean that Charlotte would have to testify in a court of law that she would have terminated the pregnancy if she had known. Her husband does not believe in abortion, and Willow would hear and know of her attitude. In addition, the ob-gyn was not just her doctor, but her best friend.

Handle with Care explores questions of personal morality and ethics. Questions erupt from many sides. At what point is the physician obligated to tell expectant parents that their child may have a disability? Should a physician ever counsel for termination of the fetus? Should the parents have a right to make that choice? When is a child too disabled? As a parent, how far can your go to take care of someone that you love? Would you be willing to lie in court and alienate friends, husband, child, and possibly community for money? And last, would you be able to live with yourself after lying in a court of law?

Type I

Type I of OI is the most common and mildest form of the condition. The child or adult with this type may appear normal, but they still have the possibility of frequent fractures. Following are the hallmarks of this type:

- Bones break easily
- Collagen produced but in insufficient quantities
- Discoloration of the whites of the eyes; blue-gray color indicates the sclera is thinner than normal
- Slight protrusion of eyes
- Poor muscle tone
- Loose or bendy joints
- Hearing loss in some children
- Teeth may be fragile and prone to cavities or cracking

Type II

Collagen is insufficient and does not form properly. Following are the hallmarks of this most serious type:

- Most die within the first year or are stillborn
- Fragile rib cage causes breathing failure; bones in skull may cause cerebral hemorrhage
- Severe bone deformity

Type III

This type is different from other types of OI in that it is more progressive. The newborn may have only mild symptoms at birth but then symptoms get worse during life. The lifespan with physical handicaps may be normal. Following are the symptoms of type III:

- Infant often born with fractures
- Bones fracture easily
- Bone deformity
- Short stature
- Curvature of the spine
- Respiratory problems
- Poor tone in muscles, especially in arms and legs
- Sclera or whites of eyes may be white, blue, gray, or purple
- Brittle teeth
- Loss of hearing possible

Type IV

Moderately severe but white of eyes may be normal. Following are the symptoms of type IV:

- Bones fracture easily especially during childhood
- Short stature
- Curvature of spine and barrel-shaped rib cage
- Early hearing loss

Type V

This type has clinical features similar to type IV; however, X-rays show the differences. The bones have a mesh-like appearance, and the bones then break down, making them calcify. The person cannot move wrist because bones of the forearms are fused.

Type VI

This type has the same clinical features as type IV, except the bones upon X-ray examination appear like fish scales.

Type VII

This is a rare recessive form of the disorder appearing only in people living in Quebec.

Type VIII

Clinical symptoms are similar to Type IV.

What Are the Genetic Causes of Osteogenesis Imperfecta (OI)?

Osteogenesis imperfecta is caused by four different genes: *COL1A1*, *COL1A2*, *CRTAP*, and *LEPRE1*. *COL1A1* and *COL1A2* are responsible for 90% of cases. *CRTAP* and *LEPRE1* are responsible for very rare and often fatal conditions.

COL1A1

The first gene is *COL1A1* or the "collagen, type I, alpha 1" gene. Normally, *COL1A1* instructs for making collagen type I, the most abundant form in the human body. *COL1A1* makes a part of type I collagen called pro-α1(I) chain. Collagens start out as procollagen molecules, Enzymes outside the cell remove certain protein segments from the end, creating a rope-like molecule with three chains: two pro-α1(I), chains made from the *COL1A1* gene, and one pro-α1(i) chain made by the *COL1A2* gene. Next, the collagen molecules arrange in long, thin fibrils of very strong type I collagen.

Over 400 mutations of *COL1A1* have been detected. Most mutations cause the milder type I form. In this form of the disease, collagen is present but the number of pro-α1(I) chains are reduced, and cells make only some of the type I collagen. This reduction of collagen leads to the characteristic features of type I of OI.

Other mutations cause types II, III, and IV of OI. Genetic changes can occur along many places in the chains replacing one amino acid such as glycine with another protein building block. Substitutions made at the end of protein chains cause the chain not to organize properly. The abnormal collagen then causes the severe forms of OI. *COL1A1* is inherited in an autosomal dominant pattern and is located on the long arm (q) of chromosome 17 at position 21.33.

COL1A2

The second gene associated with OI is *COL1A2*. *COL1A2* produces the third component of the pro-α2(I) chain that is important in the procollagen molecule. About 300 mutations are responsible for OI types. A few mutations cause Type I, but most mutations are related to the more severe types II, III, and IV. The mutations prevent normal production of type I collagen causing the severe symptoms of OI. *COL1A2* is inherited in an autosomal dominant pattern and is located on the long arm (q) of chromosome 7 at position 22.1.

CRTAP

The third gene associated with OI is *CRTAP*, or the "cartilage associated protein" gene. Normally, *CRTAP* instructs for the cartilage associated protein. The exact role is not known, but it does act in normal bone development. Cartilage associated protein works with leprecan and cyclophilin B, two proteins that aid in making certain forms of collagen. The complex process produces a substance that releases collagen molecules into spaces around cells. The process is essential for forming strong bones, tendons, and cartilage.

Five mutations in *CRTAP* are responsible for the rare OI type VII. Several mutations prevent the cartilage associated protein from functioning, causing the severe form of the disorder. *CRTAP* is inherited in an autosomal recessive pattern and is located on the short arm (p) of chromosome 3 at position 22.3.

LEPRE1

The fourth gene related to OI is *LEPRE1*, or the "leucine proline-enriched proteoglycan (leprecan) 1" gene Normally, *LEPRE1* produces an enzyme called leprecan that works with cartilage associated protein and cyclophilin B to make certain forms of collagen. This complex affects proline, a building block of collagen. This protein appears to be essential for the normal assembly of collagen and releasing collagen into the body's supportive framework.

At least four mutations in *LEPRE1* have been identified in people with type VIII OI. Production of abnormal collagen weakens the connective tissue and causes the severe symptoms of this type of disorder. *LEPRE1* is inherited in an autosomal recessive pattern and is located on the short arm (p) of chromosome 1 at position 34.1.

What Is the Treatment for Osteogenesis Imperfecta (OI)?

No cure exists for this disorder. Treatment is aimed at strengthening overall bone function and treating symptoms of the types of disorders. Certain medications called bisphosphonates (BPs) work to replace bones that are not formed correctly. BP therapy is being used more frequently to increase bone mass. In some cases, surgery may be used. For example, metal rods may be inserted to improve strength of the long bones. Surgery may correct scoliosis. Various adaptive equipment, such as wheelchairs, splints, and grabbing arms, may be used. In addition, both physical and occupational therapy may help build strength.

Further Reading

"Osteogenesis Imperfecta." 2011. OrthoInfo. American Academy of Orthopaedic Surgeons. http://orthoinfo.aaos.org/topic.cfm?topic=a00051. Accessed 5/28/12.

Osteogenesis Imperfecta Foundation. http://www.oif.org/site/PageServer. Accessed 5/28/12.

"Osteogenesis Imperfecta—Overview." 2011. University of Maryland Medical Center. http://www.umm.edu/ency/article/001573.htm. Accessed 5/28/12.

Osteopetrosis

Prevalence	Autosomal dominant form most common, affecting 1 in 20,000 people; autosomal recessive form affecting 1 in 250,000; other forms very rare
Other Names	Albers-Schonberg disease; congenital osteopetrosis; generalized congenital osteosclerosis; ivory bones; marble bone disease; marble bones; osteopetroses; osteosclerosis fragilis generalisata

In 1904, Albers-Schönberg, a German physician, used the newly acquired device called the X-ray to peer below the skin at bones. He noted a condition among his patients in which they their bones broke very easily. When he looked at the X-rays, he noted the bones were very dense, almost like stone or ivory. Sometimes, the condition is referred to as marble bone disease, but commonly called osteopetrosis from two Greek words: *osteo*, meaning "bone," and *petros*, meaning "stone" or "rock."

What Is Osteopetrosis?

Osteopetrosis is a condition in which bones become very dense and break easily. In the bone-building process, three kinds of cells are present:

- Osteoblasts: These cells manufacture new bone tissue. The word comes from the Greek terms meaning "bone bud."

- Osteoclasts: These cells consume old and worn bone matter. The word comes from the Greek meaning "bone breaker."
- Osteocytes: These cells are the actual bone cells.

In osteopetrosis, the osteoclasts fail to resorb bone, and as a consequence, bone modeling and remodeling are impaired. Several types of the disorder exist, which the pattern of their inheritance determines. The conditions have some things in common but also have some differences in their signs and symptoms. Following are symptoms that are common to the types:

- Pain: The person experiences generalize pain in the bones.
- Broken bones: Fractures are frequent, especially to the long bones, which heal very slowly or not at all.
- Nerve pressure: Due to the density of the bone, compression on the nerves can lead to headache, blindness, and deafness.
- Osteomyelitis: The bone marrow may become inflamed, a condition known as osteomyleltis.
- Enlarged spleen.
- Skull: The frontal part of the skull appears prominent, a condition called bossing.
- Teeth: Teeth may be malformed or not erupt through gums.
- Infections.
- Bleeding and other blood difficulties.
- Stroke.

Types of osteopetrosis have different origins and some different symptoms. Following are the types and their origins:

- Autosomal dominant osteopetrosis (ADO): This type is the mildest. Some may even have no symptoms and are surprised when the condition is found in an X-ray for other reasons. For those with symptoms, they may experience multiple bone fractures, curvature of the spine, arthritis, and the bone infection osteomyelitis. The problems occur in late childhood or early adolescence.
- Autosomal recessive osteopetrosis (ARO): This form is the most severe and becomes obvious in early infancy when babies have broken bones. The dense bones in the skull may pinch the cranial nerves, causing vision loss, hearing disorders, and paralysis of the facial muscles. The heavy bone also impairs the blood cell production in the bone marrow, causing bleeding, loss of red blood cells, and infections. This type is life-threatening, with children seldom living past infancy or early childhood.
- Intermediate autosomal osteopetrosis (IOA): This form can be inherited in either autosomal dominant or recessive patterns. It has signs of the other

types but does not have the bone marrow abnormalities. However, individuals may have abnormal calcium deposits in the brain that leads to intellectual disability and kidney disorders.

- X-linked osteopetrosis (OL-EDA-ID): This rare type is characterized by swelling and fluid buildup and a condition that affects the skin, hair, teeth, and sweat glands. The individuals may also have recurring infections.

What Are the Genetic Causes of Osteopetrosis?

Mutations in several genes are associated with osteopetrosis. These genes are all related to the development and functioning of the osteoclasts, the remodeling cells that break down old bone and enable the formation of new bone cells. The following genes have been studied: *CLCN7*, *IKBKG*, and *TCIRG1*.

CLCN7

Mutations in the *CLCN7* gene, known officially as the "chloride channel 7" gene, cause several of the types of osteopetrosis. Normally, *CLCN7* instructs for making a chloride channel called ClC-7, which is found in the cells of the body. The channels are especially active in the transport of chlorine atoms to help transmit electrical signals. The channels may operate within the cells or across the cell membrane. The channels are essential for the normal function of the osteoclasts. The channels help balance the acidic environment that the osteoclasts use to dissolve bone tissue.

Over 50 mutations in *CLC7* can cause different forms: the autosomal recessive form; the autosomal dominant form; and the intermediate form. Mutations impair the ClC-7 channels, disrupting the pH balance and keeping the osteoclasts from working properly. As a result bones become dense and break easily. *CLCN7* is located on the short arm (p) of chromosome 16 at position 13.

IKBKG

Mutations in the *IKBKG* gene, officially known as the "inhibitor of kappa light polypeptide gene enhancer in B-cells, kinase gamma" gene, cause OL-EDA-ID, the type inherited on the X chromosome. Normally, *IKBKG* instructs for a protein that regulates nuclear factor kappa-B. This protein is a member of a group that attaches to DNA and controls other genes. It is essential in the proper functioning of the immune system and keeping certain cells from self-destruction, a process called apoptosis.

Several mutations in *IKBKG* are related to OL-EDA-ID and the buildup of fluid. Individuals with this type also have immune system disorders as a result of the abnormal nuclear factor-kappa-B. *IKBKG* is located on the long arm of the X chromosome at position 28.

TCIRG1

Mutations in the *TCIRG1* gene, officially known as the "T-cell, immune regulator 1, ATPase, H+ transporting, lysosomal V0 subunit A3" gene, cause type II, the most serious form of osteopetrosis. Normally, *TCIRG1* instructs for making the a3 part of a protein complex known as a vacuolar H+-ATPase (V-ATPase). The V-ATPases are enzymes that pump positively charged hydrogen atoms across the cell membranes. The action controls the pH or acidity of the environment. These enzymes are essential in the activity of the osteoclasts.

Over 60 mutations in the *TCIRG1* gene cause ARO, the most serious type. Most of the mutations cause a change in only one building block in the a3 subunit causing the protein to malfunction. *TCIRG1* is inherited in an autosomal recessive pattern and is located on the long arm (q) of chromosome 11 at position 13.2.

What Is the Treatment for Osteopetrosis?

Many of the symptoms, such as pain, can be treated on an individual basis. Surgery may be necessary to relieve the pressure on the nerves of the head due to the density of bone. The serious malignant form can only be cured with a bone marrow transplant. Several experiments treatments are being tried at this time. The use of vitamin D and recombinant human interferon gamma can possible stimulate osteoclasts. Another future treatment is the use of bone stimulators for fracture non-unions.

Further Reading

"Osteopetrosis." 2011. Genetics Home Reference. National Library of Medicine (U.S.). http://ghr.nlm.nih.gov/condition/osteopetrosis. Accessed 11/26/11.

"Osteopetrosis." 2011. Medscape. http://emedicine.medscape.com/article/123968-overview #showall. Accessed 11/26/11.

Osteopetrosis Website. 2002. http://www.osteopetrosis.org. Accessed 11/26/11.

Otogenic Vertigo. *See* Ménière Disease

Oxalosis

Prevalence	About 1 to 3 per million of type 1; common in Tunisia and other Mediterranean countries; prevalence of less-common type 2 unknown
Other Names	D-glycerate dehydrogenase deficiency; glyceric aciduria; glycolic aciduria; hepatic AGT deficiency; HP1; HP2; oxaluria, primary; peroxisomal alanine: glyoxylate aminotransferase deficiency; primary hyperoxaluria

Oxalates are chemicals that are found in certain foods, such as chocolate, leafy greens, rhubarb, and berries. For a normal person, the substances in these foods are used in the body without any problems. But if a person does not have the enzymes that processes oxalate properly, it can lead to serious consequences.

Oxalates are the natural end product of food metabolism. The body does not appear to have any real need for oxalates, and more than 90% of it is excreted from the kidney in the urine. The presence of too much oxalate in the urine is called "hyperoxaluria," coming from three Greek words: *hyper*, meaning "above and beyond"; *oxa*, meaning "oxalates"; and *uria*, meaning "urine." The condition is also known as oxalosis.

What Is Oxalosis?

Oxalosis, or primary hyperoxaluria, is a condition of the kidneys and urinary system. Oxalosis happens in the kidney when oxalates, which are the salts of oxalic acid, combine with calcium to form calcium oxalate. This compound makes up the main component of kidney stones, but it can also deposit in other areas of the body. Such deposits can damage the kidney and lead to kidney failure and injure other organs.

The following two primary forms of the disorder exist, depending on the enzyme that is not present:

- *Oxalosis or primary hyperoxaluria, type 1*: In the liver, an enzyme called alanine-glyoxylate aminotransferase (AGXT) is produced that balances the amount of oxalate that builds up. If there is a shortage of AGXT, oxalic acid builds up and combines with calcium to cause kidney stones. The stones build up, causing kidney damage and possibly kidney failure.

- *Oxalosis or primary hyperoxaluria, type 2*: In the liver, an enzyme called glyoxylate reductase/hydroxypyruvate reductase (GRHPR) helps balance the amount of oxalate buildup. If the enzyme is deficient, the excess oxalate combines with calcium to form kidney stones.

Individuals with oxalosis get progressively worse over time as the oxalates become more concentrated. In addition to the kidney, the oxalates can affect several organs. Involved may be deposits in the small blood vessels that can cause painful sores that do not heal, in the bone marrow causing anemia, and in the bone causing fractures. Deposits in the heart can cause defects.

What Are the Genetic Causes for Oxalosis?

Two genes are involved in oxalosis: *AGXT* and *GRHPR*.

AGXT

Mutations in the *AGXT* gene or the "alanine-glyoxylate aminotransferase" gene, cause oxalosis or primary hyperoxaluria, type 1. Normally, *AGXT* instructs for a liver enzyme called alanine-glyoxylate aminotransferase. In the liver, the process is controlled in a high complex of chemical actions. The liver has enzymes in structures called peroxisomes that work with cells to get rid of toxic substances and break down fats. The enzymes in the peroxisomes are imported from the cytosol or internal fluid of the liver cells. The enzyme in the peroxisome, alanine-glyoxylate aminotransferase, converts a compound called glycoxalate into the basic building block glycine. Glycine is then used for making other proteins, such as other enzymes.

About 50 mutations in *AGXT* cause type 1 oxalosis. The activity of the *AGXT* enzyme is either absent or does not work Sometimes the enzyme is misplaced. As a result of the lack of enzyme activity, glycoxalate gathers, and instead of being converted to the important building block, the amino acid glycine is turned to oxalate. The element calcium, which is also processed in the liver, combines with the oxalate to form the hard calcium oxalate, which the kidney cannot get rid of. This leads to the symptoms of kidney stones and other problems related with the disorder. *AGXT* is inherited in an autosomal recessive pattern and is located on the long arm (q) of chromosome 2 at position 37.3.

GRHPR

Mutations in the GRHPR gene, or the "glyoxylate reductase/hydroxypyruvate reductase" gene, cause oxalosis, type 2. Normally, *GRHPR* instructs for the enzyme glyoxylate reductase/hydroxypyruvate reductase, found in the liver with only a small amount in the kidneys. This enzyme has two functions: to prevent the buildup of glycoxylate by converting it into glycolate, and to convert hydroxypyruvate to D-glycerate for conversion to glucose for energy.

About a dozen mutations of *GRHPR* cause oxalosis type 2. The mutations disrupt the production of the GRHPR enzyme by changing its structure. When this change occurs, the enzyme cannot properly function. The enzyme shortage causes the glycoxylate to build up and the compound oxalate forms rather than glycolate. Calcium that is processed in the liver is there to combine with the oxalate to form calcium oxalate, which forms kidney stones and deposits in other parts of the body. *GRHPR* is inherited in an autosomal pattern and is located on the long arm (q) of chromosome 9 at position 12.

What Is the Treatment for Oxalosis?

Because oxalosis gets progressively worse as the concentration builds, it is critical to diagnose and treat as early as possible. Kidney dialysis can remove oxalate, but because of the large amount of oxalate produced, dialysis may not be able to keep pace. Some people respond to treatment with vitamin B6 and potassium or sodium citrate to decrease stone formation. People with oxalosis are advised to avoid vitamin C as the body makes oxalate from this vitamin. A kidney transplant may be advised.

Further Reading

Coulter-Mackie, Marion B.; Colin T. White; R. Morrison Hurley; Ben H Chew; and Dirk Lange. 2011. "Primary Hyperoxaluria Type 1." *GeneReviews*. http://www.ncbi.nlm.nih.gov/books/NBK1283. Accessed 5/28/12.

"Primary Hyperoxaluria." 2011. Genetics Home Reference. National Library of Medicine (U.S.). http://ghr.nlm.nih.gov/condition/primary-hyperoxaluria. Accessed 5/28/12.

"What Is Hyperoxaluria and Oxalosis?" 2011. Oxalosis and Hyperoxaluria Foundation. http://www.ohf.org/about_disease.html. Accessed 5/28/12.

P

Pachyonychia Congenita

Prevalence	Unknown but probably very rare; several people with disorder
Other Names	congenital pachyonychia; Jackson-Lawler syndrome (PC-2); Jadassohn-Lewandowsky syndrome (PC-1); Pachyonychia congenita syndrome

In 1904, Müller, a German physician, noted a combination of thickened toenails and thickened skin on the feet and palms of hands, which made his patient very miserable. Several other physicians—Wilson, Jadassohn, and Lewandowsky—reported similar conditions in 1905 and 1906. Over the years, several subtypes emerged. None of the physicians were honored with the name as the medical community referred to the condition as pachyonychia congenita. The term "pachyonychia congenita" comes from the Greek words: *pachy*, meaning "thick," and *onyx*, meaning "nail." The word "congenita" is a Latinized form of the word congenital, meaning "born with." This thickening condition of the nails is a disorder the infant has at birth.

What Is Pachyonychia Congenita?

Pachyonychia congenita is a condition that primarily affects the nails and skin. The signs and symptoms of this condition usually become apparent within the first few months of life, although a rare form of the condition known as pachyonychia congenita tarda appears in adolescence or early adulthood. Following are the symptoms of pachyonychia congenita:

- Abnormal nails: A hallmark of this disorder is thickened and abnormally shaped fingernails and toenails.
- Painful blisters: The feet will have painful blisters and calluses on the soles of the feet. When this occurs, it is almost impossible for the child to walk.

Calluses and blisters also occur on the hands. This condition, which affects the hands and feet, is known as palmoplantar keratoderma.

- Bumps around hair follicles: These bumps occur in areas that get lots of friction, such as the knees, elbows, and waistline.

- Patches in the mouth: Thick, white patches may occur inside the mouth and on the tongue. This condition is known as oral leukokeratosis.

- Excessive sweating: The palms of the hands and soles of feet may sweat excessively, a condition known as palmoplantar hyperhidrosis.

- Voice box: The conditions may affect the larynx, causing hoarseness or breathing difficulties.

Two types of pachyonychia congenita exist, and these forms are determined by their genetic origin and some differences in symptoms.

- *PC-1 or pachyonychia congenita Type 1*: This type has most of the features that are listed above and are related to mutations in two genes: *KRT6A* or *KRT16*.

- *PC-2 or pachyonychia congenita Type 2*: This type has the symptoms listed above but also some additional features. In this type, the person may develop cysts, called steatocystomas, all over the body. In addition, the person may have coarse, twisted, and brittle hair. Some babies may be born with prenatal teeth. The genes *KRT6B* or *KRT 17* are associated with this type.

What Are the Genetic Causes of Pachyonychia Congenita?

Mutations in four genes cause pachyonychia congenita. *KRT6A* and *KRT16* are related to type 1, and *KRT6B* and *KRT17* are related to type 2. All these genes instruct for tough, fibrous proteins called keratins, found in the skin, hair, and nails.

KRT6A

Mutations in the *KRT6A* gene, officially called the "keratin 6A" gene, causes PC-1. Normally, *KRT6A* provides instructions for making the protein keratin 6a or K6a, which is the keratin that gives the tough framework to skin of the palms and feet, nails, and the mucous lining inside the mouth. K6a teams with another protein K16, made by the *KRT16* gene, to create the dense networks that give strength and resiliency to the skin and nails. K6A is also involved in wound healing.

Over 20 mutations in *KRT6A* have been seen in people with PC-1. Most mutations involve only a single amino acid protein building block; a few add or delete genetic material. The mutations change the structure of K6A and disrupt the creation of the keratin filament network. Without functional keratin, the skin and nails are easily damaged, leading to the painful and distorted nails, skin, and other symptoms of the disorder. *KRT6A* is inherited an autosomal dominant pattern and is located on the long arm (q) of chromosome 12 at position 13.13.

KRT16

Mutations in the *KRT16* gene, officially known as the "keratin 16" gene, cause PC-1. Normally, *KRT16* provides instruction for the protein keratin 16 or K16. K16 works with K6 to make strong, tough keratin intermediate filaments that provide strength and resiliency to the nails, skin, and other tissues.

Over 13 mutations in *KRT16* cause PC-1. In most cases, the condition is present at birth or recognized soon after. In addition, this gene is related to a rare form of the disorder that appears later in adolescence. Most of the mutations occur in the exchange of only one amino acid building block. A few mutations may delete a few amino acids. The changes disrupt the assembly pathways of keratin and lead to the symptoms of the disorder. *KRT16* is inherited in an autosomal dominant pattern and is located on the long arm (q) of chromosome 17 at position 21.2.

KRT6B

Mutations in the *KRT6B* gene, officially called the "keratin 6B" gene, cause PC-2. Normally, *KRT6B* instructs for the protein keratin 6b or K6b. The protein is found in the tough fibrous framework of cells that form skin, hair, and nails. It is also found in the sweat glands.K6b joins with keratin 17 to from keratin intermediate filaments, which will become the network for the tough structure.

Two mutations in *KRT6B* cause PC-2. One mutation changes only one amino acid building block, while the other deletes genetic material. The mutations disrupt the work with keratin 17 and the creation of the keratin network. Without the proper structure of keratin, the soles of the feet become very fragile and blister easily, and nails do not function properly. The abnormal keratin also affects the sweat glands and leads to the development of cysts. *KRT6B* is inherited in an autosomal dominant pattern and is located on the long arm (q) of chromosome 12 at position 13.13.

KRT17

Mutations in *KRT17*, officially known as the "keratin 17" gene, cause PC-2. Normally, *KRT17* instructs for the protein keratin 17 or K17, which is produced in the nails, hair follicles, skin on the palms of hands and the soles of feet, and in the sweat glands or sebaceous glands. K17 partners with K6 to make the proper keratin structures that are tough and resilient.

About 16 mutations in *KRT17* cause PC-2. The condition is usually known in the first few months of life. Most mutations are the result of only one amino acid building block. Changes in *KRT17* alter the structure of K-17, keeping it from working effectively with K-6b. The mutations disrupt the filament network, leading to a breakdown in the skin cells, the formations of cysts, and the malfunction of the nails and hair follicle cells. *KRT17* is inherited in an autosomal dominant pattern and is located on the long arm (q) of chromosome 17 at position 21.2.

What Is the Treatment for Pachyonychia Congenita?

At present, there is no treatment for PC. However, scientists are working on a new technology called gene silencing that may have promise. The technology actually inactivates the mutant keratin gene allowing the three other genes to function. Drug therapies are limited, and people must control the symptoms by cleaning the blisters and keep the skin moist and cold. Any thing that will take pressure off feet, such as special shoes, canes, or wheelchairs can be used.

Further Reading

"Pachyonychia Congenita." 2011. Genetics Home Reference. National Library of Medicine (U.S.). http://ghr.nlm.nih.gov/condition/pachyonychia-congenita. Accessed 12/15/11.

"Pachyonychia Congenita." 2011. Medscape. http://emedicine.medscape.com/article/1106169-overview. Accessed 12/15/11.

Pachyonychia Congenita Project. 2011. http://www.pachyonychia.org. Accessed 12/15/11.

Paget Disease of Bone

Prevalence	About 2% of people over age 40 in the United States; about 1 million of Western European heritage; early-onset rare
Other Names	Osseous Paget's disease; osteitis deformans; Paget disease, bone; Paget's disease of bone; PDB

Sir James Paget, who lived and worked during the nineteenth century, was known as one of the founders of medical pathology. He wrote many works that paved the pathway for the tremendous developments in medicine that were part of the last half of the 1800s. The works were called *Lectures on Tumours* (1851) and *Lectures on Surgical Pathology* (1853). Paget researched and described many diseases and had three diseases named for him. The one that most associated with his name is Paget disease of the bone.

What Is Paget Disease of Bone?

The most known disorder that Paget is remembered for and is usually just called Paget disease is the one related to the bones. In Paget disease, the bones grow larger, become weaker, and tend to break easily. Appearing in middle age or later in only one or two bones, the condition may not have a lot of symptoms and only be detected when the person goes for an X-ray for another condition.

The classic form of the disease may affect any bone and cause symptoms of other conditions depending upon which bone is affected. Following are the bone areas that may be affected:

- Legs: One of the most common places that Paget disease affects is the leg. In Paget disease the bone is distorted, causing arthritis in near-by joints. Some people with the disorder experience pain, bowed legs, and difficulty walking.

- Knees: Distortion in any of the bones in the legs can lead to extra wear and tear and cause arthritis to develop in the knee.

- Hips: If the bones in the hips are distorted, arthritis can also affect many of the pelvic joints. Fractures are also common.

- Spine: If the disorder affects the spine, the person may experience numbness, tingling due to pinched nerves, and abnormal curvature of the spine.

- Skull: Paget disease can affect the skull, causing the head to enlarge. The person may also experience dizziness, headaches, and hearing loss.

- Bone cancer: Osteosarcoma is a rare type of cancer that may occur in fewer than 1 in 1,000 people with the disorder.

Although the classic condition causes weakened and misshapened bones, it does not appear to spread from one bone to another.

Much less common is an early-onset form that occurs in the teens or the early 20s. The symptoms vary somewhat from the classic form in that it appears to affect the bones of the skull, spine, ribs, and small bones of the hands. The form may also cause hearing loss in early life.

What Are the Genetic Causes of Paget Disease?

This disease is one in which environment may play a role. Several genes increase the risk for the disorder, but other factors may trigger the disorder. For example, certain viruses may cause the disorder. The genes that create risk and possibly cause Paget disease are *SQSTM1*, *TNFRSF11A*, and *TNFRSF11B*.

SQSTM1

Mutations in the *SQSTM1* gene, officially known as the "sequestosome 1" gene, may put the person at risk for Paget disease. Normally, *SQSTM1* instructs for a protein called p62, which is essential for breaking down old bone and replacing it with new bone. Osteoclasts are specialized bone cells that break down old bone; osteoblasts are the cells that build up new bone. The p62 protein controls the formation of osteoclasts. It may also play a role in recycling the worn-out cell parts and the body's immune response.

About 20 mutations in the *SQSTM1* gene cause Paget disease. Most of the mutations occur when the amino acid leucine replaces proline. The mutations appear to overactivate the chemical signals that form the osteoclasts, thereby triggering the cells to break down bone before it is time to do so. The broken down bone is then replaced with weaker and less organized normal bone. The bones are then larger, more misshapened, and easily fractured. *SQSMT1* is inherited in an autosomal dominant pattern and is located on the long arm (q) of chromosome 5 at position 35.

TNFRSF11A

Mutations in the *TNFRSF11A* gene, officially known as the "tumor necrosis factor receptor superfamily, member 11a, NFKB activator" gene, increase the risk for Paget disease. Normally, *TNFRSF11A* instructs for the protein receptor activator of NF-kB, also called RANK. The RANK protein plays an active role in bone remodeling, the process in which old bone cells are broken down and replaced with new bones. RANK is located on the surface of immature osteoclasts and receives the signals as to when it is time for these cells to mature and do the work of breaking down unneeded cell parts.

Two mutations in *TNFRSF11A* have been found to cause the early-onset form of Paget disease. An abnormal copy of genetic material is in the gene, causing RANK to have extra protein building blocks. The extra material appears to cause RANK to send signals to the osteoclasts to start breaking down old bone. This signal then calls upon the osteoblasts to start building up bone. The new bone that was quickly produced is much weaker and less organized than normal bone and is easily fractured and misshapened. *TNFRSF11A* is inherited in an autosomal dominant pattern and is located on the long arm (q) of chromosome 18 at position 22.1.

TNFRSF11B

Mutations in *TNFRSF11B*, or the "tumor necrosis factor receptor superfamily, member 11b" gene, increase the risk of Paget disease. Normally, *TNFRSF11B* instructs for a protein called osteoprotegerin. This protein is essential for the bone remodeling process. In bone remodeling, osteoclasts break down old bone and osteoblasts from new bone. Osteoprotegerin is one of two receptors that can bind to the activator NF-κB ligand (RANKL). The other is NF-κB (RANK). Because only one receptor at a time can bind with the ligand, the two compete with each other. When RANKL binds to RANK, it triggers the immature osteoclasts to mature and begin breaking down bone cells. When osteoprotegerin binds to RANKL, the signal prevents the activation of osteoclasts, and no action takes place. The two help balance each other so that the amount of production of the osteoclasts is normal.

About six mutations in *TNFRSF11B* cause juvenile Paget disease. Each mutation disrupts the function of osteoprotegerin, letting RANL bind only to RANK and thereby producing too many osteoclasts to break down bone. When bone is broken down so rapidly, the osteoblasts work to replace quickly bone cells, causing bone cells that are weaker and less organized.

Although mostly connected with the juvenile form, one mutation of *TNFRSF11B* indicates that the gene may increase the risk of Paget disease especially in women. *TNFRSF11B* is inherited in an autosomal dominant pattern and is located on the long arm (q) of chromosome 8 at position 24.

What Is the Treatment for Paget Disease?

Because the disease is one of the more common bone disorders, several researchers have investigated and developed treatments that can slow down the rate of bone breakdown and pain. Following are several of the interventions:

- Bisphosphonates: These pharmaceuticals have been shown to be effective in several conditions relating to bone loss. The drugs target the excessive breakdown that happens in Paget disease. Examples are alendronate or Fosamax and risedronate or Actonel.
- Calcitonin: An injectable form can treat Paget disease, but it might not be as effective as bisphosphonates.
- Pain relievers: These include acetaminophen, aspirin, or ibuprofen.
- Exercise: Regular exercise is recommended to keep bone strength.
- Diet: Adequate amount of calcium and vitamin D helps bone strength.
- Surgery: Surgery may be necessary on the affected bone or joint. Hip and knee replacement may help.

See also Aging and Genetics: A Special Topic

Further Reading

"Paget's Disease." 2011. Arthritis Foundation. http://www.arthritis.org/disease-center.php?disease_id=19&df=treatments. Accessed 5/28/12.

"Paget Disease of Bone." 2011. OrthoInfo. American Association of Orthopedic Surgeons. http://orthoinfo.aaos.org/topic.cfm?topic=A00076. Accessed 5/28/12.

The Paget Foundation. 2011. http://www.paget.org. Accessed 5/28/12.

Pallister-Hall Syndrome

Prevalence   Rare; prevalence unknown

Other Names CAVE complex; cerebroacrovisceral early lethality complex; Hall-Pallister syndrome; hypothalamic hamartoblastoma syndrome; PHS

In 1980, Philip Pallister and Judith Hall noted several serious conditions in some patients. The child was born with a hamartoblastoma, a new growth in normal tissues that appears at random but does not reproduce. However, the presence near the hypothalamus created grave concerns. Several conditions were present. They wrote about it in the *American Journal of Medical Genetics*, calling it a new syndrome. The disorder was named for the two doctors—Pallister-Hall syndrome.

What Is Pallister-Hall Syndrome?

Pallister-Hall syndrome is a developmental disorder that affects many parts of the body. Following are the areas affected and the problems that are related:

- Hypothalamic hamartoma: The hypothalamus is an endocrine gland that is located deep in the brain and controls many functions of growth and reproduction. A hamartoma is a benign growth that appears in normal tissue and does not expand. Hypothalamic hamartoma is characteristic of Pallister-Hall syndrome. Some hamartomas can lead to seizures or hormone problems that can be life-threatening.

- Extra fingers or fused fingers and toes: Most people with this disorder have some digital disorders. They may have extra fingers and toes, a condition known as polydactyly. They may have fused fingers and toes where the toes are grown together, a condition known as syndactyly.

- A bifed epiglottis: The epiglottis is the small door that closes over the trachea to keep food from going down into the lungs. The word "bifed" means split in two. In this condition, a cleft of split is in the epiglottis, causing it not to work properly.

- Anal opening: The area of the anus at the end of the digestive tract has an obstruction that causes problems with proper elimination of waste.

- Kidney abnormalities.

The symptoms can vary from mild to severe. However, only a small number appear to have serious complications.

What Is the Genetic Cause of Pallister-Hall Syndrome?

Mutations in the *GLI3* gene, officially known as the "GLI family zinc finger 3" gene, cause Pallister-Hall Syndrome. Normally, *GLI3* works during embryonic development instructing for proper patterns and shaping of many organs and tissues. *GLI3* is a member of a family of *GLI* genes that attach to specific areas of DNA, turning genes on and off as needed. Thus the GLI proteins are called transcription factors.

The *GLI* family works with a protein called Sonic Hedgehog, which creates a path for early development of structures in brain and limbs. The signal from Sonic Hedgehog tells GLI3 proteins to activate or turn off other genes.

Several mutations in *GLI3* cause Pallister-Hall syndrome. The mutations are located near the middle of the gene and create early stop signals for making the protein. This abnormal protein is short and can only turn off genes but not turn them on. The disruption of the protein causes defects in the embryonic development. *GLI3* is inherited in an autosomal dominant pattern and is located on the short arm (p) of chromosome 7 at position 13.

What Is the Treatment for Pallister-Hall Syndrome?

The most urgent treatment is for endocrine abnormalities, especially cortisol deficiencies. The bifed epiglottis generally does not cause a problem necessary for treatment. Polydactyly or anal disorders may require surgery. Special education for developmental delays is available.

Further Reading

Biesecker, Leslie G. 2010. "Pallister-Hall Syndrome." *GeneReviews*. http://www.ncbi.nlm .nih.gov/books/NBK1465. Accessed 5/28/12.

"Pallister-Hall Syndrome." 2011. Genetics Home Reference. National Library of Medicine (U.S.). http://ghr.nlm.nih.gov/condition/pallister-hall-syndrome. Accessed 5/28/12.

"Pallister-Hall Syndrome." 2011. RightDiagnosis.com. http://www.rightdiagnosis.com/p/ pallister_hall_syndrome/intro.htm. Accessed 5/28/12.

Pallister-Killian Mosaic Syndrome

Prevalence　　Rare but unknown; more than 100 reported in literature; mild symptoms may go undiagnosed

Other Names　isochromosome 12p syndrome; Pallister-Killian syndrome; PKS; Teschler-Nicola/Killian syndrome; tetrasomy 12p, mosaic

In 1977, Philip Pallister described a condition in which a child was born with an unusual skin color pattern, along with other severe birth defects. Later in 1981, Maria Teschler Nicola and W. Killian noted the same condition. The disorder is most commonly named Pallister-Killian syndrome. About 100 of the most severe cases have been described in the literature. However, the actual number of cases is unknown because people with mild symptoms may go undiagnosed.

What Is Pallister-Killian Mosaic Syndrome?

Pallister-Killian mosaic syndrome is a developmental disorder that involves chromosome 12. The disorder is not inherited. The condition, which is present at birth, affects many parts of the body.

The symptoms may vary from mild to severe. The following signs generally relate to this disorder:

- Skin: The term mosaic well describes the color of the skin. People with this condition have streaks or patches of skin that are darker or lighter than surrounding pigment. The changes can be anywhere on the body and can appear at birth or later.

- Hair: Children have little hair on their heads. Hair is especially sparse around the temples but may fill in as they get older.

- Weak muscle tone: This symptom, called hypotonia, is usually very severe. Because of the weak muscle tone, the child has trouble with breathing, feeding, sitting, standing, and walking.

- Intellectual disability: This disability is usually profound.

- Facial appearance: The face is described as "coarse." The features are a very high forehead, broad nasal bridge, wide space between the eyes, droopy eyelids, epicanthic folds, rounded cheeks, wide mouth with thin upper lip, large ears that are thick and protrude outward, and a large tongue.

- Speech: The child may never learn to speak or may have limited speech.

- Several birth defects: Other defects may include a cleft or highly arched palate, heart defects, hernia involving the diaphragm, extra nipples, crossed eyes, abnormal opening in the anus, narrowing of the ear canal, and skeletal malformations.

Most of the defects in severe cases are life-threatening in early infancy. Even children with mild cases have intellectual disabilities and some physical abnormalities.

What Is the Genetic Cause of Pallister-Killian Mosaic Syndrome?

Abnormalities on chromosome 12 cause Pallister-Killian mosaic syndrome. Normally, a person has two copies of each chromosome 12 in each of the cells. One is inherited from each parent. A normal chromosome has a long arm designated as q and a short arm called p. However, people with Pallister-Killian mosaic syndrome have the two usual copies of chromosome 12, but some cells have an extra chromosome called an isochromosome. The Greek root word *iso* means "equal." Instead of having a long and short arm, an isochromosome has equal arms. The isochromosome may have two q arms or two p arms. The version of chromosome 12 has two p arms.

Chromosome 12 is fairly large, having about 132 million DNA base pairs and representing about 4% of the total DNA in cells. In the human genome, about 20,000 to 25,000 genes are present. Between 1,200 and 1,400 genes are on chromosome 12. Several genetic disorders, such as several types of cancers and tumors, are related to this chromosome.

Pallister-Killian mosaic syndrome is the result of a nondisjunction or error in cell division in the either the sperm or the egg. It is a random anomaly. There is no history of the condition in either family. As the cells divide in early embryonic development, some of the cells lose the isochromosome 12p, resulting in a condition called mosaicism. Because of the extra material from an abnormal chromosome in some cells, normal development of the fetus is affected causing the features and defects of this disorder.

What Is the Treatment for Pallister-Killian Mosaic Syndrome?

Because this is a serious chromosomal condition, no specific therapy for individuals with Pallister-Killian mosaic syndrome exists. The prognosis for those with serious defects is poor, and only supportive care is available. Children may benefit from early intervention programs and special education.

Further Reading

"Chromosome 12." 2011. Genetics Home Reference. National Library of Medicine (U.S.). http://ghr.nlm.nih.gov/chromosome/12. Accessed 5/28/12.

"Pallister-Killian Mosaic Syndrome" 2011. Genetics Home Reference. National Library of Medicine (U.S.). http://ghr.nlm.nih.gov/condition/pallister-killian-mosaic-syndrome. Accessed 6/1/11.

PKS Kids. 2010. http://www.pkskids.net. Accessed 5/28/12.

Pantothenate Kinase-Associated Neurodegeneration (PKAN)

Prevalence   Unknown: about 1 to 3 per million worldwide

Other Names   NBIA1; neuroaxonal dystrophy, juvenile-onset; neurodegeneration with brain iron accumulation type 1; PKAN

In 1922, Hallervorden and Spatz, two German physicians, noted a family with a progressive nervous system disorder that began in childhood and continued with loss of motor and intellectual functions. Upon autopsy, the doctors found the brains were loaded with iron. The disorder was called Hallervorden-Spatz syndrome (HSS). However, during World War II, the two doctors were considered to be involved in unethical practices related to the Nazi regime. The medical community recognized the term neurodegeneration with brain iron accumulation type 1 (NBIA-1). Most recently, the name of the disorder is called according to the clinical description pantothenate kinase-associated neurodegeneration.

What Is Pantothenate Kinase-Associated Neurodegeneration (PKAN)?

Pantothenate kinase-associated neurodegeneration is a disorder of the nervous system characterized by the buildup of iron in the brain. At one time, this concentration of iron was only determined at autopsy, but now it can be diagnosed with a scan of the brain with magnetic resonance imaging (MRI). When the doctor looks at the brain scan, he will note a specific change called the eye-of-the-tiger sign, which indicates the accumulation of iron in the brain.

The following symptoms characterize PKAN:

- Movement problems: The person will slowly begin involuntary muscle spasms and rigidity, and then trouble walking, which worsens over time.
- Problems with speech.
- Vision loss.
- Loss on intellectual function, called dementia.
- Psychiatric problems: The person has changes in behavior and personality, which may include depression.

Several forms of PKAN exist.

- *Classic form of PKAN*: This form begins in childhood and is associated with the severe and rapid worsening of movement.
- *Atypical form of PKAN*: The features of movement begin more slowly and progress later in childhood or adolescence. However, people with this type of PKAN have more speech defects and psychiatric problems.
- *HARP*: At one time this disorder was considered a separate syndrome but now is considered part of PKAN. HARP is an acronym for hypoprebetalipo-proteinemia, acanthocytosis, retinitis pigmentosa, and pallidal degeneration. All these conditions describe issues with movement, dementia, and vision.

What is the Genetic Cause of Pantothenate Kinase-Associated Neurodegeneration (PKAN)?

Mutations in the *PANK2* gene, officially known as the "pantothenate kinase 2" gene, cause pantothenate kinase-associated neurodegeneration. Normally, *PANK2* provides instructions for the pantothenate kinase 2 enzyme. Active in the mitochondria, the energy powerhouse of the cells, this enzyme is essential for regulating the formation of the molecule coenzyme A. Coenzyme A is necessary for producing energy from carbohydrates, fats, and some amino acids. *PANK2* is active in cells throughout the body, especially in the brain and nervous system.

About 100 mutations in *PANK2* cause PKAN. Persons with the rapid, early-onset classic type have mutations that prevent the production of any of the enzyme pantothenate kinase 2. People with the later-onset atypical form have changes in a single amino acid in the enzyme. The most common exchange in the enzyme is the replacement of arginine for the amino acid glycine. Whether the enzyme is missing or altered, both lead to the disruption of coenzyme A and harmful buildup of iron in the brain. The toxic buildup causes swelling, brain damage, and impaired energy production. *PANK2* is inherited in an autosomal recessive pattern and is located on the short arm (p) of chromosome 20 at position 13.

What Is the Treatment for Pantothenate Kinase-Associated Neurodegeneration (PKAN)?

Treating the spasms and muscular disorders with medication or injections of botulinum may be necessary. Deep brain stimulation has been used in some cases with effectiveness. Other treatments are symptomatic and on an individual basis. Adaptive aids for movement, educational services, and services for the blind are available.

Further Reading

"Hallervorden-Spatz Disease." 2011. Medscape. http://emedicine.medscape.com/article/ 1150519-overview. Accessed 12/15/11.

"*PANK2*." 2011 Genetics Home Reference. National Library of Medicine (U.S.). http:// ghr.nlm.nih.gov/gene/PANK2. Accessed 12/15/11.

"Pantothenate Kinase-Associated Neurodegeneration." 2011. Genetics Home Reference. National Library of Medicine (U.S.). http://ghr.nlm.nih.gov/condition/pantothenate -kinase-associated-neurodegeneration. Accessed 12/15/11.

Paramyotonia Congenita

Prevalence   Uncommon disorder; affects fewer than 1 in 100,000

Other Names   Eulenburg disease; paralysis periodica paramyotonia; Paramyotonia congenita of von Eulenburg; PMC; Von Eulenberg's disease

What Is Paramyotonia Congenita?

Paramyotonia congenita is a disorder of the skeletal muscles and is a form of myotonia, or muscle dysfunction. As the name implies, the child is born with the condition, and symptoms begin in early childhood. The following four major symptoms characterize this disorder:

- Muscle stiffness: The stiffness occurs in the muscles of the neck, face, arms and hands.
- Flaccid muscle paralysis: The muscles are flabby and do not move.
- Reduced reflexes.
- Difficulty relaxing muscles following contraction: The muscles become tense and do not relax normally.

Two conditions appear to make the symptoms worse: exercise and cold.

- *Exercise*: The above symptoms appear after exercise, and muscle cooling induces the stiffness. Although most people with muscle stiffness respond to exercise, those with this condition worsen with activity.
- *Cold*: People, even without muscle disorder, report that muscles do not function when they are cold. However, people with paramyotonia congenital have serious reactions. They react to brief periods of cold and respond severely to prolonged periods.

What Is the Genetic Cause of Paramyotonia Congenita?

Mutations in the *SCN4A* gene, officially called the "sodium channel, voltage-gated, type IV, alpha subunit" gene, cause paramyotonia congenita. Normally, *SCN4A* provides instructions for creating sodium channels. These channels are essential for moving sodium atoms or ions into cells, which then enable the cell to transmit electrical signals. Muscle contractions depend upon this flow of sodium into the cells. This flow is essential for muscles to move, tense, and relax.

About 13 mutations in *SCN4A* are related to paramyotonia congenita. Most of the mutations involve the exchange of only one amino acid building block. The most common change occurs when one of the other amino acids replaces arginine. The mutations change the structure and function of the sodium channels in the muscle cells. For example, a mutation may delay the closing of a channel in response to cold; the delay also keeps the channels closed longer. Extra sodium ions then move in causing the episodes of muscle stiffness. *SCN4A* is inherited in an autosomal dominant pattern and is located on the long arm (q) of chromosome 17 at position 23.3.

What Is the Treatment for Paramyotonia Congenita?

Physical therapy in this case does not appear to help. Some individuals learn to manage their symptoms without specific treatment by learning to recognize the triggers for episodes and avoid them. Others may need certain medications, such as mexiletine and acetazolamide, to treat severe stiffness.

Further Reading

"Paramyotonia Congenita." 2011. Genetic and Rare Diseases Information Center (GARD). National Institutes of Health (U.S.). http://rarediseases.info.nih.gov/GARD/Condition/7325/Paramyotonia_congenita.aspx. Accessed 12/16/11.

"Paramyotonia Congenita." 2011. Muscular Dystrophy Association. http://www.mda.org/disease/pc.html. Accessed 12/16/11.

Parkinson Disease (PD)

Prevalence　Affects more than 4 million persons worldwide; 1 million in the United States; occurs in about 13 per 100,000 people; 50,000 new cases each year

Other Names　idiopathic Parkinsonianism; paralysis agitans; Parkinson's disease; PD; Primary Parkinsonism

In 1817, a British apothecary named James Parkinson wrote an article describing people as having a condition with shaking or trembling hands or limbs, who then ended up with severe motor dysfunction. In this *Essay on the Shaking Palsy*, he

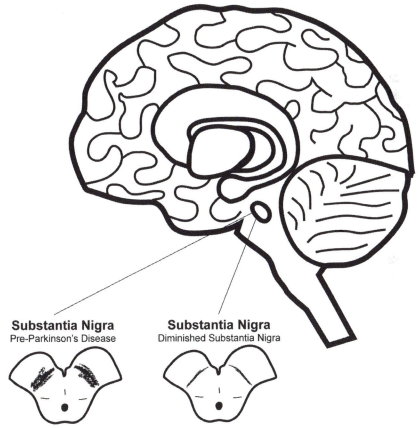

Substantia Nigra
Pre-Parkinson's Disease

Substantia Nigra
Diminished Substantia Nigra

In Parkinson disease, the disorder affects an area in the brain called the substantia nigra. Neurons in this area of the brain become dysfunctional and die. (ABC-CLIO)

called the condition "paralysis agitans." Later Jean-Martin Charcot called the condition Parkinson's disease after the apothecary.

On Parkinson's birthday each year on April 11, activists raise funds for research for a treatment of Parkinson disease. They have adopted the red tulip as the symbol of the disease. Several celebrities who have touted the cause of Parkinson include the actor Michael J. Fox and boxer Muhammad Ali, both of whom have Parkinson disease.

What Is Parkinson Disease (PD)?

Parkinson disease (PD) is a progressive disorder of the brain and its control of movement functions. The condition affects several areas of the brain, especially an area deep in the brain called the substantia nigra. Neurons in this area of the brain become dysfunctional and die. The normal function of this area is to produce a chemical messenger called dopamine, which acts to produce steady physical movements. When the neurons that produce dopamine die, movements become unsteady and erratic. Deposited in the place of the dopamine-producing neurons are deposits called Lewy bodies.

The death of the dopamine-producing cells then causes several symptoms. Following are the signs of Parkinson disease:

- Tremor: Probably one of the first symptoms is a shaking limb that occurs when the person is at rest. The tremor disappears with voluntary movements and sleep. Tremor usually begins on one side of the body and then eventually moves to arms, legs, feet, and face. Another feature of tremor is called "pill rolling." The individual appears to be rolling pills. Early pharmacists used the thumb and index fingers to make round pills for their patients; the tremor reminded some early physicians of this action. Some people do not develop tremors at the onset of PD, but as the disease progresses, tremors occur.

- Slow movement or bradykinesia: This symptom is one of the most disabling signs of the early stages of the disorder. The person cannot do things that require fine motor skills, such as writing, sewing, or getting dressed. This feature can be puzzling. The person may be barely able to walk but can ride a bicycle. Eventually, the person may develop a shuffling movement in which he or she takes little baby steps. Some people describe the movement as "the march of little feet."

- Rigidity: Rigidity is characterized by stiffness of the limbs and trunk. Increased muscle tone leads to the continuous contraction of muscles. The two types of rigidity paint a picture of the movement: uniform or lead-pipe rigidity, and ratchety or cogwheel rigidity. Pain may occur in the joints. In early stages of the disorder, the rigidity tends to affect the face and shoulders. Eventually, the stiffness spreads to the entire body, keeping it from moving freely.

- Loss of balance and frequent falls: The person cannot keep good balance, a condition called postural instability. The number of falls is related to the severity of the disease. Bone fractures may result as a secondary issue.

- Other motor signs may include rapid shuffling of feet and a forward-leaning posture when walking.

- Speech and swallowing disturbances.

- A mask-like expression.

- Some neuropsychiatric problems: Usually these occur in advanced stages of PD. These problems are with the executive function of the brain, which includes planning, abstract thinking, appropriate action, and flexibility. Memory may be affected but can improve when aided by cues. Some people develop dementia, and other individuals become anxious and disturbed.

Late-onset Parkinson begins after age 50; early-onset begins before age 50. Cases that start before the age of 20 are called juvenile-onset Parkinson disease.

What Are the Genetic Causes of Parkinson Disease (PD)?

Although many of the cases of Parkinson are sporadic with no family history, the condition may be the interaction of the environment and certain unidentified genetic factors. However, about 15% of people have a family history of PD. Mutations are related to the following eight genes: *GBA*, *LRRK2*, *PARK2*, *PARK7*, *PINK1*, *SNCA*, *SNCA1P*, and *UCHK1*.

GBA

The *GBA* gene is officially known as "glucosidase, beta, acid." Normally, *GBA* instructs for an enzyme called beta-glucocerebrosidase, which is active in the recycling centers of the lysosomes. The lysosomes break down toxic invaders, digest bacteria, and recycle worn-out cell parts. Enzymes in the lysosomes are housekeeping enzymes. Beta-glucocerebrosidase is one of these housekeeping enzymes. It breaks down a large molecule of glucocerebroside into glucose and a simple fat molecule. Mutations of *GBA* may affect movement and balance. Symptoms of PD result from loss of dopamine-producing nerve cells in the brain. The *GBA* mutations disrupt the housekeeping function of the lysosomes, causing toxic substances to build up. The toxic substances kill the neurons, leading to abnormal movement and balance. The inheritance pattern is unknown. *GBA* is located on the long arm (q) of chromosome 1 at position 21.

LRRK2

The gene *LRRK2* is officially called the "leucine-rich repeat kinase 2" gene. Normally, this gene instructs for a little-known protein called dardarin. The gene itself gives clues about dardarin's function. Part of *LRKK2* instructs for a building block called leucine. Leucine appears to play a part in assembling the cell framework. Dardarin also is involved in the cells in phosphorylation and the ultimate production of energy. Twenty mutations of *LRRK2* have been found in families with late-onset PD. In the dardarin molecule, amino acids are exchanged, disrupting the function and causing tremor of late-onset PD. *LRRK2* is inherited in an autosomal dominant pattern and is located on the long arm (q) of chromosome 12 at position 12.

PARK2

The gene *PARK2* is officially named the "Parkinson protein 2, E3 ubiquitin protein ligase (parkin) gene. Normally, *PARK2*, one of the largest human genes, instructs for making the protein parkin. Parkin is part of the housekeeping machinery that breaks down unwanted proteins and tags them with ubiquitin. Ubiquitin gives the signal to move unneeded proteins into structures known as proteosomes, which act as cell garbage disposals. More than 100 mutations of *PARK2* have been found in juvenile PD and early-onset PD. The mutations lead to very small molecules of parkin, which do not work to degrade unwanted cell material. Disruption of the ubiquitin-proteasome system could allow buildup of toxic materials in the dopamine-producing areas of the brain, leading to the characteristic symptoms of PD. *PARK2* is inherited in an autosomal recessive pattern and is located on the long arm (q) of chromosome 6 at position 25.2-q27.

PARK7

The gene *PARK7* is officially called the "Parkinson disease (autosomal recessive, early onset)" gene. Normally, *PARK7* instructs for a protein called DJ-1, which helps protect brain cells from oxidative stress. Oxidative stress happens when certain unstable molecules called free radicals roam around the body and collect in large numbers and kill cells. Also, DJ-1 may act as a chaperone molecule that helps newly formed proteins hold their shapes. It may also be active in repairing damage to proteins. Chaperone molecules are very important molecules in maintaining body health. For example, DJ-1, acting as a chaperone molecule, delivers tagged proteins to the proteasomes, the machinery that breaks down unneeded proteins. DJ-1 may also be active in the RNA process.

More than 10 mutations in *PARK7* cause early-onset Parkinson. Deletions or exchanges of building blocks cause an unstable DJ-1 to malfunction. Researchers suggest that the most disruption occurs in the protein's chaperone function, allowing toxic materials to build up. Others believe that the abnormal protein does not protect the dopamine-producing cells against the oxidative stress of the free radicals. *PARK7* is inherited in an autosomal recessive pattern and is located on the short arm (p) of chromosome 1 at position 36.23.

PINK1

The *PINK1* gene is officially called the "PTEN induced putative kinase 1" gene. Normally, this gene instructs for the protein PTEN induced putative kinase 1, which is found in the mitochondria of cells throughout the body. The enzyme appears to protect the mitochondria during periods of high energy demands. More than 20 *PINK1* mutations cause early-onset Parkinson. The mutations disrupt the function of the enzyme, leading to an inability to protect the mitochondria when overstressed. Cells die when the mitochondria function. Just how *PINK1* mutations cause the death of the dopamine-producing cells is not known. *PINK1* is inherited in an autosomal recessive pattern and is located on the short arm (p) of chromosome 1 at position 36.

SNCA

The gene *SNCA* is officially named the "synuclein, alpha (nonA4 component of amyloid precursor)" gene. Normally, *SNCA* instructs for alpha-synuclein, a small protein found especially in the brain. This protein plays a special part in the release of chemical messengers called neurotransmitters, which help impulses go from one neuron to another across a gap called the synapse. Several studies suggest that the protein plays a role in regulating the release of dopamine.

Two mutations in *SNCA* disrupt this process. One type replaces one amino acid with another, causing the alpha-nuclein protein to misfold or make a faulty 3-D shape. In another type of alteration, two *SNCA* genes duplicate incorrectly and, instead of having two copies, may have three or four copies, leading to the production of an excess of alpha-synuclein. The mutated forms may gather and disrupt the recycling of dopamine. The misfolded alpha-synucleins are a major component of Lewy bodies, the deposits that appear in the substantia nigra. *SNCA* is inherited in an autosomal dominant pattern and is located on the long arm (q) of chromosome 4 at position 21.

SNCA1P

The *SNCA1P* gene is officially named the "synuclein, alpha interacting protein." Normally, *SNCA1P* instructs for the protein synphilin-1 and another version called synphilin-1A. These proteins gather at an area called the presynaptic vesicles that are located at the tips of neurons. Here the proteins interact with alpha-synuclein to help neurotransmitters like dopamine cross the synaptic gap. Only one mutation of *SNCA1P* has been found in very few cases of Parkinson. The mutation occurs when one of the building blocks, cysteine, replaces the amino acid arginine. Disruption of the protein appears to cause faulty synphilin-1 proteins to gather together, causing the death of dopamine-producing cells in the substantia nigra. The inheritance pattern of *SNCA1P* is unknown. It is located on the long arm (q) of chromosome 5 at position 23.2.

UCHL1

The gene *UCHL1* is officially named the "ubiquitin carboxyl-terminal esterase L1 (ubiquitin thiolesterase)" gene. Normally, *UCHL1* instructs for ubiquitin carboxyl-terminal esterase L1, an enzyme found in the nerve cells of the brain. Ubiquitin is a substance that tags unwanted cells for breakdown and tells the proteasomes when to move on these defective proteins. A mutation caused by a single amino acid exchange can disrupt the process, allowing unwanted proteins to build up and kill the neurons in the brain. The inheritance pattern of *UCHL1* is unknown. It is located on the short arm (p) of chromosome 4 at position 14.

The genetics of Parkinson disease is quite complicated, and researchers continue to investigate the genetic causes of the disorder.

What Is the Treatment for Parkinson Disease (PD)?

Finding a cure for Parkinson disease has been a passion of activists and several foundations. However, at present, no cure exists. However, a variety of strategies are under investigation.

Drugs and Pharmaceuticals

Early motor symptoms respond well to the use of levodopa, a dopamine agonist that converts to dopamine in the brain. Combined with carbidopa, which delays the conversion to dopamine until it reaches its destination in the brain, levodopa enables the nerve cells to use dopamine for the brain's functions. Not all symptoms respond well to levodopa. For example, bradykinesia and rigidity respond best, but problems of balance and tremors may not be affected at all. As people use the drug over an extended period of time, the symptoms may not respond, and serious side effects may occur. A condition called tardive dykinesia in which the person develops involuntary writhing movements may occur.

Other medications mimic the role of dopamine in the brain. Some of the medications are bromocriptine, pramipexole, and ropinirole. In 2006, the FDA approved rasagaline to be used with levodopa in patients with advanced PD.

Deep Brain Stimulation

Some cases of PD do not respond to drugs, and the person may want to investigate surgery. The U.S. Food and Drug Administration has approved a procedure known as deep brain stimulation. Small electrodes are implanted in the brain and connected to an external device called a pulse generator. This generator is programmed to reduce the need for levodopa and the other drugs that cause such serious side effects. In some patients, deep brain stimulation has reduced tremors, slow movement, and shuffling gait. One issue is programming the impulses properly.

Stem Cell Research

Usually by the time the person has clinical symptoms, the majority of the cells in the substantia nigra have died. This makes PD a good candidate for stem cell replacement. Several reasons give hope to this strategy:

> ➤ The neurological damage is confined to one area of the brain, the substantia nigra.
> ➤ A specific type of cell, the dopamine-producing neurons, is needed to relieve the symptoms of the disease.

➤ Preclinical animal research has been shown to produce the needed neurons.

➤ Injection of neuronal stem cells into Parkinson animal models has relieved some of the motor deficits.

Gene Therapy

For the same reasons as listed above, gene therapy appears to be a viable option. However, locating the specific gene that causes the disorder is essential for this procedure. Investigators in the field are working on the possibilities of using gene therapy, but at this time, the potential seems remote.

See also Aging: A Special Topic; Gene Therapy: A Special Topic

Further Reading

National Parkinson Foundation. 2011. http://www.parkinson.org/Parkinson-s-Disease/ PD-101. Accessed 5/28/12.

"Parkinson Disease." 2011. MedlinePlus. National Library of Medicine (U.S.). http:// www.nlm.nih.gov/medlineplus/parkinsonsdisease.html. Accessed 5/28/12.

"Parkinson's Disease." 2011. MedicineNet.com. http://www.medicinenet.com/ parkinsons_disease/article.htm. Accessed 5/28/12.

Paroxysmal Nocturnal Hemoglobinuria

Prevalence   Rare; affects between 1 and 5 per million people

Other Names   hemoglobinuria, paroxysmal; Marchiafava-Micheli syndrome

In 1882, German physician Paul Strubing described a condition in which the patient had reoccurring symptoms of red urine and anemia. Beginning in 1911, two Italian physicians, Dr. Ettore Marchiafava and Dr. Alessio Nazari, studied the syndrome in great detail and wrote about it in several publications. However, the Dutch physician Enneking coined the long term "paroxysmal nocturnal hemoglobinuria" in 1928, and this descriptive term has become the common name for the disease.

The term "paroxysmal nocturnal hemoglobinuria" can be broken down into understandable words. "Paroxysmal" means that the symptoms keep reoccurring. Nocturnal refers to night. "Hemoglobinuria" indicates that the red blood cells, which make hemoglobin red, are found in the urine.

What Is Paroxysmal Nocturnal Hemoglobinuria?

Paroxysmal nocturnal hemoglobinuria is different from many genetic disorders in that it is acquired. Body stresses, such as overexertion or infection, may trigger the attacks. During the episodes, the body destroys the red blood cells, which carry oxygen in the hemoglobin. Hemoglobin is what gives blood its red color. The destroyed hemoglobin is passed in the urine, giving it a red color. It is usually most noticeable in the morning, thus assuming the attack was carried out during the night.

Several processes of the disorder are noted in the major blood cells:

- Red blood cells or erythrocytes: When the red blood cells are destroyed, the person may experience fatigue, weakness, very pale skin, shortness of breath, and an increased heart rate. The condition is referred to as hemolytic anemia. Hemolytic anemia and the premature destruction of red blood cells are related to a component of the immune system called complement. Complement normally protects the cells of a person from being destroyed. However, the abnormal cells in people with paroxysmal nocturnal hemoglobinuria are missing the two important GPI anchor proteins that attach them to the cell membrane. These red blood cells are destroyed, and hemolytic anemia results.

- White blood cells or leucocytes: In this disorder, the white blood cells are also affected. White blood cells are involved in the immune system. The individual then has little resistance to infection. People with this disorder are at risk for developing leukemia, a cancer of the blood.

- Platelets or thrombocytes: In this disorder, the thrombocytes, which are involved in blood clotting, may be abnormal. People with this disorder experience clotting or thrombosis, especially in the large abdominal veins. Sometimes, excessive bleeding can occur.

What Is the Genetic Cause of Paroxysmal Nocturnal Hemoglobinuria?

Mutations in the *PIGA* gene, officially called the "phosphatidylinositol glycan anchor biosynthesis, class A" gene, cause paroxysmal nocturnal hemoglobinuria. Normally, *PIGA* instructs for making the protein phosphatidylinositol glycan class A. This protein helps produce a molecule called the GPI anchor. This anchor helps attach many proteins to the cell membrane, which make these proteins ready for action when needed.

This is where *PIGA* differs from other genetic conditions. In people with this disorder, gene mutations occur during their lifetime and are called somatic mutations. They are not inherited or present at birth. Red blood cells are formed in the bone marrow in cells known as hematopoietic stem cells. The somatic mutations occur during the process of this formation, resulting in abnormal blood cells. The abnormal cells multiply along with the normal ones that are produced.

The abnormal white blood cells that are also produced can mistakenly treat normal blood-forming stem cells as an enemy. In addition, the abnormal cells may be less susceptible to self-destruction and the process of apoptosis. More than 100 mutations of *PIGA* have been found on the short arm (p) of chromosome X at position 22.1.

What Is the Treatment for Paroxysmal Nocturnal Hemoglobinuria?

This condition is difficulty to treat. Treating the symptoms is the only real strategy. Some physicians say that using steroids is effective, but this strategy is debatable.

Further Reading

"Diagnosis and Management of Paroxysmal Nocturnal Hemoglobinuria." 2005. American Society of Hematology. http://bloodjournal.hematologylibrary.org/content/106/12/3699.long. Accessed 5/28/12.

"Paroxysmal Nocturnal Hemoglobinuria." 2011. Genetics Home Reference. National Institutes of Health (U.S.). http://ghr.nlm.nih.gov/condition/paroxysmal-nocturnal-hemoglobinuria. Accessed 5/28/12.

"Paroxysmal Nocturnal Hemoglobinuria." http://www.ncbi.nlm.nih.gov/books/NBK22166. Accessed 5/28/12.

Patau Syndrome. *See* Trisomy 13 (Patau Syndrome)

Patent Ductus Arteriosus (Char Syndrome)

Prevalence   Rare; only a few families worldwide

Other Names   Char syndrome; patent ductus arteriosus with facial dysmorphism and abnormal fifth digits

Both names, patent ductus arteriosus and Char syndrome, appear in the scientific literature. The rare condition has been known since 1938, when Dr. Robert Gross performed the first successful ligation of patent ductus arteriosus on an eight-year-old girl at Children's Hospital in Boston.

What Is Patent Ductus Arteriosus (Char Syndrome)?

Patent ductus arteriosus, also known as Char syndrome, affects many body areas. Abnormal defects are seen in the face, heart, and hands. Following are characteristics of the three major defects of this disorder:

- Face: Facial appearance is flat. Person has a flat nose, flat tip of nose, and nasal bridge with flattened cheek bones. Eyes slant down and are wide-set with droopy eyelids. A very short distance between the nose and upper lip, which is an area called the philtrum, exists. The mouth is shaped like a triangle with very thick lips.

- Heart: The patent ductus arteriosus is a connection between the two major arteries, the aorta and the pulmonary artery. The connection closes soon after birth. However, if the area does not close, the infant will have breathing difficulties, eating problems, and failure to thrive. The child is at risk for infection. Hearing and vision loss may occur. If untreated, the condition is life-threatening because of the risk of heart failure.

- Hand: The middle section of the little finger is short or missing. Other abnormalities of the hands and feet may exist.

What Is the Genetic Cause of Patent Ductus Arteriosus (Char Syndrome)?

Mutations in the *TRAP2B* gene, officially called the "transcription factor AP-2 beta (activating enhancer binding protein 2 beta)" gene, cause patent ductus arteriosus. Normally, *TFAP2B* instructs for a protein called AP-2β that binds to certain regions of DNA. This transcription factor controls the action of specific genes on the DNA. These genes regulate cell division and the process of self-destruction called apoptosis. AP-2β is especially active in embryonic development. One of the early areas of development is called the neural crest. Cells in the neural crest form parts of the nervous system, endocrine glands, skin pigment, and smooth muscle especially in the heart, the face, skull, and the limbs.

About 10 mutations of the *TFAP2B* cause the condition patent ductus arteriosus. The mutations change the structure of AP-2β so that the binding ability is lost. Some of the mutations prevent any of the transcription factors from working. The structures that normally develop as part of the neural crest are disrupted, resulting in the abnormal development of many of the body tissues. *TFAP2B* is inherited in a dominant pattern and is located on the short arm (p) of chromosome 6 at position 12.

What Is the Treatment for Patent Ductus Arteriosus (Char Syndrome)?

Attention to the heart defect is the first order of treatment. The surgeon will determine the degree of opening between the two arteries. Surgical options include ligation or closing the area so that the blood will flow normally. Any other problems will be dealt with according to the symptom in a routine manner. Children can benefit from special education and training.

Further Reading

Gelb, Bruce D. "Char Syndrome." 2008. *GeneReviews*. http://www.ncbi.nlm.nih.gov/books/NBK1106. Accessed 5/28/12.

"Patent Ductus Arteriosus (PDA)." 2011. Mayo Clinic. http://www.mayoclinic.com/health/patent-ductus-arteriosus/DS00631. Accessed 5/28/12.

"Patent Ductus Arteriosus (PDA)." 2011. Medscape. http://emedicine.medscape.com/article/759542-overview. Accessed 5/28/12.

Pelizaeus-Merzbacher Disease (PMD)

Prevalence   About 1 in 200,000 to 500,000 males in the United States; rare in females

Other Names   brain, Pelizaeus-Merzbacher; Cockayne-Pelizaeus-Merzbacher disease; PMD sclerosis

In 1885, two German physicians noted five boys in one family who were very limp. All the boys had little head control, developmental delay, and involuntary eye movements. Later, in 1910, Dr. Merzbacher studied the same family, which had then progressed to 14 members including two girls. Merzbacher traced the condition to a single female ancestor and concluded the disorder was passed thought the female line and not father-to-son. The disorder was named after these two doctors—Pelizaeus-Merzbacher disease.

What Is Pelizaeus-Merzbacher Disease (PMD)?

Pelizaeus-Merzbacher disease is a condition that affects the myelin of the brain and spinal cord. Myelin is the tough, fatty white covering in the central nervous system that insulates the axons of nerve cells and also is involved in transporting nerve signals. PMD is one of a group of diseases called leukodystrophies that affect the myelin sheath. Because of the dysfunctional myelin sheath, the following symptoms may occur:

- Delayed motor skills: The individual may have difficulty with skills involving coordination, such as holding up the head, moving, and walking. This condition is referred to as hypotonia.
- Abnormal eye movements: The muscles in the eyes are weak, causing involuntary movements called nystagmus.
- Impaired intellectual skills: The individual may have poor cognitive functions such as memory and language.

Two types of conditions characterize Pelizaeus-Merzbacher disease: the classic and connatal. The two types differ in severity but may have some overlapping features. Following are the two types and their specific symptoms:

1. The classic type of PMD: This type is the most common and may occur during the first year of life. The child has really poor muscle control, delayed crawling or walking, and abnormal eye movements. As the child gets older, the eye movements stop, but other weaknesses occur. The child develops stiff movements, a condition called spasticity, balance disorders, and involuntary jerking.

2. The connatal type of PMD: This type is the most serious, with symptoms beginning almost at birth. The child has feeding problems, spasticity that limits movement, seizures, and speech difficulties. These children develop few motor or intellectual skills.

thyroid, pendrin carries the iodide ions across the membrane. In order for the thyroid to produce hormones, iodide must bond to a protein called thyroglobulin. A similar thing is happening in the inner ear. Pendrin is carrying both chloride and bicarbonate ions across the membrane, maintaining just the right level for hearing to take place. The level of fluid is important to the development of the inner ear. The same process probably shaped the vestibular canals and bony structures of the inner ear.

About 60 mutations in *SLC26A4* gene cause Pendred syndrome. Some of the mutations change only a single DNA building block. For example, in a mutation prevalent in the Japanese population, arginine replaces histidine. In Europeans, proline replaces threonine. Other mutations may include deletions or additions of building blocks. The results of all the changes make an abnormal protein that is very small and does not function. The abnormal protein disrupts the transport of the negatively charged ions across the membrane causing the problems within the inner ear and with the thyroid. *SLC24A4* is inherited in an autosomal recessive pattern and is located on the long arm (q) of chromosome 7 at position 31.

What Is the Treatment for Pendred Syndrome?

Many specialists may be involved in the treatment and care of the person. The physicians should include an audiologist, an endocrinologist, a clinical geneticist, a genetic counselor, an otolarnygologist, and a speech pathologist. Children with the disorder should take special precautions when participating in any activity that can cause head injury. Early intervention with speech therapy and special education is essential. A cochlear implant may help the person understand speech. The goiter should be checked for regularly.

Further Reading

"NIDCD Fact Sheet: Pendred Syndrome." 2011. National Institute on Deafness and Other Communication Diseases (U.S.). http://www.nidcd.nih.gov/staticresources/health/hearing/FactSheetPendredSyndrome.pdf. Accessed 6/19/11.

"Pendred Syndrome." 2011. MedicineNet. http://www.medicinenet.com/pendred_syndrome/article.htm. Accessed 5/28/12.

Periventricular Heterotopia

Prevalence   Unknown

Other Names   familial nodular heterotopia; peri ventricular nodular heterotopia

During embryonic development, many things can happen. Periventricular heterotopia is one of those random mistakes of nature. The term "periventricular" combines the Greek roots *peri*, meaning "around," and *ventriculos*, meaning "little belly" or "small cavity." The cavity that the term refers to is located near the center on either side of the brain. The word "heterotopia" comes from two Greek words: *hetero*, meaning "other" or "different," and *tapas*, meaning "place." Putting the terms together means that periventricular heterotopia is a disorder when bits of gray matter in the brain locates in the area near a cavity in the brain rather than where it should be.

What Is Periventricular Heterotopia?

Periventricular heterotopia is a rare condition that occurs during embryonic development when the neurons do not get to the right place. Between the 6th week and 24th week of pregnancy, neurons normally form in the periventricular area, the fluid-filled cavities in the center of the brain. After forming, the neurons then move to the outside layers of the brain called the cerebral cortex. In periventricular heterotopia, the neurons do not move to their proper places, and clumps of gray matter or neurons are located in the wrong place in the brain. The clumps remain in the area around the ventricles or periventricular area. The neurons appear to be normal but not in their right place.

The symptoms of the disorder vary and include the following:

- Epilepsy: Almost all individuals with this disorder have some type of epilepsy or seizures. The clumps in the ventricles are discovered when the person has magnetic resonance imaging (MRI) studies. Sometimes the seizures are severe and lead to loss of motor skills and cognitive function.

- Intelligence: Normally, the person is of normal intelligence but may have difficulty reading or spelling, a condition known as dyslexia.

- Brain malformations: This malformation is less common and may result in a small head, developmental delays, many infections, and blood vessel problems.

- Association with other disorders: Sometimes periventricular heterotopia is seen with other disorders such as Ehlers-Danlos syndrome, a condition in which the person has extremely flexible joints, stretchy skin, and weak blood vessels.

People with periventricular heterotopia generally have a normal life expectancy.

What Are the Genetic Causes of Periventricular Heterotopia?

Three genetic areas may be responsible for periventricular heterotopia: the gene *FLNA*, the gene *ARFGEF2*, and chromosome 5.

FLNA

Mutations in the *FLNA* gene, officially called the "filamin A, alpha" gene, cause most cases of periventricular heterotopia. Normally, *FLNA* instructs for the protein filamin A. Filamin A is part of the network of the cytoskeleton or framework of the cell. Filamin attaches to another protein called actin to give shape and structure to the cell and also to help other proteins regulate body functions such as skeletal and brain development.

More than 25 mutations in *FLNA* are related to periventricular heterotopia. In most mutations, the protein filamin A is too short, leading to disrupting the formation of the cytoskeleton and normal cell movement. When nerve cells are affected, they do not move or migrate to their proper place and just clump around the area of their formation in the periventricular area. A few mutations cause the exchange of one building block for another resulting in a protein that partially works. These mutations cause a milder form of the disorder. *FLNA* is inherited in an X-linked dominant pattern and is located on the short arm (q) of the X chromosome at position 28.

ARFGEF2

Mutations in the *ARFGEF2* gene, officially called the "ADP-ribosylation factor guanine nucleotide-exchange factor 2 (brefeldin A-inhibited)" gene, cause periventricular heterotopia. Normally, *ARFGEF2* instructs for a protein that helps move small sacs within the cells called vesicles. The ARFGEF2 protein is responsible for converting guanine diphosphate (GDP) to another molecule called guanine triphosphate (GTP). The two work together to set into motion a molecule that is involved in cell movement called the ADP-ribosylation factor. Movement of the neuron cells is central to normal brain development.

Only a few mutations in *ARGEF2* have been identified with periventricular heterotopia. The mutations disrupt the movement patterns of the neurons, causing them to form the clumps of gray matter in the periventricular area. Some mutations may also weaken the cells that form the lining of the ventricle or cavity. This weakening could cause some of the neurons to clump but let others migrate to the normal exterior of the brain. *ARFGEF2* is inherited in an autosomal recessive pattern and is located on the long arm (q) of chromosome 20 at position 13.13.

Chromosome 5

A few cases of periventricular heterotopia have been traced to chromosome 5. This chromosome has between 900 and 1,300 genes and has numerous roles in the body. It also has many genetic disorders associated with it. Extra genetic material in chromosome 5 may lead to the signs and symptoms of periventricular heterotopia. Finding genes on chromosome 5 is an active area of current research.

What Is the Treatment for Periventricular Heterotopia?

Management of epilepsy is the main area of concern. Many cases of periventricular heterotopia are diagnosed only when the person has recurrent seizures and the source of the seizure needs to be pinpointed in the MRI. Most cases have focal epilepsy, and the drug of choice is generally carbamezipine. However, because antiepileptic drugs may cause side effects, treatment is often cautious. If the person has difficulties in school with dyslexia, he may qualify for special education under the Individuals with Disabilities Education Act (IDEA).

Further Reading

"Periventricular Heterotopia." 2011. Genetics Home Reference. National Library of Medicine (U.S.). http://ghr.nlm.nih.gov/condition/periventricular-heterotopia. Accessed 5/28/12.

"X-Linked Periventricular Heterotopia." 2011. Genetics and Rare Disorder Information (GARD). National Institutes of Health (U.S.). http://rarediseases.info.nih.gov/GARD/ Condition/7371/Xlinked-periventricular_heteroto pia.aspx. Accessed 5/28/12.

Peters Plus Syndrome

Prevalence   Rare; fewer than 70 people worldwide

Other Names   Krause-Kivlin syndrome; Krause–van Schooneveld–Kivlin syndrome; Peters anomaly–short limb dwarfism syndrome; Peters'-plus syndrome; Peters' plus syndrome

In 1906, Peters described three brothers that had clouding and severe abnormalities of the front part of the eye, especially the cornea. The cornea is the transparent outer covering of the eyeball. The condition was referred to as Peters anomaly. In 1984, Van Schooneveld wrote about 11 cases of people with Peters anomaly but who also with had several other characteristics such as short-limbed dwarfism. He referred to it a Peters Plus syndrome because the syndrome had some more disorders not documented by Peters earlier.

What Is Peters Plus Syndrome?

Peters Plus syndrome is a disorder that affects several areas of the body especially the eyes and stature. Following are the characteristics of Peters Plus syndrome:

- Eye abnormalities: The front part of the eye has a fluid-filled structure called the anterior chamber that is located between the cornea, or clear covering of the eye, and the iris, or colored part of the eye. A condition known as Peters anomaly involves the clouding and thinning of the cornea and the attachment of the iris to the cornea. Blurred vision is the result. The condition may also be involved with clouding or the lenses, a condition known as cataracts. The severity of the condition varies from family to family.

- Short stature: All people with Peters Plus are short in stature. The condition is noted at birth when the baby is shorter than average. The average height for an adult man is about 4 feet, 7 inches; females are shorter, with average height being about 4 feet, 2 inches. The upper part of the arms are short, a condition called rhizomelia; and fingers and toes are short, a condition called brachydactyly.

- Distinct facial features: The facial features of Peters Plus include a large forehead, very narrow eyes, and a long distance between the nose and mouth, an area called the philtrum. The person also has a broad neck and a double curve on the upper lip, called a Cupid's bow.
- Clefts in lip or palate: A cleft is one of the areas seen in about half of the individuals with this condition.
- Developmental delay: Most children with Peters Plus have mild to severe intellectual disability, although some may have normal intelligence.
- Less common features: Not as usual with Peters Plus are heart defects, brain defects, hearing loss, hypothyroidism, kidney disturbances, and genital defects.

What Is the Genetic Cause of Peters Plus Syndrome?

Mutations in the *B3GALTL* gene, officially called the "beta 1,3-galactosyltransferase-like" gene, cause Peters Plus syndrome. Normally, *B3GALTL* instructs for an enzyme beta-I, 3-glucosyltransferase (B3Glc-T). This enzyme is one of the steps for adding sugar molecules to proteins, a process called glycosylation. This process then allows proteins to do a lot of different type of functions. The *B3GALTL* gene is active in most cells of the body, suggesting the enzyme it instructs for is important to cells and their functions.

About four mutations cause Peters Plus syndrome. Most common is the mutation called a splice-site mutation, in which the building block adenine replaces guanine. The mutation causes a very short enzyme that does not work properly, leading to the symptoms of Peters Plus. *B3GALTL* is inherited in an autosomal recessive pattern and is located on the long arm (q) of chromosome 13 at position 12.3.

What Is the Treatment for Peters Plus Syndrome?

A corneal transplant before the age of three to six months may prevent severe disability from clouding of the eye. The other symptoms of Peters Plus may be treated with interventions according to the symptoms. Educational interventions are needed. There are several support networks for parents of children with Peters Plus syndrome.

Further Reading

Aubertin, Gudrun Marjolein Kriek, and Saskia, A. J. Lesnik Oberstein. 2011. "Peters Plus Syndrome." *GeneReviews*. http://www.ncbi.nlm.nih.gov/booksINBK1464. Accessed 5/28/12.

"Peters Plus Syndrome." 2011. OMIM. Online Mendelian Inheritance in Man. http://www.omim.org/entry/261540?search=peters%20plus%20syndrome&highlight=syndromic%20peter%20plus%20syndrome%20peters. Accessed 5/28/12.

Peutz-Jeghers Syndrome (PJS)

Prevalence   Unknown; about 1 in 25,000 to 300,000 births

Other Names intestinal polyposis-cutaneous pigmentation syndrome; lentigino-sis, perioral; periorificiallentiginosis syndrome; Peutz-Jeghers polyposis; PJS; polyposis, hamartomatous intestinal; polyposis, intestinal, II; polyps-and-spots syndrome

In 1921 Jan Peutz, a Dutch physician, noted a condition in a Dutch family charac-terized by growths in the intestines combined with dark spots on the lips and inside the mouth. Later in 1949, Harold Jeghers, an American physician, wrote about the syndrome connecting the melanin spots with generalized intestinal polyps. In honor of the two physicians, Andre Bruwer introduced the eponym Peutz-Jeghers syndrome or PJS in 1954.

What Is Peutz-Jeghers Syndrome (PJS)?

Peutz-Jeghers syndrome is a condition in which a number of polyps or growths develop in the stomach and intestines combined with spots on various parts of the body. The symptoms are characterized as follows:

- Numerous growths or polyps: The growths that develop are hamartomas, a type of benign or noncancerous tumor, which consist of a lot of tree-like networks of smooth muscle wrapped in tissue. The polyps can be located anywhere in the gastrointestinal tract but primarily are found in the stomach and intestine. The polyps develop during childhood and adolescence and can cause recurrent bowel obstructions, bleeding, and stomach pain. Although the polyps are not malignant, a number of people with PJS develop cancer in the gastrointestinal tract, pancreas, cervix, ovary, and breast with a higher frequency than the general population.

- Small dark spots: A small amount of melanin or dark pigment cells may gather around the mouth, inside the mouth, near the eyes, around the nose, around the anus, and on the hands and feet. The spots appear in childhood and tend to fade or disappear as the child gets older.

What Is the Genetic Cause of Peutz-Jeghers Syndrome (PJS)?

Mutations in the *STKII* gene, officially known as the "serine/threonine kinase 11" gene, cause Peutz-Jeghers syndrome. Normally, *STKII* instructs for the tumor sup-pressor enzyme called serine/threonine kinase 11 A tumor suppressor keeps cells from growing uncontrollably. This enzyme also has other roles such as helping cells orient within specific tissues, determining the amount of energy the cells use, and promoting cell death. Using all these mechanisms, serine/threonine kinase 11 prevents tumors from growing wildly in certain body parts.

More than 140 mutations in *STKII* cause Peutz-Jeghers syndrome. Most of the mutations produce a very short, abnormal version of serine/threonine kinase 11 enzyme. When the enzyme is lost or does not function properly, cells continue to divide too often, causing the growths in the gastrointestinal tract. The loss of the enzyme also increases the risk of malignant cancers to grow. *STKII* is inherited in an autosomal dominant pattern and is located on the short arm (q) of chromosome 19 at position 13.3.

What Is the Treatment for Peutz-Jeghers Syndrome (PJS)?

Gastrointestinal specialists must monitor the stomach and intestines for the development of cancers. Standard treatment of malignancies in other parts of the body may be needed. Testing for cancer in all areas of the body should begin as early as eight years of age.

Further Reading

Amos, Christopher I.; Marsha L. Frazier; Chongjuan Wei; and Thomas J. McGarrity. 2011. "Peutz-Jeghers Syndrome." *GeneReviews*. http://www.ncbi.nlm.nih.gov/books/ NBK1266. Accessed 5/28/12.

"Peutz-Jeghers Syndrome." 2011. Genetics Home Reference. National Library of Medicine (U.S.). http://ghr.nlm.nih.gov/condition/peutz-jeghers-syndrome. Accessed 5/28/12.

"Peutz-Jeghers Syndrome." 2011. Medscape. http://emedicine.medscape.com/article/ 182006-overview. Accessed 5/28/12.

Pfeiffer Syndrome

Prevalence   About 1 in 100,000 births

Other Names   acrocephalosyndactyly, type V; ACS5; ACS V; craniofacial-skeletal-dermatologic dysplasia; Noack syndrome

In 1964, Rudolf Arthur Pfeiffer, a German physician, described a condition in which the newborns had a prominent forehead and bulging eyes. He also noted that instead of the fontanelle or soft spot, the bones of the skull had grown together before birth. The condition was named after this doctor—Pfeiffer syndrome.

What Is Pfeiffer Syndrome?

Pfeiffer syndrome is a disorder in which the bones of the head fuse together before birth, a condition known as craniosynostosis. The word "craniosynostosis" comes from the Greek words "*cranio*," meaning "head"; *syn*, meaning "together"; *ost*, meaning "bone"; and *osis*, meaning "condition." At birth, the skull normally has areas called sutures where the bones have not grown together. This are is referred to as the fontanelle or "soft spot." The fontanelles enable the child to pass through the birth canal in a normal head-first position with the bones having some flexibility. In Pfeiffer syndrome, these bones have fused prematurely. The following symptoms are related to the syndrome:

- Skull fusion: The premature fusion of the skull gives an abnormal shape to head and face.
- Facial features: Bulging and wide-set eyes are characteristic of the disorder. The premature fusion of the bones in the skull causes these distortions. The child has a high forehead, a small lower jaw, sunken midface, and a nose like a beak.
- Hearing loss.
- Dental problems.
- Digit disorders: Thumbs and big toes are very wide and bend away from the other digits. Fingers are very short, a condition known as brachydactyly. Digits may be fused or webbed, a condition known as syndactyly.

Pfeiffer syndrome has three subtypes as follows:

1. Type 1, the classic type: These individuals have most of the symptoms listed above but have a normal life expectancy and normal intelligence. This type is most common of the three.
2. Type 2: These individuals have much more serious problems with the brain and nervous system. Development may be delayed. Individuals with type 2 have obvious fusion of the bones in the skull that develop a cloverleaf-shaped head.
3. Type 3: Serious neurological and developmental problems exist but the bones of the head are similar in shape to type 1.

What Are the Genetic Causes of Pfeiffer Syndrome?

Two genes appear to be the causes of Pfeiffer syndrome. The genes are *FGFR1* and *FGFR2*.

FGFR1

Mutations in the *FGFR1* gene, officially known as the "fibroblast growth factor receptor 1" gene, cause Pfeiffer syndrome. Normally, *FGFR1* instructs for making the protein fibroblast growth factor receptor 1. This protein is related to cell division, blood vessel formation, and embryonic development. The protein is so constructed that one end is in the cells and the other projects out from the surface of the cell membrane. This protein is thus called a transcription factor. Growth factors that are in the bloodstream bind to the protein, causing chemical reactions that enable the cell to function. The FDFR1 protein is essential in the development of the nervous system and may help regulate the growth of long bones in the arms and legs.

The different types of Pfeiffer syndrome are caused by mutations in the *FGFR1* gene. Type 1 occurs when the amino acid arginine replaces praline. The mutation appears to send extra signals that keep the bones growing, causing the premature fusion. The changed proteins also affect the bones of the hands and feet. When a mutation causes the abnormal signaling such as that in *FGFR1*, it is called a "gain of function" mutation. *FGFR1* is inherited in an autosomal dominant pattern and is located on the short arm (p) of chromosome 8 at position 12.

FGFR2

The second gene involved with Pfeiffer syndrome is the *FGFR2* gene, officially known as the "fibroblast growth factor receptor 2" gene. Normally *FGFR2* instructs for a protein called fibroblast growth factor receptor 2. Like the FGFR1 protein, the FGFR2 protein is involved in important cell processes of growth, especially during embryonic development. Also like FGFR1 protein, this protein is a transcription factor scanning the bloodstream for specific growth factors from its position just outside the cell membrane. The FGFR2 protein is essential for bone growth, signaling immature cells in the embryo to become bone cells.

About 25 mutations of *FGFR2* cause Pfeiffer syndrome. Most of the mutations change the number of cysteine building blocks in the protein. The mutations cause the FGFR2 protein to increase the signaling, causing the premature fusion of the skull bones and problems with the bones of the hands and feet. *FGFR2* is inherited in an autosomal dominant pattern and is located on the long arm (q) of chromosome 10 at position 26.

What Is the Treatment for Pfeiffer Syndrome?

The treatment depends on the severity of the disease. Surgery may be necessary to reshape the skull. Jaw surgery and orthodontal work may straighten teeth. Surgery also may be essential for eye disorders. For type 1 especially, the prognosis is good. Children may have normal intelligence and life expectancy.

Further Reading

"A Guide to Understanding Pfeiffer Syndrome." 2011. http://www.ccakids.com/syndrome/pfeiffer.pdf. Accessed 5/28/12.

"Pfeiffer Syndrome." The Craniofacial Center. http://www.thecraniofacialcenter.org/pfeiffer.html. Accessed 5/28/12.

"Pfeiffer Syndrome." 2011. Genetics Home Reference. National Library of Medicine (U.S.). http://ghr.nlm.nih.gov/condition/pfeiffer-syndrome. Accessed 5/28/12.

Phelan-McDermid Syndrome

Prevalence   About 500 cases known worldwide

Other Names   deletion 22q13; deletion 22q13.3 syndrome; syndrome monosomy 22q13; 22q13 deletion syndrome

Interest in communication disorders and speech delays or autism has spurred search for genetic causes for the condition. Autism is not just one disease, but a collection of several conditions dubbed as autism spectrum disorders. Likewise, many genetic connections have been made. In 2001, M. C. Phelan located a deletion on chromosome 22 that was connected with communication delays and wrote about it in the *American Journal of Medical Genetics*. The condition Phelan-McDermid syndrome is also known as 22q13.3 syndrome.

What Is Phelan-McDermid Syndrome?

Phelan-McDermid syndrome, or 22q13.3 deletion syndrome, occurs when a small bit of chromosome 22 is deleted. The missing part is located on the long arm (q) at position 13.3. This deletion in this area appears to affect many different areas of the body and bodily function.

The syndrome has a wide variety of symptoms that vary greatly. Following are the characteristics:

- Communication disorders: The child has poor social interaction, little eye contact, and problems with speech.
- Autistic-like behaviors: The person may be sensitive to touch and have aggressive behaviors. They may do unusual things such as chewing on non-food items such as clothing.
- Subtle physical features: The child has droopy eyelids with puffiness around the eyes, a long, narrow head with a pointed chin, and deep-set eyes with long eyelashes. The physician may not recognize these symptoms.
- Rapid growth.
- Feet: The person may have large hands and/or feet. Fusion of second and third toes along with small and abnormal toenails is common.
- Developmental delay: Child is delayed in sitting up, rolling over, crawling, or walking. Moderate to profound intellectual disability almost always occurs.
- Decreased muscle tone: Newborns usually have floppy muscle tone, a condition known as hypotonia.
- Seizures.
- Decreased sensitivity to pain.
- Inability to sweat: This condition causes a risk for overheating and dehydration.

- Episodes of vomiting and backflow or reflux of stomach acid into the esophagus causing heartburn. Newborns especially have trouble feeding; fluids tend to leak out of the mouth.
- Emotional instability.

What Is the Genetic Cause of Phelan-McDermid Syndrome?

Absence or loss of genes on the lip of chromosome 22 causes Phelan-McDermid syndrome—also known as 22q 13 Deletion. A ring chromosome may occur that breaks off important areas of the gene. A ring chromosome results when chromosome tips break off and fuse together, forming a ring. People with this ring have a copy of the abnormal chromosome in some or all of the cells. Chromosome 22 is the second-smallest human chromosome, with about 50 million base pairs and about 500 to 800 genes.

Specifically lost with the deletion of chromosome 22 is the *SHANK3* gene, officially called the "SH3 and multiple ankyrin repeat domains 3" gene. Normally, *SHANK3* instructs for a protein that is associated with the development of the human nervous system and brain. The protein is essential for the proper working of the synapses, the gap between dendrite of one neuron and the axon of another. The impulses or nerve signals move in one direction by crossing these synapses. The SHANK3 protein creates a scaffold that enables the nerve impulses to move from one neuron to the other. *SHANK3* also plays a role in the function of these dendrites, the small tree-like structures on one side of the neuron.

In Phelan-McDermid syndrome, the part of chromosome 22 that houses *SHANK3* is missing or deleted. Without this gene, cell-to-cell communication across the synapse does not function. This lack of communication leads to the developmental delay, intellectual deficits, and other characteristics of this disorder. Most cases of this deletion are caused by a random event during the formation of egg or sperm or early in fetal development. However, affected people can pass on the disorder to offspring. It is thought to be inherited in an autosomal dominant pattern. The deletion is located on the long arm (q) of the chromosome at position 13.3.

What Is the Treatment for Phelan-McDermid Syndrome?

The treatment for Phelan-McDermid syndrome varies with different children and different families. Because this is a genetic condition, physicians can only treat the symptoms. This treatment depends on the display of symptoms. However, physical and occupational therapy, and speech and vision therapy, can help. The child qualifies for free education under the Individuals with Disabilities Education Act.

Further Reading

"22q13 Deletion Syndrome." 2011. Genetics Home Reference. National Library of Medicine (U.S.). http://ghr.nlm.nih.gov/condition/22q133-deletion-syndrome. Accessed 5/28/12.

"Chromosome 22." 2011. Genetics Home Reference. National Library of Medicine (U.S.). http://ghr.nlm.nih.gov/chromosome/22. Accessed 5/28/12.

"Phelan-McDermid Syndrome." 2011. Disabled World News. http://www.disabled-world.com/health/pediatric/phelan-mcdermid-syndrome.php. Accessed 5/28/12.

Phenylketonuria (PKU)

Prevalence   About 1 in 10,000 to 15,000 newborns in the United States; known worldwide; one of the disorders screened in newborns

Other Names   deficiency disease, phenylalanine hydroxylase; Folling disease; Folling's disease; P AH deficiency; phenylalanine hydroxylase deficiency disease; PKU

In 1909, Dr. Archibald Garrod in England noted the dark urine in a family of patients and acclaimed he had found an inborn error of metabolism. The disorder was alkaptonuria. "Uria" in a word indicates found in the urine.

A second inborn error of metabolism was discovered by a Norwegian physician, Ivar Asbjorn Folling, in 1934. Folling noted hyperphenylalaninemia (HPA) was associated with mental retardation. He also studied the urine of these patients and found chemicals that were in all the patients of certain children that were blonde and had a tendency toward eczema. Folling was the first to apply chemical analysis to the study of disease. The disorder called phenylketonuria in the United States is still called Folling disease in Norway.

What Is Phenylketonuria?

Phenylketonuria (PKU) is a condition in which the child has an elevated level of phenylalanine in the blood. Normally, phenylalanine is one of the amino acid protein building blocks and is found in all food with protein and in some artificial sweeteners. Phenylalanine builds up when certain inherited metabolic conditions do not function. If PKU is not treated, it is very toxic and can cause serious health problems and mental disability.

Untreated children are normal at birth but develop the symptoms when given food. Classic PKU, the most severe form of the disease, has the following symptoms if the children are not treated with a special low-phenylalanine diet:

- Permanent intellectual disability
- Delayed development
- Seizures
- Behavioral problems
- Psychiatric disorders

PKU: Dr. Folling and Another Inborn Error of Metabolism

In 1934, a desperate mother with two retarded children nervously waited for Dr. Ivar Folling. She had come to ask the doctor for help, although she thoroughly expected none. But this woman would not take "no" for an answer. The field of biochemistry was not advanced, but was just opening up. At the University of Norway where Dr. Folling was a professor of nutrient research, there was no well-equipped laboratory; his laboratory was in the attic of the medical ward.

Folling agreed to examine the children. The little girl was 6.5 years old but could say only a few words and walked with a labored spastic gait. A little four-year-old boy could not walk, speak, or eat on his own. His bathroom habits were those of a baby.

The urine had a unique smell, but ne found no protein or glucose in the urine. When he added ferric chloride, he was puzzled by a deep green color. He wondered if he would get the same reaction another time with all forms of medication stopped. Both children had something in their urine that was not found in normal urine.

Now he had to identify the substance. Chemists today have all kinds of techniques to help them, but in 1934, he had to isolate and purify the substance. After using traditional chemical methods, he finally found crystals of phenylpyruvic acid. The children had this in their urine; normal children did not.

But did this have any connection with the mental disability? He collected samples from 430 patients in different institutions and found the green color in 8. These children were fair-complexioned with eczema. He published his results in a German journal in 1934.

He continued to work on the topic and found later after studying many families that it fit the pattern of an autosomal recessive disorder. Later, he found that something in the food that the children were eating was causing the illness.

This discovery of PKU not only helped people with the disease, but opened up the whole field of medicine to the biochemistry and inborn error of metabolism. PKU and its discovery has been a model in the field of clinical genetics.

- Distinct mouse-like smell when phenylalanine is excreted through the skin and in the urine
- Albino-like light skin and hair
- Eczema

A mild form of PKU is called variant PKU or hyperphenylalaninemia. These children have fewer symptoms and may not have to follow the low-phenylalanine diet.

Another condition occurs when people with PKU have children. If the adults with PKU have abandoned their low-phenylalanine diets, the babies may have intellectual disability at birth because they are exposed to high doses of phenylalanine. These children may have several health problems including heart defects, low birth weight, small head, and behavioral problems. Women with PKU are at risk during pregnancy for loss of the child.

In most states, children are screened at birth. The blood is taken by a tiny prick on the heel and given a test called the HPLC test. Newborn screening is typically 6 to 14 days after birth. A repeat test is done within two weeks. In most cases, this condition can be prevented with the test and following the diet.

What Is the Genetic Cause of Phenylketonuria?

Mutations in the *PAH* gene, officially known as the "phenylalanine hydroxylase" gene, cause PKD. Normally, *PAH* instructs for making the enzyme phenylalanine hydroxylase. The enzyme is a first step in processing the building block phenylalanine obtained from the diet. Phenylalanine hydroxylase converts phenylalanine to another amino acid, tyrosine. To carry out the chemical reaction, the enzyme must work with another molecule, tetrahydrobiopterin (BH4). Tyrosine is active in the creation of certain hormones, the pigment that gives skin and hair color, and the creation of energy.

More than 500 mutations in the *PAH* gene cause PKU. A mutation can cause only a single amino acid change in the protein. The most common is when tryptophan replaces

arginine. Other mutations delete small amounts of the gene. The mutations disrupt the functioning protein, causing the buildup of excessive amounts of phenylalanine and the symptoms of phenylketonuria. *PAH* is inherited in an autosomal recessive pattern and is found on the long arm (q) of chromosome 12 at position 22-q24.2.

What Is the Treatment for Phenylketonuria?

Restricting the diet of items containing phenylalanine is essential. These include most protein-rich foods. The following foods should be avoided: milk, eggs, cheese, nuts, soybeans, beans, chicken, beef, fish, chocolate, and peas. A warning label on certain sodas with artificial sweeteners reads: "Phenylketonuriacs: Contains phenylalanine." Some medicines are made with aspartame, an artificial sweetener, which converts to phenylalanine when digested. Doctors today recommend sticking to the diet throughout life although at one time it was thought that people with PKU could stop in the teenage years. The FDA has approved the use of the drug sapropterin (Kuvan) for treatment in addition to continued diet.

See also Newborn Screening: A Special Topic

Further Reading

"Phenylketonuria." 2011. Genetics Home Reference. National Library of Medicine (U.S.). http://ghr.nlm.nih.gov/condition/phenylketonuria. Accessed 5/28/12.

"Phenylketonuria." 2011. Mayo Clinic. http://www.mayoclinic.com/health/phenylketonuria/ds00514/dsection=treatments-and-drugs. Accessed 5/28/12.

PKU News. 2011. http://www.pkunews.org. Accessed 5/28/12.

Polycystic Kidney Disease (PKD)

Prevalence   One of the most common disorders caused by mutations in a single gene; affects 500,000 individuals in the United States; dominant type affects 1 in 500–1,000; recessive type affects 1 in 20,000–40,000.

Other Names   PKD; polycystic renal disease

Foundations have been so helpful to people with genetic disorders. The PKD Foundation is one of those organizations that have helped people with polycystic kidney disease. Dr. Jared Grantham and Joseph Bruening, a Kansas City real estate developer, started the foundation in 1982. At that time, no one had identified the gene mutations responsible for one of the most numerous of all single gene disorders, and only a few researchers were involved in studying PKD. Now this foundation has funded over 32 projects in five countries for a total of nearly $2 million. Today, two treatments—kidney transplants and dialysis—have extended the lives of people with PKD, and hopefully someday, research will lead to a cure of polycystic kidney disease.

What Is Polycystic Kidney Disease (PKD)?

Polycystic kidney disease is a condition that affects not only the kidneys, but other organs as well. The kidneys are the organs that filter waste products from the blood. The kidneys have small tubules where the work of filtering the waste products is done. The products are then passed through the ureters into the bladder until it is full. If the kidneys do not work right, then toxic chemicals build up that can affect many parts of the body. As the waste products elevate, they cause a condition known as uremia, and the person becomes very sick.

In polycystic kidney disease, small cysts form in the kidneys and stop that organ from filtering waste from the blood. A cyst forms when cells begin to divide abnormally and then fluid fills in around the cells, forming a small sac. Over time, the cysts grow and become so large that they block the function of the kidney. Other organs—especially the liver—can develop cysts.

Following are the symptoms of PKD:

- Urinating often, especially at night
- Blood in the urine, which gives it a dark-red or brown color
- Tenderness or pain in the abdomen or side
- Recurrent urinary infections and kidney stones
- Drowsiness
- Elevated blood pressure
- Menstrual pain in women
- Joint pain
- Abnormal nails

Bleeding in a cyst, which may occur in the kidney, liver, or testes, can cause excruciating pain. In addition, people with PKD have an increased risk of aneurysm either in the brain or aorta, the large artery leading out of the heart. An aneurysm is an abnormal ballooning or bulging of the blood vessels; if the balloon bursts, it can be life-threatening. People with PKD tend to develop very severe high blood pressure.

What Are the Genetic Causes of Polycystic Kidney Disease (PKD)?

Mutations in three genes cause the two major forms of PKD. The three are distinguished by the age of onset and pattern of inheritance. The most common form has an autosomal dominant inheritance pattern and is sometimes referred to as ADPKD. ADPKD is divided into two forms referred to as type 1 and type 2, depending upon the mutation of one of the genes. ADPDK usually begins in adulthood, but cysts may have formed much earlier. The much rarer form of PKD is autosomal recessive and appears early in infancy or early childhood. The three genes are *PKD1*, *PKD2*, and *PKHD*.

PKD1

Mutations in the *PKD1* gene, officially known as the "polycystic kidney disease 1 (autosomal dominant)" gene, cause type 1 of PKD. Normally, *PKD1* provides instructions for making the protein polycystin-1. Before birth, polycystin-1 is very active in the embryonic kidney, but the production slows in the adult kidney. Polycystin-1 appears to interact with a smaller protein polycystin-2.

Here is the way polycystin-1 works: In the kidney cell, it is positioned so that one end of the protein is inside the cell and the other end is outside the cell. The end that is outside the surface of the cell can interact with all types of other components such as proteins, carbohydrates, and fats. A molecule binds to the polycystin-1 on the cell surface and then interacts with pholycystin-2 to begin a series of chemicals inside the cell. Polycystin-1 and polycystin-2 are partners. They work together to help the cell grow, move, and interact with other cells. In the tubules where urine is formed, tiny hairlike projections called primary cilia are present. The cilia are essential to the movement of urine through the tube thereby maintaining the size and structure of the cells.

More than 250 mutations in the gene *PDK1* cause the disease PKD. About 85% of cases of autosomal dominant polycystic kidney disease are caused by this gene. The mutations may include insertions or deletions of DNA building blocks or alterations of one or more base pairs. The mutations disrupt the action of the primary cilia, and as a result, cells lining the tubules may begin to divide abnormally and form numerous cysts. *PKD1* is inherited in an autosomal dominant pattern and is located on the short arm (p) of chromosome 16 at position 13.3.

PKD2

Mutations in the *PKD2*, officially called the "polycystic kidney disease 2 (autosomal dominant)" gene, cause type 2 of PKD. Normally, *PKD2* provides instructions for the protein polycystin-2. This protein is regulated by the larger polycystin-1 protein. With its partner polycystin-1, this protein spans the membrane of the kidney cell, transports calcium ions into the cell, and tells the cell to act and take on specialized functions.

Polycystin-2 is also active in other parts of the cell, including cellular structures called primary cilia. Primary cilia are tiny, fingerlike projections that line the small tubes where urine is formed (renal tubules). Researchers believe that primary cilia sense the movement of fluid through these tubules, which appears to help maintain the tubules' size and structure. The interaction of polycystin-1 and polycystin-2 in renal tubules promotes the normal development and function of the kidneys. Polycystin-2 is also active in the primary cilia.

More than 75 mutations in *PKD2* are related to PKD and account for about 15% of individuals with the dominant type. Mutations may occur in a single DNA base pair or deletions or insertions. The mutation makes a small abnormal version of the protein polycystin-2 that interferes with polycystin-1. Thus the cells lining the tube grow rapidly, causing a cyst. *PKD2* is inherited in an autosomal dominant pattern and is located on the long arm (q) of chromosome 4 at position 22.1.

PKHD1

Mutations in the *PKHD1* gene, officially known as the "polycystic kidney and hepatic disease 1 (autosomal recessive)" gene, cause the recessive type of PKD. Normally, *PKHD1* provides instructions for the protein fibrocystin, which is present in fetal and adult kidney cells, as well as in the liver and pancreas. Like the other two related genes, fibrocystin has one end in the cell and the other end outside to respond to things outside the environment. The protein may help cells connect or help them remain apart, thus controlling the rate of growth and division. Fibrocystin is also found in the primary cilia and may have a role in cell maintenance.

About 270 mutations in *PKHD1* cause the autosomal recessive form of PKD. This is a very severe type and is usually lethal. Mutations can be the result of changes in one building block pair or insertions or deletions, which create an abnormal fibrocystin protein. Disruptions in the protein lead to the numerous cysts of PKD. *PKHD1* is inherited in an autosomal dominant pattern and is located on the short arm (p) of chromosome 6 at position 12.2.

What Is the Treatment for Polycystic Kidney Disease (PKD)?

At one time, people with autosomal dominant polycystic kidney disease (ADPKD) seldom lived past the age of 53. ADPKD reflected about 6% of end-stage renal disease cases in the United States. People did not know about inheritance but did know that about half of the children born to a parent with kidney disease had the disorder. Few treatments were available. In the 1960s, two lifesaving therapies were developed—hemodialysis and kidney transplants.

Today genetic tests have identified 85% of cases with ADPKD. Early diagnosis is available through genetic testing, magnetic resonance imaging, and better kidney cyst imaging. Now the average age of developing the kidney disorder has increased to age 57. Kidney transplantation is now widely available, with nearly 13% of ADPDK patients receiving a transplant. With care, the life expectancy is expanded.

Researchers at the National Institutes of Health are continuing to test new potential therapies for kidney disorders, especially ADPKD.

What is Dialysis?

People with autosomal dominant polycystic kidney disease may need dialysis because the cysts are interrupting the formation of urine. Dialysis takes over when the kidneys are no longer working. About 200,000 people in the United States use dialysis to help their bodies perform the function of their failed kidneys.

Patients require dialysis when the waste product accumulation begins to make them ill. Two tests indicate when to begin dialysis. Results of blood tests show an elevated creatinine level and a blood urea nitrogen or BUN level. These two chemicals are indicators that the kidneys are not cleansing the body of waste products. Another indicator is bloating in the abdominal area because of the body's inability to get rid of excess water.

Following are the two types of dialysis:

- Hemodialysis: The Greek root *hemo* means blood. In hemodialysis, the blood goes from the patient's body to a filter in the dialysis machine called a dialysis membrane. A tube is placed between an artery and a vein in the arm or leg and a direct connection is made between the artery and vein. This connection is called a "Cimino fistula." Needles are then placed in the graft or fistula and connected to a tube that moves the blood into the dialysis machine. In the machine, a solution takes out the waste products, and the clean blood is returned to the patient. The procedure takes from two to four hours and is done three times a week. The person is in a chair and can read, sleep, or watch television during the procedure.

- Peritoneal dialysis: The peritoneal area is the section around the abdomen. This procedure uses the body's own tissues inside the abdominal cavity to act as a filter. A dialysis catheter or plastic tube is inserted through the abdominal wall into the intestines. A special fluid is then washed into the cavity around the intestines. The walls of the intestine become the filter between the fluid and the bloodstream. Waste products are removed in this way. In this procedure, patients take a more active role in their treatment and may clean, fix, and remove the bags themselves at home. It must be done four to five times a day. Some patients may use a machine called a "cycler" every night. Five to six bags of dialysis fluid is used here, and the machine changes the fluid while the patient sleeps.

An excellent resource for patients is: American Association of Kidney Patients; 3505 E. Frontage Rd., Ste. 315; Tampa, FL 33607; 800-749-2257; http://www .aakp.org.

Further Reading

"Autosomal Dominant Polycystic Kidney Disease." 2011. National Institutes of Health (U.S.). http://report.nih.gov/NIHfactsheets/ViewFactSheet.aspx?csid=29&key=A#A. Accessed 5/28/12.

PKD Foundation. 2010. http://www.pkdcure.org. Accessed 5/28/12.

"Polycystic Kidney Disease." 2011. National Institute of Diabetes and Digestive and Kidney Diseases (U.S.). http://kidney.niddk.nih.gov/kudiseases/pubs/polycystic. Accessed 5/28/12.

Polycythemia Vera

Prevalence   About 1 in 200,000 people worldwide

Other Names Osler-Vaquez disease; polycythemia ruba vera; primary polycythemia; PRV; PV

Like most long medical terms, polycythemia vera can be readily understood by breaking down the words into their root meanings. The word "polycythemia" comes from three Greek root words: *poly*, meaning "many"; *cyt*, meaning "cell"; and *heme*, meaning "blood." Thus, polycythemia is a condition with many blood cells. *Vera* is the Latin word meaning "true."

What Is Polycythemia Vera?

Polycythemia vera is a disorder of the way red blood cells are made. The condition is the result of an increased number of red blood cells, white blood cells, and blood platelets, which are the clotting factors. All these extra cells cause the blood to thicken and to block the flow of blood through the blood vessels.

The following symptoms characterize polycythemia vera:

- Headaches.
- Dizziness.
- Fatigue.
- Ringing in the ears, a condition called tinnitus.
- Impaired vision.
- Itchy skin, especially after a hot bath or shower.
- Abnormal blood clots: Polycythemia vera may cause deep vein thrombosis (DVT), a condition in which blood clots form in the veins and then break away, traveling through the bloodstream into the lungs. The person may die of a heart attack or stroke.
- Redness of skin.
- Black and blue spots on skin.
- Breathing difficulties: The person may have shortness of breath or trouble with breathing when lying down.
- Problems with extremities: The person may feel numbness, tingling, burning, or weakness in the hands, feet, or legs.
- Organ dysfunction: The person may have a feeling or fullness or bloating in the upper abdomen because of the enlarged spleen. As a result of polycythemia vera, he may also develop stomach ulcers, gout, heart disease, and leukemia.

The condition usually appears in late adulthood, but rare cases have occurred in children and young adults. It may progress very slowly and may not even be recognized until severe symptoms occur.

What Are the Genetic Causes of Polycythemia Vera?

Mutations in two genes—*JAK2* and *TET2*—are associated with polycythemia vera. Most of the cases are not inherited but are somatic, meaning that the condition is acquired during a person's lifetime and is found only in some cells. Rare cases may appear to run in some families, inherited in a dominant pattern.

JAK2

Mutations in the *JAK2* gene, officially known as the "Janus kinase 2" gene, are associated with polycythemia vera. *JAK2* provides instruction for a protein that is part of the signaling pathway called JAK/STAT, which transports chemical signals from outside the cell to the nucleus of the cell. The JAK2 protein is essential for controlling the production of blood from the stem cells in the bone marrow.

The mutations are usually acquired and are only in certain cells. The most common mutation, which accounts for about 96% of the cases, occurs when phenylalanine replaces valine in the protein. The mutations cause the JAK2 protein to be constantly turned on, increasing the production of red blood cells. When so many extra cells are present in the bloodstream, the normal flow through the arteries and veins is blocked, and many organs do not receive the oxygen that they need. *JAK2* is located on the short arm (p) of chromosome 9 at position 24.

TET2

Mutations in the *TET2* gene, officially known as the "tet oncogene family member 2" gene, are associated with polycythemia vera. Normally, *TET2* instructs for a protein, whose specific functions have not been determined. Researchers do believe that the TET2 protein regulates the transcription process of cells in the bone marrow and plays an essential role in producing red blood cells from stem cells. *TET2* protein may act as a tumor suppressor, preventing cells from growing wildly to form cancers.

The mutations of this gene appear to be acquired during a person's lifetime and only in specific cells. All the mutations, which are not inherited, result in a nonworking protein that results in polycythemia vera. *TET2* is found on the long arm (q) of chromosome 4 at position 24.

What Is the Treatment for Polycythemia Vera?

Several goals are essential treatment: reduce the thickness of the blood; prevent the blood from clotting; and prevent bleeding. The following treatments may be used:

- Phlebotomy: Phlebotomy is a process in which blood is taken from a person. These people may be considered blood donors. They will need to give blood, weekly if necessary, to reduce the thickness of the blood. The physician measures the level in a process called hematocrit.
- Chemotherapy: The person may be given a chemotherapy agent called hydroxyurea to reduce the number of red blood cells made in the bone marrow.
- Interferon: Administering interferon may reduce the number of blood cells.
- Medication: The drug anagrelide may reduce blood platelet count.
- Aspirin: Some patients may be advised to take aspirin, although it may increase the risk for stomach bleeding.
- Ultraviolet-B light therapy: This treatment may reduce some of the skin problems, especially itching.

Further Reading

"Polycythemia Vera." 2011. Genetics Home Reference. National Library of Medicine (U.S.). http://ghr.nlm.nih.gov/condition/polycythemia-vera. Accessed 12/17/11.

"Polycythemia Vera." 2010. Mayo Clinic. http://www.mayoclinic.com/health/polycythemia-vera/DS00919. Accessed 12/17/11.

"Polycythemia Vera." 2012. National Institutes of Health (U.S.). http://www.ncbi.nlm.nih.gov/pubmedhealth/PMH0001615. Accessed 12/17/11.

Pompe Disease

Prevalence   Affects about 1 in 40,000; varies among ethnic groups

Other Names   acid maltase deficiency; alpha-1,4-glucosidase deficiency; AMD; deficiency of alpha-glucosidase; GAA deficiency; glycogenosis Type II; glycogen storage disease type II; GSD2; GSD II; Pompe's disease

In 1932, Johann Pompe, a Dutch pathologist, described a condition in which glycogen accumulated in the muscle tissue. The disorder was puzzling because all the enzymes involved in the usual body metabolism were working well. The condition still baffled scientists until Christian de Duve discovered lysosomes in 1955, and his colleague Henri Hers found that a deficiency in an enzyme prevalent in the lysosomes explained the breakdown of glycogen. The disorder was one of the first glycogen storage diseases to be identified and was given the name Pompe disease after the Dutch pathologist.

What Is Pompe Disease?

Glycogen is a complex sugar whose breakdown is essential for body metabolism. When that breakdown does not occur, glycogen may accumulate in certain tissues like muscles and disrupt their function. Pompe disease is a condition in which an enzyme in the lysosomes of cells is defective, does not allow for the breakdown of glycogen, and leads to accumulation of the substance in body tissues and organs.

Following are three types of Pompe disease:

- *Infantile-onset Pompe disease classic form*: Within a few months after birth, infants experience muscle weakness, poor muscle tone, an enlarged liver, and obvious heart defects. The individual does not gain weight and has breathing problems. If the infant is not treated immediately, he or she will usually die of heart failure within the first year.

- *Infantile-onset Pompe disease, non-classic form*: This form usually appears by age one. The child does not roll over or sit up and has obvious delayed

---

One Man's Fight against Pompe Disease

John Crowley was the son of a New Jersey police officer who was killed when John was eight. However, he was able to attend law school and to forge a distinguished career in the health care industry. But in 1998, his world changed. Two of his children, Megan and Patrick, were diagnosed with Pompe disease, and he watched as they progressively deteriorated.

Discouraged by the slow pace of research on the disorder, he knew what to do—quit his job and form his own biotech company. He joined a research company called Novazyme Pharmaceuticals in Oklahoma City that was doing research on a new experimental treatment for the disease. In 2001, Genzyme, the world's third-largest biotech company, acquired Novazyme. In 2003, Megan and Patrick received the enzyme replacement therapy for Pompe disease, which Genzyme had marketed.

Crowley's dedication and work were chronicled by Pulitzer Prize winner Geeta Anand in a 2006 book, *The Cure: How a Father Raised $100 Million—and Bucked the Medical Establishment—in a Quest to Save His Children.* Harrison Ford bought rights to the book and produced it as a film entitled *Extraordinary Measures*, released in 2010. Today, Crowley is continuing his work fighting for cures for rare genetic disorders.

---

motor skills and muscle weakness. The heart may be enlarged although the person does not have heart failure. Because the muscle weakness affects breathing, the child with this type of disorder does not live past early childhood.

- *Late-onset type*: This form of the disease becomes apparent in later childhood, adolescence, or even adulthood. The form is usually milder and probably does not involve the heart. Muscle weakness is usually progressive beginning in the legs and trunk but then moving to the muscles that control breathing. If untreated, breathing issues can lead to respiratory failure.

What Is the Genetic Cause of Pompe Disease?

Mutations in the *GAA* gene, officially known as the "glucosidase, alpha; acid" gene, cause Pompe disease. Normally, *GAA* provides instructions for the enzyme acid alpha-glucosidase (also known as acid maltase). In the cells are structures called lysosomes, which are recognized as the recycling and garbage disposal centers of the cells. The lysosomes have digestive enzymes that break down complex molecules such as glycogen into simpler sugar known as glucose. For most cells, glucose is the main source of energy.

Over 200 mutations of *GAA* cause Pompe disease. Most of the mutations involve a simple change of one amino acid used to make the enzyme acid alpha-glucosidase. The mutations disrupt the action of the enzyme and keep it from breaking down glycogen. Thus, the complex sugar builds up in the lysosomes and has a toxic effect on the muscles and other organs. *GAA* is inherited in an autosomal recessive pattern and is located on the long arm (q) of chromosome 17 at position 25.2-q25.3.

What Is the Treatment for Pompe Disease?

Treatment of the disorder must be individual. Careful attention must first be given to the heart defect. Some of the drugs used to treat cardiac illness may not work in individuals with Pompe disease, and the risk of sudden death is high. Physical therapy is essential for treatment of muscle weakness, and respiratory support may include necessary life support or surgery.

As soon as a diagnosis is established, the individual should receive enzyme replacement therapy or ERT with Myozyme or Lumizyme. Infants receiving ERT improved survival rates, reduced cardiac defects, and enhanced motor skills when compared to controls that were not treated. ERT also has been effective with late-onset cases.

Further Reading

Pompe Community. http://www.pompe.com/en/healthcare-professionals.aspx. Accessed 5/28/12.

Tinkle, Brad T., and Nancy Leslie. 2010."Glycogen Storage Disease Type II (Pompe Disease)." *GeneReviews*. http://www.ncbi.nlm.nih.gov/books/NBK1261. Accessed 5/28/12.

United Pompe Foundation. 2011. http://www.unitedpompe.com. Accessed 5/28/12.

Porphyrias

Prevalence   Probably one in 500 to 1 in 50,000 worldwide; acute intermittent porphyria most common form of acute porphyria occurring in northern European countries such as Sweden and UK

Other Names Hematoporphyria; porphyrin disorder

At one time in medical history, medical practitioners studied urine and feces as an important tool for diagnosis. In the fourth century BC, the Greek doctor Hippocrates described a disorder in which the urine looked purplish red. However, it was not until 1871 that biochemist Felix Hoppe-Seyler analyzed the blood chemistry involved in the condition. In 1889, the Dutch physician Bared Stokvis described the pigments in the blood and referred to them as porphyrins. The term comes from the Greek *porphyra*, meaning "purple pigment."

Porphyria in History

The infamous King George III, who ruled at the time of the American Revolution, was thought to have had acute intermittent porphyria. It is well known that he was plagued with abdominal pain, rashes, reddish urine, and psychotic episodes. Written about as the "mad king," George III was depicted as the victim of neurological symptoms, trances, seizures, and hallucinations that lasted for weeks. In 1788, he experienced a crisis that was well known in Great Britain. Although it is very difficult to look at diseases in retrospect, several studies have sought to make the connection. The question of the nature of the illness still remains.

What Are Porphyrias?

The porphyrias are group of disorders of the production of heme, an important element in blood. Heme is one of a group of organic compounds called porphyrins. When a problem exists in the production of these organic compounds or their precursors, the porphyrins build up, causing a variety of health problems.

The word *heme* comes from the Greek work for blood, and heme is the substance that gives blood its red color. Heme makes up the iron-containing proteins that include hemoglobin. A vital molecule for every organ of the body, heme is very abundant in the blood, bone marrow, and liver.

Two general classes of porphyrias or disorders of the production of heme exist: the erythropoietic type, which originates in the blood-producing cells of the bone marrow; and the hepatic type, which originates in the liver. Actually, there are eight different kinds of porphyrias, each with a distinct and specific genetic cause.

Problems of red cell production in the bone marrow cause the erythropoietic porphyrias. The person will have a very low number of red blood cells, a condition called anemia, and an enlarged spleen, a condition known as splenomegaly.

Following are the eight kinds of porphyrias, which have distinguishing signs and symptoms and are classified by their genetic causes:

- *Porphyria cutanea tarda*: This type of porphyria is the most common kind of the disorder. It affects the skin. When sun beats down on exposed skin, the area becomes fragile and blisters. The blisters burst, leading to infection, scarring, and changes in the skin coloring. An abundant amount of hair may grow on the area. This type may be more prevalent than thought because many people with this type of condition never get a diagnosis.

- *Erythropoietic protoporphyria*: This is another of the skin porphyrias, which result in the blistering and infection of exposed skin.

- *Hereditary coproprophyria*: This form can be both acute with nervous system implications and cutaneous.

- *Variegate porphyria*: This form is common among the Afrikaner population of South Africa and may be present in about 3 in 1,000 people in this population.

- *Acute intermittent porphyria*: This form affects the nervous system and is acute because the symptoms appear suddenly and then go away just as rapidly. The individual may experience severe abdominal pain, vomiting, constipation, diarrhea, and mental problems such as hallucinations. This form is the second-most common form and is more frequent in the United Kingdom and in northern European countries such as Sweden. Some medical historians believe that King George III of England had this type of porphyria.

- *ALAD deficiency porphyria*: This rare form is caused by a defective *ALAD* gene and is one of the acute forms of porphyria.

- *Congenital erythropoietic porphyria*: This form is rare. The person is born with the signs and symptoms of the condition.

- *Hepatic porphyrias*: These disorders originate in the malfunction of the porphyrins and porphyrin precursors in the liver. The liver ceases normal function and may develop cancer.

The lifestyle of the individual may greatly affect the symptoms of porphyria. Certain things such as alcohol, certain drugs, hormones, smoking, illness, fasting, stress, and smoking can trigger the disorder. Exposure to sunlight is threatening to those with the cutaneous type.

What Are the Genetic Causes of the Porphyrias?

The porphyrias are complex diseases caused by mutations in several genes. The genes are all related to the instructions for making the enzymes that produce heme. Heme is the main component of the iron-containing proteins that carry oxygen to the body's cells. Here is generally what happens in people with mutations in the genes: Porphyrins or porphyrin precursor cells are formed during the production of heme. The porphyrins build up in the liver or other organs, causing damage. If the buildup is in the skin, the damage comes from sunlight. If the buildup is in the nervous system, damage can cause the acute form of the disorder.

Mutations in one of eight different genes are related to porphyria. The genes are the following: *ALAD, ALAS2, CPOX, FECH, HMBS, PPOX, UROD,* or *UROS.*

ALAD

Mutations in the *ALAD* gene, officially known as the "aminolevulinate dehydratase" gene, cause ALAD deficiency porphyria. Normally, this gene provides instructions for the enzyme delta-aminolevulinate dehydratase, which produces the molecule heme. This complex production of heme requires eight different enzymes. Delta-aminolevulinate dehydratase is responsible for the second step in the process.

Ten or more mutations in *ALAD* cause a rare from of the disorder called ALAD deficiency porphyria. Most of the mutations involve a change in a single protein building block in delta-aminolevulinate dehydratase. Because this enzyme is essential in the second stop of the production of heme, the change can disrupt the formation of heme and cause toxic levels of porphyrins to accumulate, causing abdominal pain and the other symptoms of ALAD deficiency porphyria. ALAD is inherited in an autosomal recessive pattern and is located on the long arm (q) of chromosome 9 at position 33.1.

ALAS2

Mutations in the *ALAS2* gene, officially known as the "aminolevulinate, delta-, synthase 2" gene, causes erythropoietic protoporphyria. Normally, *ALAS2* provides instructions for an enzyme 5-aminolevulinate synthase 2 or erythroid ALA-synthase. There are two genes that cause several versions of the enzyme. The first gene, *ALAS1*, is active in many cells throughout the body; however, *ALAS2* is responsible for developing the prototype of red blood cells called erythroblasts. Of the eight steps involved in the production of heme, ALA-synthase begins the process with the first step of forming a compound called delta-aminolevulinic acid (ALA).

Two mutations in *ALAS2* are related to erythropoietic protoporphyria. The mutations are the result of deletions of a small amount of genetic material at the end of the gene. The changes disrupt the function of the ALA-synthase and increase ALA in the red blood cells, which convert to porphyrins. If these build up in the precursors to the red blood cells, they will be carried in the bloodstream to the skin and other tissues. The high level of porphyrins in the skin causes the reaction to the sunlight, which is characteristic of this form of the disorder. Males may experience the severe form of the disorder. *ALAS2* is inherited in an X-linked dominant pattern and is located on the short arm (p) of the X chromosome at position 11.21.

CPOX

Mutations in the *CPOX* gene, officially known as the "coproporphyrinogen oxidase" gene, can cause the form of the disorder known as hereditary coproporphyria. Normally, *CPOX* provides instructions for making the enzyme coproporphyrinogen oxidase, an enzyme that is responsible for the sixth step in the production of heme. At least 45 mutations in *CPOX* appear to cause changes in the creation of the enzyme. The mutations disrupt the activity of coproporphyrinogen oxidase, allowing the porphyrins to build up in the body. *CPOX* is inherited in an autosomal dominant pattern and is located on the long arm (q) pf chromosome 3 at position 12.

FECH

Mutations in the *FECH* gene, officially known as the "ferrochelatase" gene, can be another cause of the erythropoietic protoporphyria. Normally, *FECH* provides instructions for the enzyme ferrochelatase. This enzyme is responsible for the last or eighth step in the process of making heme. In this step, an iron atom (Fe) is inserted into a molecule called protoporphyrin IX, the product of the seventh step.

Over 110 mutations in *FECH* cause erythropoietic protoporphyria. The mutations reduce the production of the enzyme ferrochelatase by about one-half. The person must have two copies of the gene in order for the condition to develop. This gene is referred to as a low expression allele, which reduces but does not eliminate the amount of ferrochelatase in cells. However, the two mutated copies together are enough to allow the porphyrins to build up in developing red blood cells. The buildup can cause skin damage, gallstones, and liver disease. *FECH* is inherited in an autosomal dominant pattern and is located on the long arm (q) of chromosome 18 at position 21.3.

HFE

Mutations in the *HFE* gene, officially known as the "hemochromatosis" gene, cause the most common form of porphyria, porphyria cutanea tarda. Normally, *HFE* provides instructions for a protein located in many body cells, including the liver, intestine, and immune cells, and it interacts with other proteins to detect the amount of iron in the cell. The *HFE* protein controls the production of an important hormone called hepcidin, which acts as a master iron regulator. Produced in the liver, hepcidin determines how much iron is absorbed from food and then stored in the body. The average amount of iron absorbed from the diet is about 10%.

Mutations in *HFE* increase the risk for porphyria cutanea tarda. In this type of porphyria, the absorption of iron is increased, interfering with the production of heme. Other genetic and nongenetic factors may also be involved in the production of toxic levels of iron. Some cases of porphyria cutanea tarda cannot be traced to genetics but may be acquired from environmental factors. *HFE* is inherited in an autosomal dominant pattern and is located on the short arm (p) of chromosome 6 at position 21.3.

HMBS

Mutations in the *HBMS* gene, officially known as the "hydroxymethylbilane synthase" gene, cause the form of porphyria known as acute intermittent porphyria. Normally, *HMBS* provides instructions for the enzyme hydroxymethylbilane synthase. This enzyme is involved in the third of eight steps in the production of heme.

More than 300 mutations in *HMBS* are involved in acute intermittent porphyria. The mutations, which may be a change in a single protein or in additions or deletions, disrupt the function of the enzyme. Porphyrins are allowed to build up in the liver and other organs, causing a toxic accumulation. Environmental factors, such as drugs and alcohol, may combine with the toxic level buildup and cause serious abdominal pain and other symptoms. *HMBS* is inherited in an autosomal dominant pattern and is located on the long arm (q) of chromosome 11 at position 23.3.

PPOX

Mutations in the *PPOX* gene, officially called the "protoporphyrinogen oxidase" gene, cause variegate porphyria, a form that is prevalent in South Africa. Normally, *PPOX* provides instructions for making protoporphyrinogen oxidase, an enzyme responsible for the seventh of the eighth steps in the production of heme.

Over 130 mutations appear to be involved in variegate porphyria. The mutations involve changing in a single protein building block and proceeding to disrupt the activity of the enzyme, allowing porphyrins to accumulate to toxic levels. Other environmental, nongenetic factors such as drugs, alcohol, or diet may also contribute to the disorder. *PPOX* is inherited in an autosomal dominant pattern and is located on the long arm (q) of chromosome 1 at position 22.

UROD

Mutations in the gene *UROD*, officially known as the "uroporphyrinogen decarboxylase" gene, cause two forms of porphyria, porphyria cutanea tarda and hepatoerythropoietic porphyria. Normally, *UROD* provides instructions for the enzyme uroporphyrinogen decarboxylase. This enzyme is responsible for the fifth of the eight steps in the process of making heme.

More than 50 mutations of UROD may be present. The mutations disrupt the activity of the enzyme uroporphyrinogen decarboxylase. If a mutation is on one copy of the *UROD* gene in each cells, the common type of porphyria, porphyria cutanea tarda, may present mild symptoms. Mutations in both copies of the gene can cause the more serious hepatoerythropoietic porphyria. *UROD* is inherited in an autosomal dominant pattern and is located on the short arm (p) of chromosome 1 at position 34.

UROS

Mutations in the *UROS* gene, officially known as the "uroporphyrinogen III synthase" gene, cause a form of porphyria known as congenital erythropoietic porphyria. Normally, *UROS* is responsible for making the enzyme uroporphyrinogen III synthase." This enzyme is responsible for the fourth step in the eight-step process of producing heme.

More than 35 mutations involve a change in only one building block of the protein. When the mutations disrupt the process, prophyrins leak out of the red blood cells and accumulate in the skin, causing the sensitivity to the sunlight. UROS is inherited in an autosomal recessive pattern and is located on the long arm (q) of chromosome 10 at position 25.2-q26.3.

The porphyrias are a group of disorders that affect the skin, nervous system, or both. There are some points to remember:

- The disorders are the result of a deficiency in the production of heme, which results in an excessive build up of porphyrins.

- The production of heme results from eight different steps using different enzymes instructed by different genes. Any problem in any of the steps may cause a buildup and disrupt the process.

- Most porphyrias are inherited disorders, but environmental factors may enter evoking signs and symptoms.

What Is the Treatment for Porphyria?

Each form must be treated differently. Some forms are treated with medications or drawing blood to relieve symptoms. If the attacks are severe, people may have to be hospitalized. The National Institute of Diabetes and Digestive and Kidney Disease has many research projects involving porphyria that are in clinical trials. Cutting-edge research involves gene therapy, seeking to replace the defective gene with a normal gene.

Further Reading

"Learning about Porphyria." National Human Genome Research Institute. http://www.genome.gov/19016728. Accessed 5/28/12.

"Porphyria." Better Medicine. http://www.bettermedicine.com/article/porphyria. Accessed 5/28/12.

"Porphyria." National Digestive Diseases Information Clearinghouse (NDDIC). http://digestive.niddk.nih.gov/ddiseases/pubs/porphyria/index.aspx. Accessed 5/28/12.

The Porphyrias Consortium. http://rarediseasesnetwork.epi.usf.edu/porphyrias/index.htm. Accessed 5/28/12.

Prader-Willi Syndrome (PWS)

Prevalence   Affects 1 in 10,000 to 30,000 people worldwide

Other Names   Prader-Labhart-Willi syndrome; PWS; Willi-Prader syndrome

Several researchers especially in Switzerland noted a syndrome in which children with mild to moderate mental retardation and learning disabilities had abnormally insatiable appetites leading to morbid obesity. Andrea Prader, a Swiss pediatric endocrinologist, first described the condition in 1956 along with Heinrich Willi, also a Swiss physician. Three other Swiss doctors—Andrew Ziegler, Alexis Labhart, and Guido Fanconi—also were interested in the condition. However, the first two contributed their names—Prader-Willi syndrome.

What Is Prader-Willi Syndrome (PWS)?

Prader-Willi syndrome is a complex of conditions that affect many parts of the body. However, the hallmark of the condition is the overeating and consequent morbid obesity. The mother will sometimes note some irregularities during pregnancy. The fetus will not move often and will sometimes be in an abnormal fetal position. The birth will often be breech or require Cesarean section. The following are symptoms of PWS:

- In infancy, feeding difficulties and excessive sleeping
- Weak muscle tone
- Poor growth and delayed development
- In childhood, beginning about age two, child will overeat, a condition called hyperphagia
- Excessive weight gain leading to morbid obesity
- Speech delay
- Sleep disorders
- Behavioral and learning disabilities
- In adolescence and adulthood, infertility in both males and females, poor muscle tone

The overeating or hyperphagia continues throughout life. Adults with the disorder have a distinct appearance that includes a prominent nose bridge, small hands and feet, soft skin that is easily bruised, and excess fat especially in the midsection. The person also may develop type 2 diabetes related to morbid obesity. Many people with PWS have light skin and hair relative to other members of their families.

What Is the Genetic Basis for Prader-Willi Syndrome (PWS)?

Missing pieces of chromosome 15 cause PWS. Mutations in the *OCA2* gene or "oculocutaneous albinism II" gene are associated with Prader-Willi syndrome. Normally, *OCA2* instructs for making the P protein, which is essential for normal pigmentation but also regulates the relative acidity of the melanin-producing bodies. Control of pH or acidity is very important for most biological process.

People with PWS are missing a section of chromosome 15, which contains the *OCA2* gene. Normally, people inherit two copies of this chromosome, one from each parent. In certain genes, only one copy will be turned on. This turn-on of the gene is known as genomic imprinting. For example, the gene that is from the father or from the mother may be turned on; the other gene is still there but is inactive. The *OCA2* gene from the father is turned on. However, people with this missing copy do not make the P protein and the individual will have the light coloring of hair and skin. Researchers are still working to connect the other symptoms of PWS with *OCA2*. *OCA2* is located on the long arm (q) of chromosome 15.

In about 70% of people with PWS, the whole segment of chromosome 15 containing the paternal gene is missing, and the individual will have two copies of the maternal gene. In about 25% of cases, the person has two copies of the maternal chromosome instead of a copy from each parent. Rarely, PWS is caused by rearranging chromosome 15, a condition called translocation or by a mutation that inactivates the genes on the paternal chromosome. Most cases of PWS are not inherited but occur as random events in embryonic development. However, the main thing in Prader-Willi syndrome is that there is a loss of the important gene function in chromosome 15.

What Is the Treatment for Prader-Willi Syndrome (PWS)?

Although PWS has no cure, some strategies can lessen the symptoms of the conditions. Early in infancy, the child should undergo therapy to improve muscle tone. As they grow into childhood, speech and occupational therapy are essential. The children qualify under the Individuals with Disabilities Education Act (IDEA) and can receive special instruction in school.

However, obesity and overeating is still the most serious symptom. Parents have to be active in monitoring the diet of the person. Some prescriptions of growth hormones may assist children to support muscle mass and proper growth. Because of obesity, sleep apnea may be a problem requiring treatment with an airway pressure machine.

Further Reading

"Prader-Willi Syndrome." 2011. Mayo Clinic. http://www.mayoclinic.com/health/prader-willi-syndrome/DS00922. Accessed 5/28/12.

"Prader-Willi Syndrome." 2011. MedlinePlus National Institutes of Health (U.S.). http://www.nlm.nih.gov/medlineplus/praderwillisyndrome.html. Accessed 5/28/12.

Prader-Willi Syndrome Association. 2010. http://www.pwsausa.org. Accessed 5/28/12.

Presenile and Senile Dementia. *See* Alzheimer Disease (AD)

Primary Carnitine Deficiency

Prevalence   About 1 in 100,000 newborns; affects 1 in 40,000 in Japan

Other Names   carnitine transporter deficiency; carnitine uptake defect; carnitine uptake deficiency; CUD; renal carnitine transport defect; systemic carnitine deficiency

Carnitine deficiency is one of the metabolic disorders that keeps the body from using certain fats for energy. Dr. Susan Winter, medical geneticist with Children's

Hospital of Central California, and Dr. Neil Buist have been instrumental in discovering the role of carnitine and the problems of deficiency. Winter was one of the first doctors to give intravenous carnitine to treat this error of metabolism.

What Is Primary Carnitine Deficiency?

Carnitine is a compound that is made from the amino acids lysine and methionine. Cells use it for moving fatty acids from the cytosol, which is the intracellular fluid, into the mitochondria to break down fats or lipids. Carnitine is acquired mostly in the diet. However, if the process is not working properly, the body cannot use fats for energy, especially during a period of stringent dieting or fasting.

Following are the signs of primary carnitine deficiency that can appear during infancy or early childhood:

- Vomiting
- Confusion
- Brain dysfunction called encephalopathy
- A weak and enlarged heart
- Muscle weakness
- Low blood sugar
- Liver problems
- Coma
- Sudden death

The severity of the symptoms varies with the individual; yet, all persons are at risk for coma and sudden death. Illnesses and periods of fasting may trigger the symptoms. Occasionally, this condition is mistaken for Reye syndrome that can develop after a period of viral infections such as chicken pox or influenza.

What Is the Genetic Cause of Primary Carnitine Deficiency?

Mutations in the *SLC22A5* gene, known officially as the "solute carrier family 22 (organic cation/carnitine transporter), member 5" gene, cause primary carnitine deficiency. Normally, *SLC22A5* provides instructions for making the protein OCTN2, which is located in tissues such as the heart, liver, muscles, and kidneys. OCTN2 is found in the cell membrane where it carries carnitine into the cells.

About 60 mutations of *SLC22A5* cause primary carnitine deficiency. Some mutations cause a very short, nonfunctional protein, and others change only one building block. As a result, the work of carnitine is disrupted, and fatty acids cannot get into the mitochondria to produce energy. Without the process, the symptoms appear and fatty acids may build up and damage the heart, liver, and other organs. *SLC22A4* is inherited in an autosomal recessive pattern and is located on the long arm (q) of chromosome 5 at position 23.3.

What Is the Treatment for Primary Carnitine Deficiency?

As soon as primary carnitine deficiency is determined in the newborn screening, the child should being to receive oral supplements and then plasma carnitine. Guidelines for management of the deficiency, as well as other fatty acid mitochrondrial disorders, have been established. If other symptoms such as heart disorders develop, these should be treated separately. The person will have to also follow a special diet with supplements such as riboflavin, glycine, or biotin.

Further Reading

"Carnitine Deficiency Treatment and Management." 2011. Medscape. http://emedicine .medscape.com/article/942233-treatment. Accessed 5/28/12.

"Primary Carnitine Deficiency." 2011. Genetics Home Reference. National Library of Medicine (U.S.). http://ghr.nlm.nih.gov/condition/primary-carnitine-deficiency. Accessed 5/28/12.

Primary Carnitine Deficiency Support Group. 2010. http://www.dailystrength.org/c/ Primary-Carnitine-Deficiency/support-group. Accessed 5/28/12.

Prion Disease

Prevalence   Very rare; affects about one per million each year; in the United States, about 300 cases annually

Other Names   inherited human transmissible spongiform encephalopathies; prion-associated disorders; prion-induced disorders; prion protein diseases; transmissible dementias; transmissible spongiform encephalopathies; TSEs

In the middle of the twentieth century, an unusual epidemic broke out among the Fore people on the island of Papua New Guinea. People had body tremors but also would have spells of uncontrolled laughter. Investigators traced the disorder to a condition caused by a prion, which was thought to be passed through the practice of cannibalism and eating the brains of infected individuals.

Dr. Daniel Gajdusek took samples of an 11-year-old girl who had died of kuru and injected the infected material into two chimpanzees. Within two years, one of the chimps developed kuru, showing that this unknown disease was transmitted through infected tissue. Gajdusek and a colleague received the Nobel Prize in physiology and opened up investigation into the relationship of this disease with other mysterious human disorders. These disorders became designated as transmissible spongiform encephalopathies or TSEs and led to the development of diagnostic criteria for other prion-related diseases. The word "spongiform" refers to the soft, spongy appearance of the brain.

What Is Prion Disease?

Prion diseases are a group of disorders caused by a prion. A prion is a small abnormal and transmissible agent that can cause normal brain cells proteins to fold abnormally. In 1982 after years of research on this small protein, Prusiner coined the term *prion* from "proteinaceous infection particles" to distinguish it from viruses or viroids. The protein part was called PrP, or prion protein. X-ray crystallography showed that scientists were dealing with a very small particle, even smaller than a virus.

Prion diseases affect the nervous system in both humans and animals. They are usually progressive over time and affect brain function, memory, personality, and movement. When the symptoms begin, usually in adulthood, they progress rapidly and lead to death in a few months.

Following are the human prion diseases:

- Kuru: Study of this condition in New Guinea led to understanding of other TSEs.
- Creutzfeldt-Jakob disease (CJD): This disease is a TSE characterized by dementia and progressive deterioration.
- Variant Creutzfeldt-Jakob disease: This form of the disorder begins much earlier, around age 29. It is slower in progressing, and the person may live longer than the classic form of CJD.
- Fatal familial insomnia (FFI): This type was described in 1986. The person has progressive sleepiness and loss of autonomic functions.
- Gerstmann-Straussler-Scheinker syndrome: This very rare form affects people from age 20 to 60 and appears to run in families. It is a transmissible spongiform encephalopathy.

Animals may have prion diseases. One form in cattle called bovine spongiform encephalopathy, or BSE, is known as "mad cow disease." In the early 1990s in the United Kingdom, the fear that BSE was related to human cases of vCJD caused thousands of cattle to be slaughtered. Other forms may be seen in scrapie in sheep, which has been recognized for over 230 years, and chronic wasting disease (CWD) in mule deer and elk. The condition may also infect in domestic cats, zoo animals, and the mink population.

The prion diseases in humans are unique in that they can be divided as follows:

- Infectious: About 5% of cases are transmitted in some fashion from other infected animals. For example, it is thought that BSE in Europe was caused when cattle ate other materials made from infected cattle, and then the humans ate the infected meat.
- Sporadic: About 80% of cases are sporadic, meaning that a conformational change simple occurred from an abnormal folding or unknown origin.
- Inherited: About 15% of human TSEs are inherited mutations of the prion gene.

What Is the Genetic Cause of Prion Disease?

Mutations in the *PRNP* gene, officially known as the "prion protein" gene, cause the inherited form of prion disease. Normally, *PRNP* provides instructions for making the prion protein PrP. PrP is active especially in the brain and is thought to have roles in cell signaling, protection, and forming synapses in order for cells to communicate.

About 30 mutations in *PRNP* cause familial protein disease, in addition to classic Creutzfeldt-Jakob disease, Gerstmann-Sträussler-Scheinker syndrome, and fatal insomnia. For some reason the *PRNP* mutations change one amino acid in PrP or insert some additional material to make an atypical short version of the protein. The abnormal protein then builds up in the brain, making clumps that destroy neurons. When these cells are lost, the brain under the microscope appears like a sponge with holes in the brain. Those holes lead to the mental and behavioral features of the prion diseases. *PRNP* is inherited in an autosomal dominant pattern and is located on the short arm (p) of chromosome 20 at position 13.

What Is the Treatment for Prion Disease?

Several scientists are doing research on treatments for prion disease. However, to this date, no really effective drug has been found to treat or to prevent the disease.

Further Reading

Agamanolis, Dimitri. 2009. "Prion Diseases (Transmissible Spongiform Encephalopathies)." *Neuropathology*. Akron Children's Hospital. http://www.neuropathologyweb.org/chapter5/chapter5ePrions.html. Accessed 5/28/12.

"Prion Disease." 2011. Centers for Disease Control and Prevention. Department of Health and Human Services (U.S.). http://www.cdc.gov/ncidod/dvrd/prions. Accessed 5/28/12.

"Prion Disease." 2011. National Institute of Allergy and Infectious Disease. National Institutes of Health (U.S.). http://www.niaid.nih.gov/topics/prion/Pages/default.aspx. Accessed 5/28/12.

Progeria. *See* Hutchinson-Gilford Progeria Syndrome (HGPS)

Progeria-Like Syndrome. *See* Cockayne Syndrome

Progressive Cardiomyopathic Lentiginosis.
See LEOPARD Syndrome

Progressive Osseous Heteroplasia (POH)

Prevalence   Rare; the exact incidence is unknown

Other Names   cutaneous ossification; ectopic ossification; ectopic ossification, familial, heterotopic ossification; osteodermia; Osteoma cutis; Osteosis cutis; POH

Fibrodysplasia ossificans progressiva, or FOP, is a condition in which muscle turns to bone. Dr. Frederick Kaplan and colleagues at the University of Pennsylvania, specialists in this rare disease, found that some patients did not fit the mold for FOP. The bone appeared to form in the skin and fat that is under the skin. In 1994, the team decided this was a separate but sister disease to FOP and called it progressive osseous heteroplasia or POH.

What Is Progressive Osseous Heteroplasia (POH)?

Progressive osseous heteroplasia is a rare condition in which bone forms within skin and muscle tissue. The term heterotopic or ectopic is given to bone that forms in areas outside the normal skeleton. In POH, bone begins to form in the deep layers of the skin called the dermis and in subcutaneous layers of fat. The bone then forms in other tissues such as tendons and skeletal muscle. Painful ulcers or open sores may develop near the skin where the bone is pushing through. Ultimately, the condition may spread to the joints, limiting mobility. The signs may appear in early infancy, childhood, or early adulthood.

What Is the Genetic Cause of Progressive Osseous Heteroplasia (POH)?

Mutations in the *GNAS* gene, officially known as the "GNAS complex locus" gene, cause POH. Normally, *GNAS* provides instructions for making one part of the protein guanine nucleotide-binding protein (G protein). The guanine nucleotide-binding protein (G protein) is made of three component proteins called alpha, beta, and gamma subunits. *GNAS* is related to the alpha subunit. These G proteins work a complex network of signaling, which regulate hormone activity by stimulating an enzyme called adenylate cyclase. Although adenylate cyclase is involved in regulating the endocrine glands, it also regulated the development of bone and placing bone in the right place.

Most genes relate to a dominant or recessive genetic pattern of inheritance, but *GNAS* is controlled by a genetic phenomenon known as genomic imprinting. In this process, the copy from one parent—either the father or mother—is active. But *GNAS* is different. Some body cells have an active maternal copy, and other cells have the paternal copy. In POH, only the paternal copy of the gene is active, and if mutations are present, they disrupt the function of the G protein and help it from regulating the creation of and deposition of bone. The disruptions cause the deposition of abnormal bony tissue in the skin and muscles. *GNAS* is located on the long arm (q) of chromosome 20 at position 13.3.

What Is the Treatment for Progressive Osseous Heteroplasia (POH)?

At present, there is no cure for this unusual disorder. Treating is mostly symptomatic, such as pain for bones that are erupting through the skin or mobility issues related to movement of the joints.

Further Reading

"Progressive Osseous Heteroplasia." 2011. Genetics Home Reference. National Library of Medicine (U.S.). http://ghr.nlm.nih.gov/condition/progressive-osseous-heteroplasia. Accessed 6/7/11.

"Progressive Osseous Heteroplasia (POH) Causes, Symptoms and Treatment and Related Disorders." 2011. EverydayHealth.com. http://www.everydayhealth.com/health-center/progressive-osseous-heteroplasia-poh.aspx. Accessed 5/28/12.

Progressive Osseous Heteroplasia Association. http://www.pohdisease.org. Accessed 5/28/12.

Propionic Acidemia

Prevalence	Affects about 1 in 100,000 people in the United States; common among Inuits of Greenland, some Amish people, and Saudi Arabians
Other Names	hyperglycinemia with ketoacidosis and leucopenia; ketotic glycinemia; ketotic hyperglycinemia; PA; PCC deficiency; PROP; propionicacidemia; propionyl-CoA carboxylase deficiency

Garrod called them "inborn errors of metabolism." Many inherited disorders fit in this category. In 1961 Childs, and a team published a clinical report in which some children had severe episodes when they ate certain proteins, especially methionine and threonine. Other researchers noted other proteins and fats caused similar reactions. At first, the disorder, propionic acidemia, was considered to be two different diseases, but later researchers determined it was one disorder caused by genes in two different areas of the world.

What Is Propionic Acidemia?

Propionic acidemia is an organic disease. The individual is not able to process certain proteins and fats. When this occurs, certain organic acids build up in the blood, urine, and other body tissues and cause a toxic reaction with serious health complications.

Following are the features of propionic acidemia, which may be apparent only a few days after birth:

- Poor feeding and loss of appetite
- Vomiting
- Loss of muscle tone
- Lack of energy
- Heart abnormalities
- Seizures
- Coma and possibly death

In some children, the signs appear during childhood and may come and go. However, these children may experience delayed development and intellectual disability. In this late-onset form, the symptoms may be triggered by periods without food or by infections.

What Is the Genetic Cause of Propionic Acidemia?

Mutations in two genes, *PCCA* and *PCCB*, cause propionic acidemia. The two genes provide instruction for making two subunits of an enzyme called propionyl-CoA carboxylase.

PCCA

The *PCCA* gene, officially known as the "propionyl CoA carboxylase, alpha polypeptide" gene, is related to propionic acidemia. Normally, *PCCA* instructs for the alpha part of the enzyme propionyl-CoA carboxylase. This enzyme is made from six alpha subunits and six beta subunits. The enzyme is important in the processing of proteins and works to break down the protein building blocks or amino acids isoleucine, methionine, threonine, and valine. Propionyl-CoA carboxylase also helps break down certain types of lipids or fats and cholesterol. Eventually, the breakdown will lead to molecules that are used to create energy for the body.

About 45 mutations in *PCCA* cause propionic acidemia. Most mutations are changes in a single DNA building block or in insertions or deletions of small amounts of genetic material. The *PCCA* mutations disrupt the function of the propionyl-CoA carboxylase enzyme and make it unable to do its part in processing proteins and lipids. The enzyme along with other toxic compounds builds up, damaging the brain and nervous system and causing the health problems related to propionic acidemia. *PCCA* is inherited in an autosomal recessive pattern and is located on the long arm (q) of chromosome 13 at position 32.

PCCB

Mutations in the *PCCB* gene, officially known as the "propionyl CoA carboxylase, beta polypeptide" gene, cause propionic acidemia. Normally, this gene also provides instructions for making the parts of the enzyme propionyl-CoA carboxylase, called the beta subunits. The six beta parts made by *PCCB* join the six alpha parts made by *PCCA* to make the working enzyme, which is a second step in the breakdown of certain amino acids and fats to release energy.

About 55 mutations in *PCCB* cause propionic acidemia. The mutations are the result of changes in one building block or in the insertion or deletion of genetic material. The mutations disrupt the function of the propionyl-CoA carboxylase enzyme, causing toxic levels to build up and the serious health problems of propionic acidemia. *PCCB* is inherited in an autosomal recessive pattern and is located on the long arm (q) of chromosome 3 at position 21-q22.

What Is the Treatment for Propionic Acidemia?

Some states include propionic acidemia in their newborn screening; it would be wise for all states to have this screening, because knowing about the disorder is the first step of treatment. First, there must be dietary restrictions of total protein, the four amino acids valine, isoleucine, methionine and threonine, and odd-chained fats. A dietician and physician will supervise the diet and supplement it with levocarnitine to remove any toxic by-products. Some children do learn to eat, but most require a feeding tube because the child has little appetite and aversion to foods.

Further Reading

"Genetics of Propionic Acidemia." 2011. Medscape. http://emedicine.medscape.com/article/948084-overview. Accessed 5/28/12.

"Propionic Acidemia." 2011. Genetics Home Reference. National Library of Medicine (U.S.). http://ghr.nlm.nih.gov/condition=propionicacidemia. Accessed 5/28/12.

"Propionic Acidemia." 2011. STAR-G. http://www.newbornscreening.info/Parents/organicaciddisorders/PA.html. Accessed 5/28/12.

Protein C Deficiency

Prevalence   Mild protein C, about 1 in 500 people; severe cases, rare, occurring in 1 in 4 million newborns

Other Names   hereditary thrombophilia due to protein C deficiency

Thrombophilia is a term used to describe abnormal clotting of the blood. The word comes from two Greek words: *thrombus*, meaning "clot," and *philia*, meaning "to love." Many people have the condition, but problems usually arise only if there is an additional risk, such as sitting for long periods on planes or long car rides. In 1981, J. H. Griffin and colleagues found that one of the conditions that caused blood clots was related to a deficiency of protein C.

What Is Protein C Deficiency?

Protein C deficiency is a disorder in which the blood does not coagulate normally. People with this condition are at a great risk for developing blood clots, which can be life-threatening. The following two types of the deficiency exist:

- *Mild protein C deficiency*: Although many people with this condition never experience a blood clot, they have the potential for risk. The risk condition is known as deep vein thrombosis or DVT. With DVT, a clot forms in a limb and then travels to the lungs, causing a life-threatening pulmonary embolism. Risk factors include surgery, immobility of limbs such as in extended travel, pregnancy, or advancing age.

- *Severe protein C deficiency*: This life-threatening disorder occurs soon after birth. Tiny blood clots form in blood vessels throughout the body. Called purpura fulminans, the small clots use up all the clotting proteins in the body and disrupt the normal blood flow. Widespread bleeding occurs that appear as large purple skin lesions. Although the newborn may survive, the child may have later occurrences.

What Is the Genetic Cause of Protein C Deficiency?

Mutations in the *PROC* gene, officially known as the "protein C (inactivator of coagulation factors Va and VIIIa)" gene, cause protein C deficiency. Normally, *PROC* provides instructions for making protein C. Protein C is responsible for blocking two factors, called Va and factor VIIIa, that are involved in forming blood clots. Made in the liver and then released into the bloodstream, protein C is inactive until it attaches to a protein called thrombin, when it becomes an activated protein called APC. APC cuts factor Va and partially or completely activates it to a substance called factor V. APC then works with factor V to inactivate factor VIIIa.

About 270 mutations *in PROC* cause protein C deficiency. Most mutations result from changes in one building block. Mutations are of the following two types:

- Type I: Mutations cause reduced levels of protein C. Problems with blood clotting may arise because there is not have enough protein C to control clotting. These people are at increased risk for blood clots if the conditions are right.

- Type II: Mutations cause changes in protein C activity. People may have normal levels of protein C, but it cannot act with other proteins that are involved in blood clotting, causing abnormal blood clots.

PROC is inherited in an autosomal dominant pattern and is located on the long arm of chromosome 2 at position 13-q14.

What Is the Treatment for Protein C Deficiency?

General management of risk factors includes prevention of deep vein thrombosis. A common-sense approach for all people involves walking and seat exercises during long airplane rides and frequent stops during automobile trips. Fresh frozen plasma that has been viral inactivated in place of protein C can help newborns with severe protein C deficiency.

Further Reading

"Protein C Deficiency." 2008. National Blood Clot Alliance—Stop the Clot. http://www.stoptheclot.org/News/article136.htm. Accessed 5/28/12.

"Protein C Deficiency." 2011. Medscape. http://emedicine.medscape.com/article/205470-overview. Accessed 5/28/12.

"Protein C Deficiency." 2011. University of Iowa. http://www.medicine.uiowa.edu/labs/lentz/Information_For_Patients/PDF/Protein%20C%20Deficiency%20Brochure.pdf. Accessed 5/28/12.

Protein S Deficiency

Prevalence   Occurs in about 1 in 500 individuals; severe condition rare

Other Names   hereditary thrombophilia due to protein S deficiency

Not often does the name of a disorder get its name from the city of its discovery. However, protein S deficiency discovered by researchers in Seattle, Washington, in 1979 was arbitrarily named after that city. Like protein C, protein S is thromobophilic, meaning that it is a clotting disorder. Protein S depends upon vitamin K.

What Is Protein S Deficiency?

Protein S deficiency is a disorder in which the blood does not coagulate normally. People with this condition are at a great risk for developing blood clots, which can be life-threatening. Like protein C deficiency, two types exist—one mild and the other severe:

- *Mild protein S deficiency*: People with this type of deficiency are at risk for a clot known as deep vein thrombosis or DVT. The clot forms in one of the limbs and then moves through the bloodstream, lodging in the lungs, where it is called a pulmonary embolism. Not all people with this type will develop the deficiency, but having the disorder increases the risk. Other factors may also add to the risk: increasing age, surgery in which the person is immobile for long periods of time, long rides in planes or cars, or pregnancy.

- *Severe protein S deficiency*: This rare type of clotting disorder occurs soon after birth and is called purpura fulminans. In purpura fulminans, clots form within the tiny blood vessels in the infant. These clots keep blood from circulating properly and cause the tissue to die. Such widespread clotting causes normal blood clotting proteins to be used up, letting bleeding occur throughout the child's body. Large purple nodules form on the skin. Even if the infant survives, he or she may have recurring symptoms.

What Is the Genetic Cause of Protein S Deficiency?

Mutations in the *PROS1* gene, officially known as the "protein S (alpha)" gene, cause protein S deficiency. Normally, *PROS1* instructs for making the protein S, a necessary component for the control of blood clotting. Made mostly in the liver, protein S cannot act by itself but must attach itself to other enzymes to become a cofactor for another enzyme called activated protein C (APC). APC is activated by a similar protein called protein C. Now, APC acts to turn off blood-clotting proteins known as Va and VIIIa.

Of the 220 mutations in *PROS1*, most are caused by the mutation of a single protein building block, which keeps protein S from acting as a cofactor. Three types of mutations are found in *PROS1*:

- *Type I*: Mutations that cause reduced levels of protein S. People with this type do not have enough to protein to control blood clotting, leading to abnormal clots.

- *Type II*: Protein S is altered, causing reduced activity. Persons with this type have normal levels of protein S, but the protein cannot form the proper cofactor, allowing blood clots to form.

- *Type III*: The person has a very low amount of free protein S, but the overall amount of protein S is normal. Free protein S forms the cofactor more readily than bound protein S; thus, reduced levels disrupt the inactivation of clotting proteins.

PROS1 is inherited in an autosomal dominant pattern and is located on the long arm (q) of chromosome 3 at 11.2.

What Is the Treatment for Protein S Deficiency?

General management of risk factors includes prevention of deep vein thrombosis. Walking and seat exercises during long airplane rides and stopping often when taking automobile trips are common-sense ideas for all people. For newborns, replace protein S with fresh frozen plasma that has been viral inactivated.

If the person has an acute attack of DVT, heparin, an anticoagulant treatment, should be used for about five days, followed by a transition to warfarin, which may be used for longer periods of time.

Further Reading

"*PROS1*." 2011. Genetics Home Reference. National Library of Medicine (U.S.). http://ghr.nlm.nih.gov/gene/PROS1. Accessed 5/28/12.

"Protein S Deficiency." 2011. Genetics Home Reference. National Library of Medicine (U.S.). http://ghr.nlm.nih.gov/condition/protein-s-deficiency. Accessed 5/28/12.

"Protein S Deficiency." 2011. Medscape. http://emedicine.medscape.com/article/205582-overview. Accessed 5/28/12.

Proteus Syndrome

Prevalence   Exceedingly rare, with fewer than 500 confirmed cases worldwide; less than one in 1,000,000 live births.

Other Names   Elephant Man disease; Wiedemann syndrome

A disfiguring condition of overgrowth of bones and body tumors has been described in the world medical literature for many years. American physician Michael Cohen first identified the disorder in 1979. In 1983, Hans-Rudolf Wiedemann, a German pediatrician who discovered many rare genetic disorders, was the first to describe and name the condition called Proteus syndrome. Proteus was a Greek god of the sea, who could morph into different forms and shapes. Wiedemann had four unrelated patients who had similar overgrowths of bone that reminded him of the various shapes that the god Proteus manifested. Joseph Merrick, known as the "Elephant Man," is believed to have had Proteus syndrome.

What Is Proteus Syndrome?

Proteus syndrome is a disfiguring disorder involving overgrowth of bones, skin, and body tumors. The condition may be extremely variable, ranging from mild to severe. Mild cases may not be identified as Proteus syndrome.

Following are common signs of the syndrome:

- Overgrowth of one side of face, body, and limbs, a condition known as hemi-hypertrophy; lesions noted at birth in about 17% of patients
- Asymmetric growth of hand and/or feet
- Rough and raised pigment on the skin, often dark and discolored in spots
- Tumors under the skin and on the skin surface
- Lipomas of collections of fat under skin; however, trunk and limbs may not have subcutaneous fat
- Abnormal skull shape; head enlarged or asymmetrical
- Deep lines and overgrowth of soft tissue on the soles of feet
- Partial gigantism with overgrowth of fingers or toes

Joseph Merrick: Elephant Man

Joseph Merrick was born in Leicester, England, in August 1862. During the first few years of his life, he developed a bony growth on his forehead, lumpy and thick skin, and very large lips. At some point during childhood, one of his arms and both feet grew large; because of problems walking, he fell and broke his hip and could not walk well. When he was 11, his mother died, and the father's new wife refused to accept him. He dropped out of school at the age of 12. In 1884, he contacted a showman named Sam Torr and suggested that Torr exhibit him. They dubbed him the Elephant Man.

Merrick traveled with several different shows in England and on the continent and ended up back in London. He became friends with a doctor at London Hospital named Frederick Treves. People of London society, including the princess of Wales, came to visit him. He stayed at the hospital until he died on April 11, 1890, at the age of 27.

People in the medical profession were interested in his rare condition, and Treves dissected the body and made cast of his head and limbs. His skeleton is now in the collection of the Royal London Hospital. Doctors of the time determined that Merrick suffered from neurofibromatosis type 1. However, during the 1980s, evidence arose that this was another conditions called Proteus syndrome.

Merrick's story formed the basis of several works of art. In 1979, British playwright Bernard Pomerance wrote a play called *The Elephant Man* based on his life. The play was later made into a film and a television show.

As a follow-up to the study released by NHGRI, researchers intend to test DNA from the skeleton to test to see if Merrick really had Proteus disease. According to the researchers, this will not be an easy task. The nature of the mosaic disorder is very complex in that the person's cells may or may not have the mutation. Essentially, with a mosaic, scientists are dealing with more than one genome.

- Premature dental eruption and root resorption
- Eye issues such as strabismus and cysts
- Scoliosis, a curved spine, and caved-in chest may hamper breathing
- Kidney and bladder problems in some cases
- Blood-clotting problems, leading to deep vein thrombosis
- Learning disabilities or mental retardation

People with this disorder have challenging medical problems such as orthopedic complications and heart and blood vessel disorders. In addition, the severe disfigurement may lead to social problems.

What Is the Genetic Cause of Proteus Syndrome?

Mutations in the *AKT1* gene, officially called the "v-akt murine thymoma viral oncogene homolog 1" gene, cause Proteus syndrome. Normally, *AKT1* provides instructions for serine-threonine protein kinase, an enzyme that is inactive in certain fibroplasts. However, activation occurs in the presence of another enzyme, phosphatidylinositol 3-kinase. The gene is important for the normal turning on and off of machinery that controls apoptosis or cell death.

On July 27, 2011, a team of researchers at the National Human Genome Research Institute (NHGRI) identified the genetic mutation that causes Proteus syndrome. A single letter exchange in the DNA code in the *AKT1* gene causes tissue to grow erratically. This mutation occurs during embryonic development. Timing is important here because it appears that the genetic mistake occurs in single cells in specific parts of the body. Only cells that are from the ones with the original *AKT1* mutation display the disease. Thus, there is a mixture of normal and mutated cells in one individual. This condition is called mosaicism.

The baby appears normal at birth; however, during the first two years, symptoms may appear. The mutation in *AKT1* keeps the affected cells from regulating growth, causing some parts to become abnormal while others remain normal. As the person ages, the irregular growth becomes more intense. Until the discovery of the gene, diagnosis of the disease had been strictly by observation of symptoms.

AKT1 is an oncogene, a gene that is related to the uncontrolled growth of cancer cells. If the mutation was part of a cascade of events that causes wild growth of cells that locate in one part of the body, then the cancer would be localized but then spread to other parts of the body. In Proteus syndrome, the mutation occurs in embryonic development, causing more tissues to be affected by the mutated gene. According to researchers, if the mutated gene was in all body cells, the person would not live. *AKT1* is located on the long arm (q) of chromosome 14 at position 32.32.

What Is the Treatment for Proteus Syndrome?

Treating the symptoms requires a team of doctors. The orthopedic and cardiovascular challenges are enormous. Other medical experts must attend the skin and the multitude of lesions. A team of psychologists and special educators will work with the person on self-esteem and learning problems.

Researchers from NIH believe that finding the *AKT1* gene as the cause can lead to developing drugs to treat the overgrowth sooner. In the cancer field a number of potential drugs are in the pipeline to inhibit the pathway of the gene. This same treatment could be a possibility for affecting Proteus syndrome.

Further Reading

"Genetics of Proteus Syndrome." 2011. Medscape. http://emedicine.medscape.com/article/948174-overview. Accessed 5/28/12.

"NIH Researchers Identify Gene Variant in Proteus Syndrome." 2011. *National Institutes of Health News*. http://www.nih.gov/news/health/jul2011/nhgri-27.htm. Accessed 5/28/12.

Proteus Syndrome Foundation. 2011. http://www.proteus-syndrome.org. Accessed 5/28/12.

Proteus Syndrome Patient Brochure. 2011. University of Kansas Medical Center. http://www.kumc.edu/gec/support/protwww.html. Accessed 5/28/12.

Prothrombin Deficiency

Prevalence   Affects 1 in 2 million people in population; inherited form very rare

Other Names dysprothrombinemia; factor II deficiency; hypoprothrombinemia

When a person cuts himself and is bleeding, the body begins a series of reactions to stop that bleeding. A clot is formed through the process called the coagulation cascade. The cascade is made possible by a special group of proteins. If any one of the clotting factors is missing, the clot does not form properly, and there is a higher risk of bleeding. Prothrombin or factor II is one of these proteins.

What Is Prothrombin Deficiency?

Prothrombin deficiency is a disorder that occurs when the blood does not clot properly. The body does not have enough of the protein factor II, one of the essential components of clotting. The genetic disorder is very rare, and both parents must carry the trait in order for it to be passed on to their children. Often, the disorder is acquired in the following ways:

- Long-term use of antibiotics causes lack of vitamin K, a vitamin essential for clotting
- Poor absorption of vitamin K from the intestines
- Deficiency of vitamin K from birth
- An obstruction in the bile duct
- Severe liver disease
- Use of drugs that prevent clotting, such as Coumadin or warfarin

Whether inherited or acquired, people with the deficiency do have problems with prolonged bleeding. Women may have abnormally heavy menstrual bleeding. Simple procedures such as having a tooth pulled or minor traumas cause abnormal bleeding. Sometimes the bleeding is internal occurring in the joints, muscles, brain, or other organs. Milder forms may exist and only show up when one has surgery or an injury.

What Is the Genetic Cause of Prothrombin Deficiency?

Mutations in the *F2* gene, officially known as the "coagulation factor II (thrombin)" gene, cause prothrombin deficiency. Normally, *F2* provides instructions for the protein prothrombin or coagulation factor II. When one cuts him- or herself, a group of related proteins that cause blood clotting or hemostasis go into action. The clot seals the cut so that no more blood will escape.

Prothrombin or factor II is made in the liver and circulates in the bloodstream in an inactive form until an injury occurs. The prothrombin then becomes an active from called thrombin, which acts on a protein called fibrinogen to make fibrin. Fibrin is one of the major components of blood clots.

More than 50 mutations in *F2* cause prothrombin deficiency. Only one change in the protein building block in prothrombin reduces the activity of the protein, leading to severe bleeding. Some mutations allow for a moderate amount and cause a milder form of the disorder. If prothrombin is completely absent, a person cannot live. *F2* is inherited in an autosomal dominant pattern and is located on the short arm (p) of chromosome 11 at position 11.

What Is the Treatment for Prothrombin Deficiency?

Knowing about the condition and its risks is important for those with prothrombin deficiency. Being very careful to avoid accidents can prevent problems for those with mild conditions. If moderate bleeding occurs, it can be treated with fresh frozen plasma. Another treatment is the use of Prothrombin complex concentrates or PCCs. However, a problem may occur with PCCs because the factor II amount can vary with different products. Certain PCCs have known to cause too much clotting, leading to blood clots that locate in the lungs.

Further Reading

"Factor II Deficiency." 2006. National Hemophilia Foundation. http://www.hemophilia .org/NHFWeb/MainPgs/MainNHF.aspx?menuid=185&contentid=48&rptname=bleeding. Accessed 5/28/12.

"Hypoprothrombinemia." 2012. Medscape. http://emedicine.medscape.com/article/ 956030-overview. Accessed 5/28/12.

Kujovich, Jody L. 2011. "Prothrombin-Related Thrombophilia." *GeneReviews*. http:// www.ncbi.nlm.nih.gov/books/NBK1148. Accessed 6/9/11.

Prune Belly Syndrome

Prevalence   Affects about 1 per 30,000 to 40,000 births; about 4% of cases of the syndrome are twin pregnancies; 95% of cases occur in males

Other Names   Eagle-Barrett syndrome; triad syndrome; urethral obstruction malformation sequence

In 1839, the German physician Frölich first saw children with stomachs that looked wrinkled and had several other kidney and bladder problems; the Canadian Osler named the condition "prune belly syndrome" because of the appearance of the abdominal area.

What Is Prune Belly Syndrome?

Prune belly syndrome is a congenital condition, meaning it is present at birth. While carrying the infant, the mother may not have enough amniotic fluid, but the baby's abdomen may swell with a small amount of fluid. When the baby is born, the abdomen muscles shrink, leaving a wrinkled or prune-like appearance. The mortality rate of children with this disorder is about 20%.

The children with the syndrome have a number of symptoms that are present at birth, including the following:

- Appearance like Buddha, with a large stomach
- Urinary tract infections caused by obstructions in the urinary tract
- Constipation because of weak or absent abdominal muscles
- Coughing difficulties because of increased lung secretions
- Delay in sitting and walking
- Heart problems
- Bone and muscle problems
- Two undescended testicles

Prune belly is a very serious and life-threatening problem. Many infants are stillborn or die within the first few weeks of life because of the many abnormal conditions. Newborns that do survive continue to have problems throughout life.

What Is the Cause of Prune Belly Syndrome?

Prune belly is not a genetic disease, but one that occurs during embryonic development. Scientists think that the problems come when the mesodermal layer is forming. Some noxious event between the 6th and 10th weeks could explain why the abdominal wall did not develop and explain the problems with the genito-urinary tract. The condition is associated with trisomy 18 and 21.

What Is the Treatment for Prune Belly Syndrome?

Treatment for the symptoms of the condition is controversial. Some doctors call for conservative management of the urinary tract, while others recommend an aggressive therapy beginning about 10 days after birth. Likewise, some surgeons do not recommend any surgery on the abdominal wall, while others recommend reconstruction.

Infections are inevitable, and even with aggressive antibiotic therapy, prune belly syndrome has many complications.

Further Reading

"Prune Belly Syndrome." 2011. MedlinePlus. National Institutes of Health (U.S.). http://www.nlm.nih.gov/medlineplus/ency/article/001269.htm. Accessed 5/28/12.

"Prune Belly Syndrome." 2011. Medscape. http://emedicine.medscape.com/article/447619-overview. Accessed 5/28/12.

Prune Belly Syndrome Network. http://www.prunebelly.org. Accessed 5/28/12.

Pseudoachondroplasia

Prevalence   About 1 in 30,000 individuals; exact prevalence unknown

Other Names   PSACH; pseudoachondroplastic dysplasia; pseudoachondroplastic spondyloepiphyseal dysplasia syndrome

Achondroplasia is a condition in which a defect in the cartilage of the long bones results in a condition called dwarfism. The word comes from three Greek roots: *a*, meaning "without"; *chondro*, meaning "cartilage"; and *plasia*, meaning "form" or "mold." Add the prefix *pseudo*, meaning "false," and a new condition emerges. This condition called pseudoachondroplasia at one time was thought to be a form of achondroplasia. Now researchers consider it a separate disorder with some symptoms of achondroplasia but some distinct features.

What Is Pseudoachondroplasia?

Pseudoachondroplasia is a disorder of bone growth. The cartilage at the epiphysis where bone is created does not form the bone properly. This is similar to the condition known as achondroplasia. However, in pseudoachondroplasia, the face and head are normal. The individual also has normal intelligence.

The average height of males with the condition is about 3 feet, 11 inches; females are a few inches shorter. The children are not short at birth but by about age two, the growth rate begins to slow. Following are features of individuals with pseudoachondroplasia:

- Short arms and legs
- Waddling gait noted at the onset of walking
- Early-onset osteoarthritis
- Joint extensibility in hands, knees, and ankles
- Limited range of motion at elbow and hips
- Scoliosis or abnormal curvature of the spine
- Lordosis or swayed back, and kyphosis or caved-in chest
- Very long trunk with prominent abdomen
- Short, stubby fingers

What Is the Genetic Cause of Pseudoachondroplasia?

Mutations in the *COMP* gene, officially known as the "cartilage oligomeric matrix protein" gene, cause pseudoachondroplasia. Normally, *COMP* provides instructions for the making of the COMP protein, which is found in the extracellular matrix. This structure is a complex intertwining of proteins that fills the spaces between cells, holding the cells in place. The COMP protein surrounds the cells that make ligaments, tendons, and cartilage. Cartilage-forming cells are called chondrocytes and are critical in forming bone, a process called osteogenesis. Bone formation starts with cartilage that then becomes bone. This is especially found in the bones of the spine, hips, arms, and legs. Although the role of *COMP* is not completely known, it is believed to play a role in cell growth, cell division, regulation of cell movement, and in cell death, a process known as apoptosis. *COMP* also binds strongly to calcium.

About 60 mutations in *COMP* cause pseudoachondroplasia. Most of the mutations involve the substitution of one amino acid or another in the COMP protein. The mutations cause a buildup of the protein in the endoplasmic reticulum of the chondrocytes, the maze that channels materials through the cell. This buildup causes the cells to die and not form the bone properly, leading to the features of pseudoachondroplasia. *COMP* is inherited in an autosomal dominant pattern and is located on the short arm (p) of chromosome 19 at position 13.1.

What Is the Treatment for Pseudoachondroplasia?

Like in all skeletal dysplasias, individuals with pseudoachondroplasia need an interdisciplinary team for medical assessment and treatment. Regular attention by an orthopedist, geneticist, pediatrician, neurologist, and physical therapist is essential to provide comprehensive treatment.

Further Reading

"*COMP*." 2011. Genetics Home Reference. National Library of Medicine (U.S.). http://ghr.nlm.nih.gov/gene/COMP. Accessed 6/9/11.

"Pseudoachondroplasia." 2011. Genetics Home Reference. National Library of Medicine (U.S.). http://ghr.nlm.nih.gov/condition/pseudoachondroplasia. Accessed 5/28/12.

"Pseudoachondroplasia." Nemours Children's Hospitals. http://www.nemours.org/service/medical/orthopedics/dysplasia/pseudo.html. Accessed 5/28/12.

Pseudoxanthoma Elasticum (PXE)

Prevalence   Estimated 1 case per 25,000–100,000 in the United States; twice as frequent in females as males; prevalent in South Africa Afrikaner population

Other Names   Gronblad-Strandberg syndrome; PXE

In 1881, a French dermatologist, Rigal, described a condition that had the characteristics of common xanthomas. The word "xanthoma" comes from two Greek words: *xanthos*, meaning "yellow," and *oma*, meaning "tumor." A xanthoma is a lesion that appears yellow. Later in 1896, Darier named the condition by adding the words *pseudo*, meaning "false," and elasticum, which relates to the deterioration of the connective tissue. The lesions appear in the skin usually during late childhood or early adolescence.

What Is Pseudoxanthoma Elasticum (PXE)?

Pseudoxanthoma elasticum is a disorder of the connective tissues that gives strength and flexibility to organs throughout the body. The progressive condition is characterized by deposits of calcium and other minerals in the elastic fibers that make up the connective tissue. The deposits may appear as yellowish bumps on the skin. As a result of the calcification, the following symptoms may occur:

- Yellow papules on neck, underarm, and elsewhere on skin
- Angioid streaks on the eyes in the light-sensitive layers of the retina; bleeding can lead to blindness
- Calcification in arteries lead to cramps in arms and legs
- Calcification in arteries of the heart, which can lead to chest pain or heart attack
- Calcification of the blood vessels in the digestive tract, which can lead to bleeding

The average age of onset is 13 years. However, the condition may appear any time from infancy to the 70s or older. People with pseudoxanthoma elasticum can have a normal life span. Early diagnosis can prevent some damage that may be irreversible.

What is the Genetic Cause of Pseudoxanthoma Elasticum (PXE)?

Mutations in the *ABCC6* gene, officially known as the "ATP-binding cassette, sub-family C (CFTR/MRP), member 6" gene, cause pseudoxanthoma elasticum. Normally, *ABCC6* provides instruction for making the ATP-binding cassette, sub-family C, member 6 protein, made in the liver, kidneys, and other tissues such as skin. ABCC6 protein carries important molecules across the cell membrane. Researchers believe that it plays a role in the connective tissue and also aids in regulating the deposit of calcium in tissues.

Over 150 mutations of *ABCC6* cause pseudoxanthoma elasticum. The mutations disrupt the proper function of the ABCC6 protein. It is generally thought that within the liver and kidney, the malfunctioning protein keeps the proper substances from reaching the fibers of the connective tissues. Others surmise that calcium is allowed to accumulate in the elastic fibers such as the skin. *ABCC6* is inherited in an autosomal recessive pattern and is found on the short arm (p) of chromosome 16 at position 13.1.

What Is the Treatment for Pseudoxanthoma Elasticum (PXE)?

Early diagnosis is essential because many of the changes to the body are irreversible. Preventive care must be taken to minimize the disease course. Some of the skin lesions can be cut out, but problems exist with surgery because of delayed healing and scarring. Collagen injections are an option. To control heart and blood vessel lesions, diet and exercise can minimize some of the effects. If bleeding occurs in the intestines, the individual may need iron supplements, blood transfusions, or partial removal of the stomach. Certain drugs may help bleeding in the retina. Vitamins A, C, E, and zinc supplements may reduce risk of hemorrhage.

Further Reading

"Pseudoxanthoma Elasticum." 2011. Medscape. http://emedicine.medscape.com/article/ 1074713-overview#showal. Accessed 5/28/12.

PXE International. 2011. http://www.pxe.org. Accessed 5/28/12.

"Pseudoxanthoma Elasticum." 2012. Genetics Home Reference. National Library of Medicine (U.S.). http://ghr.nlm.nih.gov/condition/pseudoxanthoma-elasticum. Accessed 5/28/12.

Psoriasis. *See* Psoriatic Arthritis

Psoriatic Arthritis

Prevalence   Affects an estimated 24 in 10,000 people

Other Names   arthritis psoriatica; arthropathic psoriasis; psoriatic arthropathy

This condition psoriatic arthritis has two disorders that are seemingly unrelated. Psoriasis is a common skin condition affecting 2% to 3% of the population worldwide. Arthritis is a condition involving inflammation of joints. Seventy percent of individuals who develop psoriatic arthritis first have the symptoms appear on the skin. About 5% to 10% of people with psoriasis will develop a crippling form of arthritis similar to rheumatoid arthritis. The word "psoriasis" comes from the Greek word *psor*, meaning "an itching." Arthritis comes from the Greek roots "*arthr*," meaning "joint" and the suffix *itis*, meaning "inflammation of." Thus, psoriatic arthritis combines joint inflammation with the skin condition psoriasis.

What Is Psoriatic Arthritis?

Psoriatic arthritis is a type of disorder of the joints that combines two conditions: psoriasis and arthritis.

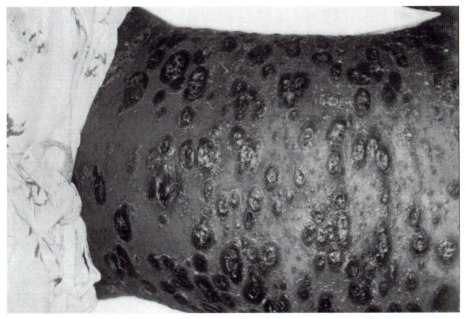

Psoriasis is a common skin disorder, ranging in severity or distribution from just a few spots on the body to large areas. Psoriatic arthritis combines this skin disease with arthritis. (CDC - Center for Disease Control and Prevention)

Psoriasis

Psoriasis is a common skin condition characterized by red, irritated patches that are covered with flaky white scales. In addition, fingernails and toenails of the person may become split, pitted, ridged, or separated from the nail bed. Psoriasis may develop during adolescence or early adulthood, although the arthritis conditions may not develop until between the ages of 30 and 50. About 30% of people with the skin condition psoriasis will develop psoriatic arthritis. In rare instances, a person will not have noticeable skin changes.

Arthritis

The person can have mild or severe arthritis, which progresses to massive joint destruction. Following are the arthritic symptoms of the disorder:

- Pain swelling in one or more joints.
- Stiffness in joints.
- Redness around joints: The area may be warm to touch.
- Swollen fingers and toes: The fingers and toes appear as sausage-like, a condition called dactylitis.
- Pain in the feet and ankles: The person may have tendinitis especially in the Achilles tendon of the heel. In addition, he may have plantar fasciitis in the soles of the feet.
- Pain in the lower vertebral area called the sacrum.

There are five types of psoriatic arthritis:

1. Asymmetric oligoarticular form: This mild type affects about 70% of patients. The joints involved are on different sides of the body and usually fewer than three on side of the body.
2. Symmetric polyarthritis type: This type affects joints on both sides of the body at the same time and is found in about 25% of all cases. It is the most similar to rheumatoid arthritis and is crippling in about 50% of all cases.
3. Spondylitis: Symptoms of this type include stiffness of the neck and spine and call also affect the hands and feet. It is similar to symmetric arthritis.
4. Distal interphalangeal predominant: The word "distal" indicates far, away from the center of the body. This type mainly affects the ends of the fingers and toes. These joints are the closest to the nails, and nail changes are frequent with this form.
5. Arthritis mutilans: This type is the most serious but least common type. Fewer than 5% of people with psoriatic arthritis have this form. This is the

type that involves severe damage to joints and loss of bone in the joints of hands, fingers, and toes.

People with any level of mild to very severe skin disease can develop psoriatic arthritis. If the skin conditions are not present, psoriatic arthritis is difficult to tell from other forms of arthritis.

What Are the Genetic Causes of Psoriatic Arthritis?

Although researchers have not pinpointed the exact cause of psoriatic arthritis, they do have evidence that it runs in families. Cases may appear at random with no family history. Evidence does exist that it is related to the immune system and inflammation of both skin and joints. Mutations in six genes may be related to psoriatic arthritis. These genes are briefly discussed.

HLA-B Gen

The *HLA-B* gene, officially called "major histocompatibility complex, class I, B" gene, is a member of a family of genes that are related to the immune system function. Several disorders of the spondyloarthropathies are related to this gene, including ankylosing spondylitis. Some of these disorders are associated with the common skin condition psoriasis. Psoriasis is a disorder in which the skin cells overproduce. *HLA-B* is located on the short arm of chromosome 6 at position 21.3.

HLA-C

Genetic variations in the *HLA-C* gene, officially the "major histocompatibility complex, class I, C" gene, are a cause of a type of psoriasis. Mutations in this gene may lead to psoriasis and the abnormal growth of skin cells. *HLA-C* is also located on the short arm (p) of chromosome 6 at position 21.3.

HLA-DRB1

Genetic variations in the *HLA-DRB1* gene, officially known as the "major histo-compatibility complex, class II, DR beta 1" gene, may pose a risk to psoriatic arthritis and juvenile arthritis. This gene is also located on the short arm of chromosome 6 at position 21.3.

IL12B

The *IL12B* gene, officially called the "interleukin 12B (natural killer cell stimu-latory factor 2, cytotoxic lymphocyte maturation factor 2, p40)" gene, encodes for a cytokine that acts on T and natural killer cells and has a wide range of bio-logical activities. Mutations in this gene are known to increase the risk of a type of psoriasis. *IL12B* is located on the long arm of chromosome 5 at position 31.1-q33.1.

IL13

Mutations in the *IL13* gene, officially called the "interleukin 13" gene, may be associated with psoriatic arthritis. The *IL13* gene is located on the long arm of chromosome 5 at position 31.

IL23R

Genetic variations in the *IL23R* gene, officially called the "interleukin 23 receptor" gene, provides information for interleukin 23 receptor that is related to several immune system responses. Variations have been associated to the risk of developing psoriasis. This gene is located on the short arm (p) of chromosome 1 at position 31.1.

TRAF3IP2

Genetic variations of *TRAF3IP2*, officially known as the "TRAF3 interacting protein 2" gene, are the cause of a type of psoriasis. This gene is located on the long arm (q) of chromosome 6 at position 21.

What Is the Treatment for Psoriatic Arthritis?

Researchers are working to find a cure and treat psoriatic arthritis. Treatment involves relieving pain, reducing swelling, keeping joints working, and preventing joint damage. Early diagnosis is very important to preserve function and movement. Following are some of the treatments approved for both psoriasis and psoriatic arthritis:

- Nonsteroidial anti-inflammatory drugs or NSAIDS, such as ibuprofen and naproxen. These drugs are effective for controlling inflammation in some patients.
- Disease-modifying antirheumatic drugs or DMARDs. These drugs are called biologic response modifiers and aim to prevent joint destruction. Examples are methotrexate and leflunomide.
- Biological response modifiers or biologics. These are injections derived from recombinant DNA technology. They target specific parts of the immune system. Examples are infliximab and etanercept.
- Joint injections with corticosteroids.
- Orthopedic surgery to correct joint damage.
- Self-care. Regular exercise can maintain flexibility. Devices, such as jar openers, can help people with everyday tasks.

Further Reading

National Psoriasis Foundation. 2012. http://www.psoriasis.org/learn02. Accessed 12/19/11.

"Psoriatic Arthritis." 2011. Genetics Home Reference. National Library of Medicine (U.S.). http://ghr.nlm.nih.gov/condition/psoriatic-arthritis. Accessed 12/19/11.

"Psoriatic Arthritis." 2012. MedicineNet.com. http://www.medicinenet.com/psoriatic_arthritis/article.htm. Accessed 12/19/11.

Pyridoxine-Dependent Epilepsy

Prevalence   Occurs about 1 in 100,000 to 700,000 cases; 100 cases reported worldwide

Other Names   AASA dehydrogenase deficiency; EPD; epilepsy, pyridoxine-dependent; PDE; pyridoxine dependency; pyridoxine dependency with seizures; pyridoxine-dependent seizures; vitamin B6–dependent seizures

Pyridoxine or vitamin B6 is one of the eight B vitamins. It assists in red blood cell production, heart health, and balancing hormonal changes in women. It also is required for the production of certain neurotransmitters. If the certain mutated genes are present, lack of the vitamin can cause a serious type of epilepsy, which can only be treated with large daily supplement of the vitamin.

What Is Pyridoxine-Dependent Epilepsy?

Pyridoxine-dependent epilepsy is a disorder in which the child has seizures beginning in infancy and sometimes before birth. Symptoms of the disorder include the following:

- Seizures that last several minutes and involve convulsions, rigidity, and loss of consciousness; these types of seizures are called tonic-clonic seizures
- Very low body temperature, a condition known as hypothermia
- Very poor muscle tone soon after birth
- Irritability especially before a seizure
- Drugs that normally treat seizures do not work

Some mothers have reported that they experienced strange movement sensations before birth, possibly indicating a prenatal seizure. The seizures begin soon after birth, although in rare instances, the child may not have seizures until he is one to three years old.

The child must be treated immediately with large doses of pyridoxine or vitamin B6 or will possibly develop a severe brain condition called encephalopathy. Most cases of seizures are controlled with the vitamins, but the person may still have problems such as developmental delay or learning disabilities.

What Is the Genetic Cause of Pyridoxine-Dependent Epilepsy?

Mutations in the *ALDH7A1* gene, officially known as the "aldehyde dehydrogenase 7 family, member A1" gene, cause pyridoxine-dependent epilepsy. Normally, *ALDH7A1* provides instructions for an important enzyme that changes chemical molecules called aldehydes. The enzyme α-aminoadipic semialdehyde (α-AASA) dehydrogenase, also known as antiquitin, is found in the nucleus of cells and also in the cell fluid, called cytosol. This enzyme is essential for breaking down the amino acid lysine in the brain, a process necessary for energy production.

Several mutations in *ALDH7A1* cause pyridoxine-dependent epilepsy. In one of the mutations, glycine replaces glutamine in the antiquitin protein. All the mutations cause pyridoxine-dependent epilepsy and produce a dysfunctional enzyme. The deficiency causes a substance called α-aminoadipic semialdehyde to build up and disrupt the normal processing of pyridoxine, or vitamin B6, from food. The neurotransmitters in the brain are affected, possibly causing the characteristic seizures. *ALDH7A1* is inherited in an autosomal recessive pattern and is located on the long arm (q) of chromosome 5 at position 31.

What Is the Treatment for Pyridoxine-Dependent Epilepsy?

All anticonvulsants must be withdrawn, and the individual is treated with large daily supplements of pyridoxine. The amount must be constantly monitored so that the individual will not have an overdose, which may cause nerve endings in the extremities to die, a condition known as neuropathy. Children with the disorder qualify for special education.

Further Reading

"*ALDH7A1*." 2011. Genetics Home Reference. National Library of Medicine (U.S.). http://ghr.nlm.nih.gov/gene/ALDH7A1. Accessed 5/28/12.

Gospe, Sidney M. 2009. "Pyridoxine-Dependent Epilepsy." *GeneReviews*. http://www.ncbi.nlm.nih.gov/books/NBK1486. Accessed 5/28/12.

"Pyridoxine-Dependent Epilepsy." 2011. Genetics Home Reference. National Library of Medicine (U.S.). http://ghr.nlm.nih.gov/search?query=pyridoxine+dependent+epilepsy. Accessed 5/28/12.

Pyruvate Dehydrogenase Complex Deficiency

Prevalence   Rare; several hundred cases reported worldwide; exact number is not known because there may be unreported mild cases

Other Names   pyruvate dehydrogenase complex (PDC) deficiency

Many rare genetic conditions are the result of metabolic disorders. These conditions mean that the person cannot process certain body chemicals. Pyruvate dehydrogenase complex deficiency is a disorder in which the person cannot process carbohydrates. The condition of pyruvate dehydrogenase deficiency, designated PDCD, is caused by one of three enzymes in the pyruvate dehydrogenase complex

(PDC). Understanding this action is difficult because it involves complex biochemical actions in the citric acid cycle in the mitochondria.

What Is Pyruvate Dehydrogenase Complex Deficiency?

Pyruvate dehydrogenase complex (PDC) deficiency (PDCD) is a neurological disorder that results from problems in processing carbohydrates. The condition is related to the deficit in the mitochondria, the powerhouses or the cell. In the mitochondria, a cycle called the citric acid cycle works to derive energy from carbohydrates. If this cycle malfunctions in any way, the body cannot get the energy it needs.

The key feature of this condition is the degeneration of the gray matter in the cerebrum of the brain, causing certain cells to die and abnormal capillary action to develop in the brain stem. An abnormal buildup of lactate causes progressive neurological symptoms that may be evident at birth or emerge in later childhood.

The symptoms of the condition may vary but generally they are the following:

- General nonspecific symptoms of any metabolic illness, include poor feeding, lethargy, and rapid breathing.
- Developmental signs include mental delays, psychomotor delays, and slow growth.
- Progressive symptoms develop as the child ages.
- Poor muscle tone combines with periods of poor coordination
- Seizures may begin.
- Eye problems develop along with poor response to visual stimuli.
- Respiratory symptoms develop that include sleep apnea and dyspnea or difficulty breathing.

If the children have early-onset or infantile-onset of pyruvate dehydrogenase complex deficiency, they usually die during the first years of life. If the onset is in later childhood, the person may survive into adulthood.

What Are the Genetic Causes of Pyruvate Dehydrogenase Complex Deficiency?

Three components make up the large pyruvate dehydrogenase complex, called E1, E2, and E3. Likewise three genes are responsible for the complex and relate to various ages and severity. The genes are *PDP1*, *PDHX*, and the E1 alpha enzyme subunit on the X chromosome.

PDP1

Mutations in the *PDP1* gene, officially known as the "pyruvate dehyrogenase phosphatase catalytic subunit 1" gene, cause a form of the deficiency. Normally,

PDP1 provides instructions for the enzyme pyruvate dehydrogenase (E1) that breaks down certain residues in the citric acid cycle.

Mutations in the gene cause pyruvate dehydrogenase phosphatase deficiency and lead to the symptoms noted in the condition. This form is quite rare. *PDP1* is inherited in an autosomal dominant pattern and is located on the long arm (q) of chromosome 8 at position 22.1

PDHX

Mutations in the *PDHX* gene, officially known as the "pyruvate dehydrogenase complex, component X" gene, cause certain forms of the pyruvate dehydrogenase complex deficiency. Normally, *PDHX* works in the complex to bind essential proteins for the conversion of energy in the mitochondria. Mutations in the gene disrupt this process and lead to a deficiency in the enzyme. These mutations are quite rare. PDHX is inherited in an autosomal recessive pattern and is found on the short arm (p) of chromosome 11 at position 13.

X Chromosome

Mutations in the E1 alpha enzyme subunit gene may result in gender differences. Normally the gene regulates stability and efficiency in the operation of the enzyme. About 90 mutations of E1 alpha enzyme subunit disrupt the process in the citric acid cycle. The gene is located on the short arm (p) of chromosome X at position 22.2-p22.1.

What Is the Treatment for Pyruvate Dehydrogenase Complex Deficiency?

Some therapies may extend the lives of individuals who are severely affected with pyruvate dehydrogenase complex deficiency; however, the progressive nature of the neurological deterioration results in significant morbidity.

Further Reading

"Pyruvate Dehydrogenase Complex Deficiency." 2011. Healthline. http://www.healthline .com/galecontent/pyruvate-dehydrogenase-complex-deficiency. Accessed 5/28/12.

"Pyruvate Dehydrogenase Complex Deficiency." 2011. Medscape. http://emedicine .medscape.com/article/948360-overview#showall. Accessed 5/28/12.

R

Rachischisis. *See* Spina Bifida

Recklinghausen Disease, Nerve. *See* Neurofibromatosis Type 1 (NF1)

Recombinant 8 Syndrome

Prevalence   Rare condition with unknown incidence

Other Names   rec(8) syndrome; recombinant chromosome 8 syndrome; San Luis Valley syndrome

The San Luis Valley is an area in southern Colorado and northern New Mexico with a large percentage of people of Hispanic heritage. A condition became evident in the population, which involves abnormalities in many body parts. The condition, which has been traced to abnormalities in chromosome 8, is sometimes called the San Luis Valley syndrome. It is rarely found outside this population.

What Is Recombinant 8 Syndrome?

Recombinant 8 syndrome is a condition that involves a rearrangement of chromosome 8, resulting in a distinct facial appearance and several body abnormalities. Following are some of the symptoms of Recombinant 8 syndrome or San Luis Valley syndrome:

- Distinct appearance with a wide, square face
- Down-turned mouth with a thin upper lip and small chin and jaw
- Diseases of the gums and abnormal tooth development
- Wide-set eyes with low-set, unusually shaped ears
- Chronic ear infections and hearing loss
- Heart and urinary tract deformities
- In males, undescended testicles

Many children with this disorder do not live past early childhood because of the complications of the heart and urinary deformities.

What Is the Genetic Cause of Recombinant 8 Syndrome?

Chromosome 8 is one of the 23 pairs of chromosomes. Like other chromosomes, it has a short arm and long arm. About 8% of the genes are involved in brain development and function. There is a unique feature on the short arm of the chromosome in which about 15 megabases or groups of base pairs of amino acids have a very high mutation rate.

Recombinant 8 syndrome is caused by a rearrangement in this short arm (p), resulting in the deletion of one piece of the arm and a duplication of a piece on the long arm (q). This recombination appears to cause the many symptoms of the disorder. Scientists are investigating the genes involved in the deletion and duplication.

The condition appears to be inherited in an autosomal dominant pattern with at least one parent having an inversion or break in the chromosome in two places. The inversion may not result in the loss of genetic material, nor will the parent have health problems. The issues arise when the inversion is passed to the next generation. People with these inversions are at risk for having a child with recombinant 8 syndrome.

What Is the Treatment for Recombinant 8 Syndrome?

Because this disorder involves a change in the chromosome, treatment of the basic genetic factors is not possible at this time. However, the heart disorders, which appear to be the major cause of death, may be monitored and treated with surgery. Other symptoms may also be treated with drugs or surgery. The child qualifies for special education and services.

Further Reading

"Chromosome 8." 2011. Genetics Home Reference. National Library of Medicine (U.S.). http://ghr.nlm.nih.gov/chromosome/8. Accessed 5/28/12.

"Recombinant 8 Syndrome." 2011. Genetics Home Reference. National Library of Medicine (U.S.). http://ghr.nlm.nih.gov/condition/recombinant-8-syndrome. Accessed 5/28/12.

Refsum Disease

Prevalence   Rare; only about 60 cases known worldwide

Other Names   adult Refsum disease; ARD; classic Refsum disease; CRD; hereditary motor and sensory neuropathy Type IV; heredopathia atactica polyneuritiformis; HMSN IV; HMSN type IV; phytanic acid storage disease; Refsum's disease; Refsum syndrome

In 1946, Norwegian neurologist Sigvald Bernhard Refsum noted a condition in which several body systems were involved. The person had neuropathy in the arms and legs, a condition of the retina, dysfunction of the cerebellum, and a skin condition known as ichthyosis. He noted the symptoms slowly progressed through childhood and adolescence, into adulthood. The rare disorder was named after the doctor—Refsum disease.

What Is Refsum Disease?

Refsum disease is a rare genetic caused by a deficiency enzyme in the peroxisomes that break down phytanic acid, a type of fat found in foods. Peroxisomes are small bodies found in animal and human cells, which have a large number of enzymes that are essential in cell metabolism.

The disorder usually begins in childhood or early adulthood and may have the following symptoms:

- Beginning in childhood or adolescence with night blindness due to the degeneration of the retina, a condition known as retinitis pigmentosa
- Person may develop cataracts and nystygamus, when the eyeballs move in erratic concentric circles
- Deafness
- Loss of sense of smell, a condition called anosmia
- Disorders of balance and coordination, showing impact on the cerebellum
- Dry and scaly skin
- Abnormal heartbeats with possible life-threatening cardiomyopathy
- Some with shortened bones in fingers or toes and an abnormally short fourth toe
- Liver and kidney disorders
- Adults develop a neurological condition called peripheral neuropathy

The condition may appear in childhood, but the symptoms do not develop until the person is in the 40s or 50s.

What Are the Genetic Causes of Refsum Disease?

Mutations in two genes—*PEX7* and *PHYH*—cause Refsum disease.

PEX7

Mutations in the *PEX7* gene, or the "peroxisomal biogenesis factor 7" gene, cause Refsum disease. Normally, *PEX7* provides instructions for a protein called peroxisomal biogenesis factor 7. This enzyme is part of a group of proteins that make up the peroxisomal assembly. The PEX proteins are in cells and are responsible for importing other enzymes into peroxisomes. These enzymes are in little sacs and break down many substances, especially fatty acids. These processes are necessary for the synthesis of fats or lipids that are used in digestion and in the nervous system.

Three mutations in the *PEX7* disrupt the import of the critical enzymes, causing toxic substances then to build up over time. It is speculated that the buildup affects the vision and the senses and causes other symptoms of the disorder. *PEX7* is inherited in an autosomal recessive pattern and is located on the long arm (q) of chromosome 6 at position 23.3.

PHYH

The second gene involved is the *PHYH* gene, or "phytanoyl-CoA 2-hydroxylase." Normally, this gene provides instruction for making the enzyme phytanoyl-CoA hydroxylase, which is essential for the work of peroxisomes in the cells. This enzyme breaks down many substances, including the fatty acid phytanic acid. This enzyme performs one of the first in the breaking down process of fats.

About 30 mutations in *PHYH* cause more than 90% of all cases of Refsum disease. The mutation disrupts the production of the enzyme and consequently affects the breakdown of phytanic acid. Phytanic acid builds up in the cells, and the toxic effect leads to the many symptoms of the disorder. *PHYH* is inherited in an autosomal recessive pattern and is located on the short arm (p) of chromosome 10 at position 13.

What Is the Treatment for Refsum Disease?

People with the disorder must avoid foods that contain phytanic acid. These foods include dairy products, beef and lamb, and fatty fishes such as tuna, cod, and haddock. Some severe cases may need to have a plasma exchange to control the buildup of the phytanic acid.

When the person develops the disorder as an adult, treatment is possible for muscle weakness, numbness, and skin disorders. However, the vision and sensory problems may persist.

Further Reading

"NINDS Refsum Disease Information Page." 2011. National Institute of Neurological Disorders and Stroke. http://www.ninds.nih.gov/disorders/refsum/refsum.htm. Accessed 5/28/12.

"*PEX7*." 2011. Genetics Home Reference. National Library of Medicine (U.S.). http://ghr.nlm.nih.gov/gene/PEX7. Accessed 5/28/12.

"Refsum Disease." 2011. Genetics Home Reference. National Library of Medicine (U.S.). http://ghr.nlm.nih.gov/condition/refsum-disease. Accessed 5/28/12.

Renpenning Syndrome

Prevalence Unknown

Other Names mental retardation, X-linked Renpenning type; mental retardation, X-linked, syndromic 8; MRXS3; MRXS8; RENS1; SHS; Sutherland-Haan syndrome; Sutherland-Haan X-linked mental retardation syndrome; X-linked mental retardation syndromic 3; X-linked mental retardation with spastic diplegia

In 1962, Hans Renpenning noted a condition in a large Mennonite family living in Manitoba, Canada. The term "Renpenning syndrome" came to be used to indicate any X-linked mental retardation; however, when the exact location was mapped, it became the name of a specific disorder.

What Is Renpenning Syndrome?

Renpenning syndrome is an X-linked disorder that is characterized by mental retardation and other body disorders. Being an X-linked syndrome, it is present mostly in males. Following are the symptoms of the disorder:

- Mental retardation
- Very short stature
- Long, narrow face
- Smaller-than-normal head, a condition known as microcephaly
  - Distinct facial appearance with upturned lip, abnormal nose, cupped ears, and short philtrum
  - Eye deformities
  - Cleft palate
  - Small testes
  - Anal anomalies

What Is the Genetic Cause of Renpenning Syndrome?

Mutations in the *PQBP1* gene, officially known as the "polyglutamine tract binding protein 1" gene, causes Renpenning syndrome. Normally, *PQBP1* provides instructions for the nuclear polyglutamine-binding protein that is active in cell transcription activation. This protein binds two highly conserved tryptophans to a certain protein with the amino acid proline. The mutations disrupt the action of the protein causing the many symptoms of Renpenning syndrome. *PQBP1* is inherited in an X-linked recessive pattern and is located on the short arm (p) of chromosome X at position 11.23.

What Is the Treatment for Renpenning Syndrome?

Because this is a serious X-linked condition, no cure is available. Treatment is supportive and possibly cosmetic. The individual qualifies for special education.

Further Reading

"PQBP1." 2011. Genetics Home Reference. National Library of Medicine (U.S.). http://ghr.nlm.nih.gov/gene/PQBP1. Accessed 5/28/12.

"Renpenning Syndrome 1." 2011. Genetics and Rare Disease Information Center (GARD). National Institutes of Health (U.S.). http://rarediseases.info.nih.gov/GARD/Condition/9509/Renpenning_syndrome_1.aspx. Accessed 5/28/12.

Retinitis Pigmentosa (RP)

Prevalence   One of the most common diseases of the retina; affects about 1 in 3,500 to 1 in 4,000 people in the United States and Europe

Other Names   cone-rod retinal dystrophy; pigmentary retinopathy; rod-cone dystrophy; RP; tapetoretinal degeneration

Retinitis pigmentosa or RP is a very common from of retinal degeneration. Actually, RP is not one disease but a large group of conditions caused by several genes inherited in several different ways. The term "retinitis" is really a misnomer because it implies there is some kind of infection or inflammation. This is not true, as causes of the condition are genetic. However, the name has stuck.

What Is Retinitis Pigmentosa (RP)?

Retinitis pigmentosa consists of a group of serious eye conditions that cause vision loss. All the disorders affect the retina, the screen at the back of the eye where images are formed. In RP, the light-sensing cells that enable one to see the pictures begin to deteriorate.

Following is the progression of RP:

- Night vision is poor, becoming evident in childhood.
- Any dim light causes a problem.
- Blind spots begin to develop in the side or peripheral vision.
- As the person gets older, the blind spots merge to form tunnel vision.
- Over the years, central vision is affected; person has trouble reading, driving, and even recognizing faces.
- The person is then legally blind.

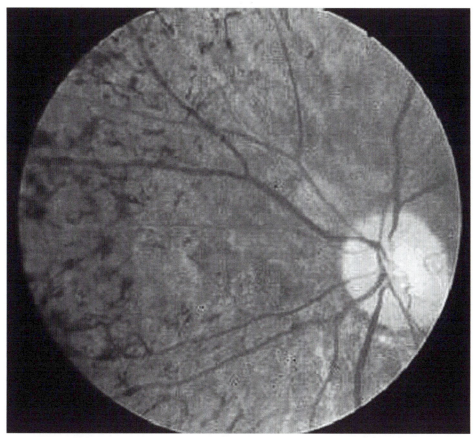

Enlarged image of eye with retinitis pigmentosa, an eye disease that damages the retina. (Getty Images News / Getty Images)

When a person has only the above symptoms, the disorder is described as nonsyndromic RP. However, some types may combine the disorder with a syndrome. Following are syndromes that may include retinitis pigmentosa:

- Usher syndrome—a combination of vision and hearing loss that begins early in life
- Bardet-Beidl syndrome
- Refsum syndrome
- NARP or neuropathy, ataxia, and retinitis pigmentosa

What Are the Genetic Causes of Retinitis Pigmentosa (RP)?

Several genes are related to RP. However, this article will present four of the most common: *RHO*, *RP2*, *RPGR*, and *USH2A*.

RHO

Mutations in the gene *RHO*, officially known as the "rhodopsin" gene, cause retinitis pigmentosa. Normally, *RHO* encodes for making the protein rhodopsin, a substance necessary for normal vision. Rhodopsin is found in the rods, specialized light receptor cells found on the retina. Rods are essential for vision in dim light. The other receptors on the retina are called cones, which help the person see in bright light. In the retina a molecule called 11 cis retinal, which is a form of vitamin A, is bound to rhodopsin. Light activates the rhodopsin and creates reactions that cause electrical signals to be transmitted to the visual centers of the brain.

About 140 mutations in *RHO* cause RP. Most of the mutations change the folding of the rhodopsin protein. Whatever the cause, the altered versions disrupt the cell functions, causing the rods to self-destruct and leading to the gradual loss of both rods and cones. This most common form of the disorder, which is found in 20% to 30% of all cases, is inherited in an autosomal dominant pattern. A few rare cases may have an autosomal recessive pattern. *RHO* is located on the long arm (q) of chromosome 3 at position 21 to q 24.

RP2

Mutations in the *RP2* gene, or the "retinitis pigmentosa 2 (X-linked recessive)" gene, cause RP. Normally, *RP2* encodes for a protein that is active in all body cells including the cells at the back of the eye. Researchers suggest that it may be involved in transporting important proteins to the light receptor cells.

More than 70 mutations in *RP2* are associated with the X-linked form of RP. The condition especially is found in early childhood. Most of these mutations cause a very short version of the RP2 protein, which then disrupts the processes of the cells in the retina. *RP2* is inherited in an X-linked recessive pattern and is located on the short arm (p) of the X chromosome at position 11.3.

RPGR

Mutations in the *RPGR* gene, officially known as the "retinitis pigmentosa GTPase regulator" gene, cause RP. Normally, *RPGR* encodes for a protein that is essential for normal vision. It appears to play an important role in structures called cilia, tiny hairlike projections that stick out of several types of cells and are involved in cell movement. Cilia are very important for all types of senses, including hearing, smelling, and seeing. There are several different isoforms or versions of the *RPGR* gene.

More than 300 mutations of *RPGR* are found in the X-linked form of RP. These mutations account for about 70% of these cases. *RPGR* is inherited in an X-linked recessive form and is located on the short arm (p) of chromosome X at position 21.1.

USH2A

Mutations in the *USH2A* gene, officially called the "Usher syndrome 2A (autosomal recessive, mild)" gene, cause some kinds of RP. Normally, *USH2A* encodes for a protein known as usherin, which is an essential part of the basement membranes. The basement membrane lines the basic structures of the sense organs. In the ear, it lines the entire ear structure, and in the eye, the basement membrane is located in the back of the eye in the retina. Usherin is found in these structures. Some scientists think that the protein usherin may have a role related to the synapses of nerves in these structures.

Several mutations in *USH2A* cause RP. They are responsible for about 10% to 15% of all cases of RP. *USH2A* is inherited in an autosomal recessive pattern and is located on the long arm (q) of chromosome 1 at position 41.

What Is the Treatment for Retinitis Pigmentosa (RP)?

No specific now exists for RP. However, people with the condition may find vitamins A and E are helpful. Another study found that an omega-3-rich diet may slow disease progression. Low-vision services can help people cope with the disease. In addition to nutritional therapy, gene therapy and implantable microchips may offer hope.

Further Reading

Foundation Fighting Blindness. http://www.blindness.org/what-is-retinitis-pigmentosa. Accessed 5/28/12.

Pagon, Roberta A., and Stephen P. Daiger. 2005. "Retinitis Pigmentosa Overview." *GeneReviews*. http://www.ncbi.nlm.nih.gov/books/NBK1417. Accessed 5/28/12.

"Retinitis Pigmentosa." 2011. Medscape. http://emedicine.medscape.com/article/1227488-overview#showall. Accessed 5/28/12.

Retinoblastoma

Prevalence About 250 to 350 cases per year in the United States; accounts for 4% of all cancers in children under 15 years; worldwide, 11 cases per million children

Other Names glioma, retinal; RB

In the early 1800s, Peter Pawius, a physician in Amsterdam, first described a tumor that invaded the back orbit of the eye. He described the growth as filled with a substance that looked like a mixture of brain tissue, thick blood, and crushed stone. Later, William Hey, an English surgeon, called the growth a mass fungus haematodes that affected the globe of the eye. Later, Paris physicians with a microscope determined the tumor arose from the retina and the optic nerve. In 1864, Rudolph Virchow, a famous German scientist, named the tumor glioma of the retina, and Flexner in 1891 noticed tiny rosettes within the tumor. Later research showed that the cells were like the undeveloped cells of the retina of the embryo called retino-blasts. This is the origin of the name "retinoblastoma." It combines the Greek term *oma*, meaning "tumor," with the name of those cells that looked like the immature eye cells of the embryo, the retinoblasts.

What Is Retinoblastoma?

Retinoblastoma is a type of eye cancer that develops generally before the age of five. The retina of the eye is a structure at the back of the eye where images are formed and then go to the brain where it is interpreted as a picture. The retina has sensitive tissues called the rod and cones that detect light and color.

Most children develop the condition only in one eye, but one out of three may develop it in both eyes. Following are the signs and symptoms of retino-blastoma:

- Visible whiteness in the pupil of the eye, called "cat's eye reflex"; this is especially noticeable if one takes a picture with a flash camera
- Crossed eyes or eyes that do not point in the same direction
- Eye pain
- Redness in eye
- Blindness or poor vision in the affected eye or eyes

What Is the Genetic Cause of Retinoblastoma?

Mutations in the *RB1* gene, known officially as the "retinoblastoma 1" gene, cause retinoblastoma. Normally, *RB1* encodes for a protein called pRB, which acts as a tumor suppressor. Tumor suppressors regulate cell growth and keep the cells from dividing wildly. The protein pRB stops the DNA replication and tightly regulates what is going on within the cells. In this way, cancerous growths do not develop.

Hundreds of mutations in *RB1* cause retinoblastoma. Researchers believe that about 40% of cases are hereditary and can be passed on to the next generation. This type is called germinal retinoblastoma. Germinal retinoblastoma readily spreads to other body parts and may lead to pineal gland (a gland in the brain) cancer, a type of bone cancer called osteosarcoma, and skin cancer called mela-noma. The other 60% occur as spontaneous mutations and cannot be passed to the next generation. People are born with two normal copies of *RB1*, but some-thing happens in early childhood when both copies of the gene acquire mutations leading to the cancerous tumors

One group of germinal mutations in the gene are related to several genes on chromosome 13, causing intellectual disability, distinctive facial features, and impaired growth *RB1* is inherited in an autosomal recessive pattern and is located on the long arm (q) of chromosome 13 at position 14.2.

What Is the Treatment for Retinoblastoma?

If diagnosed early and treated promptly, retinoblastoma is curable. However, if it spreads to other parts of the body, it is life-threatening. Several treatments are available including chemotherapy, given in pill form to kill cancer cells. Shrinking the tumor may be possible using brachytherapy, cryotherapy, and laser therapy. If the tumor is large, surgery may be used to remove the eye and fitting it with an artificial eye. Several clinical trials are ongoing investigating the best treatments.

Further Reading

"A Parent's Guide to Understanding Retinoblastoma." http://retinoblastoma.com/retinoblastoma/frameset1.htm. Accessed 5/28/12.

"Retinoblastoma." 2011. Mayo Clinic. http://www.mayoclinic.com/health/retinoblastoma/DS00786. Accessed 5/28/12.

"Retinoblastoma." 2011. National Cancer Institute. http://www.cancer.gov/cancertopics/types/retinoblastoma. Accessed 5/28/12.

Rett Syndrome

Prevalence   Affects 1 in 10,000 to 15,000 live female births worldwide

Other Names   autism-dementia-ataxia-loss of purposeful hand use syndrome; cerebroatrophic hyperammonemia; Rett disorder; Rett's disorder; Rett's syndrome; RTS; RTT

In 1966, Dr. Andreas Rett, an Austrian physician, noted a condition that occurred mostly in girls that occurred when normal children began to progressively deteriorate. He described the condition in a journal article that was soon forgotten. In 1983, a Swedish researcher, Dr. Bengst Hagberg, published a second article describing the condition. It was not until then that the disorder was generally recognized and given the name of Rett syndrome.

What Is Rett Syndrome?

Rett syndrome is a neurological disorder found almost exclusively in girls. The girls seemingly have normal development, and then symptoms begin to occur that are puzzling to parents. Although there is a rare form that begins in infancy, most of the cases show the following four stages of development:

- *Stage I—Early onset*: This stage begins between the ages of 6 months and 18 months and may display vague symptoms, which can be interpreted as normal. The child may not make eye contact and display no interest in toys. Sitting or crawling may be delayed, and loss of other gross motor skills is noted. The child may begin by wringing hands, but parents are usually are not alarmed. The stage can continue for about a year.

- *Stage II—Period of rapid decline*: Lasting between ages one and four, this second stage may begin with hand movements such as wringing, washing, clapping, tapping, or moving the hands repeatedly to the mouth. Another movement is clasping the hands behind the back or to the sides while randomly grasping and releasing. Movements may disappear during sleep. Breathing symptoms, such as apnea or hyperventilation, may also improve during sleep. The girls may show autistic-like symptoms, such as loss of communication or social interaction. There may be a slowing of the growth of the head. Walking becomes labored and difficult.

- *Stage III—Period of a plateau or pseudo-stationary stage*: This stage begins between the ages of 2 and 10 and lasts for several years. Many problems erupt. The child may develop seizure with more motor problems. However, she may begin to show more interest in her surroundings; parents are hopeful about improvement.

- *Stage IV—Late motor deterioration stage*: This stage can last for decades and is accompanied by many problems: movement issues, scoliosis, rigidity, heart problems, feeding problems, and spasticity. Intellectual and developmental disorders continue. Girls who were walking may stop. Gaze improves and hand movements may stop.

Girls with Rett syndrome usually have a normal life span, but they will need special care and assistance during their lives.

What Are the Genetic Causes of Rett Syndrome?

Rett syndrome is a genetic disorder; however, fewer than 1% of cases are passed from one generation to the other. Most of the cases are the result of a random mutation. Two genes are related to Rett syndrome: *CDKL5* and *MEPC2*.

CDKL5

Mutations in the *CDKL5* gene, known officially as the "cyclin-dependent kinase-like 5" gene, cause Rett syndrome. Normally, *CDKL5* instructs for making an essential protein for brain development. The protein acts as a kinase, an enzyme that affects other proteins by adding a cluster of oxygen and phosphorus called a phosphate groups to DNA.

About 10 mutations in *CDKL5* have been found in girls with a type of Rett syndrome that has the early onset–seizure variant. This form usually begins in infancy and has all the classic developmental symptoms and language loss. These mutations involve a change in one single protein building block; other mutations lead to short, nonfunctional versions of the protein. *CDKL5* is inherited in an X-linked dominant pattern and is located on the short arm (p) of the X chromosome at position 22.

MECP2

Mutations in the *MECP2* gene, known officially as the "methyl CpG binding protein 2 (Rett syndrome)" gene, cause Rett syndrome. Normally *MECP2* instructs for making the protein MECP2, which appears to be related to the function of nerve cells in the brain and is present is high levels in nerve cells. MECP2 protein plays a role in the proper working of the synapses, the communication gaps between neurons. The protein also is active in turning off other genes from working when they are not supposed to. These all-important proteins also may serve as genetic blueprints for making other proteins in a process known as alternative splicing. This process is necessary for the cells to communicate normally.

About 200 mutations in *MECP2* cause Rett syndrome. This gene is responsible for most cases. The mutations, which may be a change in a single DNA building block or either addition or deletions of DNA, change the structure of the protein. The mutated protein disrupts the process of turning off the other genes, and thus they make proteins that are not wanted and possibly cause the symptoms of Rett syndrome. *MECP2* is inherited in an X-linked dominant pattern and is located on the long arm (q) of chromosome X at position 28.

What Is the Treatment for Rett Syndrome?

No cure exists for Rett syndrome. A multidisciplinary approach is used to treat the various symptoms and to provide support. Medication may be needed for convulsions or breathing difficulties. The girl should be monitored for scoliosis and heart difficulties. Children will need help in learning self-directed skills such as feeding and dressing. The children qualify for special education services, along with physical and occupational therapy.

Further Reading

"Rett Syndrome." Eunice Kennedy Shriver National Institute of Child Health and Human Development (U.S.). http://www.nichd.nih.gov/health/topics/rett_syndrome.cfm. Accessed 5/29/12.

"Rett Syndrome." 2011. Mayo Clinic. http://www.mayoclinic.com/health/rett-syndrome/DS00716. Accessed 5/29/12.

"Rett Syndrome Fact Sheet." 2011. National Institute of Neurological Disorders and Stroke (U.S.). http://www.ninds.nih.gov/disorders/rett/detail_rett.htm. Accessed 5/29/12.

Rhizomelic Chondrodysplasia Punctata

Prevalence　　Rare; fewer than 1 in 100,000 people worldwide

Other Names　CDPR; chondrodysplasia punctata, rhizomelic; RCDP; RCDP 1; RCP chondrodystrophia calcificans punctata; rhizomelic chondrodysplasia punctata Type 1

The Little People of America or LPA is an organization that serves people of that are short as a result of a medical or genetic condition. The term "dwarfism" is used to describe the condition. There are about 200 different kinds of dwarfism defined as an adult height of 4 feet, 10 inches or under. The condition rhizomelic chondrodysplasia punctata is one of those conditions.

What Is Rhizomelic Chondrodysplasia Punctata?

Rhizomelic chondrodysplasia punctata affects many body systems. The long term broken down into its root words means the following: The term "rhizomelic" comes from the Greek words *rhizo*, meaning "root," and *melos*, meaning "limb," relating to the hip joint or shoulder; "chondrodysplasia" comes from the words *chondro*, meaning "cartilage," *dys*, meaning "with difficulty" and *plasia*, meaning "form" or "mold"; punctata refers to dots or points. The words describe a disorder of cartilage formation, especially at certain points in the limbs. This condition impairs the normal body functioning of the individual. Following are the many body areas affected by rhizomelic chondrodysplasia punctata":

- Skeletal abnormalities: This condition results in shortened bones of the rhizomelia, the upper bones of the arms and thighs. Points develop on the bones that can be seen on X-rays. These problems develop into joint deformities or contractures that make the joints stiff and painful.

- Distinctive facial features: The individual may have a very prominent forehead with wide-set eyes and a sunken appearance in the middle of the face, framed by full, fat cheeks. The nose is small with upturned nostrils.

- Eyes: Almost all children have cataracts that are apparent at birth or in early infancy.

- Developmental delay: The motor abilities develop more slowly. The children are late in sitting, feeding, and talking.

- Intellectual disability.

- Seizures: Many of the children have seizures.

- Breathing problems: Respiratory problems are recurrent.

Children rarely live past the age of 10, although some with milder characteristics have lived into early adulthood. There are three types of rhizomelic chondrodysplasia punctata: type 1 (RCDP1), type 2 (RCDP2), and type 3 (RCDP3). Their genetic cause distinguishes them.

What Are the Genetic Causes of Rhizomelic Chondrodysplasia Punctata?

Mutations in three genes cause rhizomelic chondrodysplasia punctata. *PEX7* causes RCDP1, *GNPAT* causes RCDP2, and *AGPS* causes RCDP3.

PEX7

Mutations in the *PEX7* gene, known officially as the "peroxisomal biogenesis factor 7" gene, cause RCDP1. Normally, *PEX7* encodes for the protein called peroxisomal biogenesis factor 7. This protein is part of a group of proteins—the PEX proteins—which import certain enzymes into peroxisomes. Peroxisomes are in little sacs and break down fatty acids and other compounds. They are essential for the synthesis of fats for digestion and in the nervous system.

The normal assembly of peroxisomes depends on the enzymes that peroxisomal biogenesis factor 7 carry. The enzymes include alkylglycerone phosphate synthase, required for synthesis of certain lipids, and phytanoyl-CoA hydroxylase, required for processing a fatty acid called phytanic acid. These enzymes then produce a multistep process to create energy.

More than 36 mutations in *PEX7* are related to rhizomelic chondrodysplasia punctata type 1 (RCDP1). These mutations lead to a completely dysfunctional protein. Thus the important enzymes that are changed prevent the synthesis of body materials for energy. Researchers are still investigating how the lack of the enzymes leads to the severe features of RCDP1. *PEX7* is inherited in an autosomal recessive pattern and is located on the long arm (q) of chromosome 6 at position 23.3.

GNPAT

Mutations in the *GNPAT* gene, known officially as the "glyceronephosphate O-acyltransferase" gene, cause rhizomelic chondrodysplasia punctata type 2 (RCDP2). Normally, *GNPAT* encodes for an enzyme known as glyceronephosphate O-acyltransferase (GNPAT) or dihydroxyacetonephosphate acyltransferase (DHAPAT). The enzyme is found in the sac-like structures known as peroxisomes that are important in the production of fats. The DHAPAT enzyme is responsible for the first step in the production of lipid molecules called plasmologens. The plasmologens make up part of the myelin, the white protective covering of nerve cells. Researchers believe they may protect the cells from the assault of free radicals that collect and attack cells. Plasmologens may also function to help interactions between lipids and proteins.

About five mutations in *GNAT* cause RCDP2. The mutations make a dysfunctional DHAPAT enzyme and disrupt the production of plasmologens in the cells. *GNPAT* is inherited in an autosomal dominant pattern and is located on the long arm (q) of chromosome 1 at position 42.

AGPS

Mutations in the *AGPS* gene, known officially as the "alkylglycerone phosphate synthase" gene, cause RCDP3. Normally, *AGPS* encodes for making the enzyme alkylglycerone phosphate synthase, which is also found in the peroxisomes. This enzyme is critical for the production of the lipid molecules called the plasmologens. Researchers believe these substances are essential for interactions between proteins, as well as other cell functions.

At least three mutations in *AGPS* cause RCDP3. The mutations change only a single building block in the enzyme, altering its structure and activity. The disruption reduces the amount of plasmologens in cells leading to the many characteristic features of RCDP3. *AGPS* is inherited in an autosomal recessive pattern and is located on the long arm (q) of chromosome 2 at position 31.2.

What Is the Treatment for Rhizomelic Chondrodysplasia Punctata?

Because this serious genetic condition has many manifestations, there is no cure. Treating the symptoms may alleviate some discomfort. The cataracts may be removed to restore vision. Physical therapy may improve some of the orthopedic pressures. Monitoring of vision, hearing, and other complications is a must.

Further Reading

Braverman, Nancy E.; Ann B. Moser; and Steven J. Steinberg. 2010. "Rhizomelic Chondrodysplasia Punctata Type 1." *GeneReviews*. http://www.ncbi.nlm.nih.gov/books/NBK1270. Accessed 5/29/12.

"Rhizomelic Chondrodysplasia Punctata Type 1." 2011. Counsyl. https://www.counsyl.com/diseases/rhizomelic-chondrodysplasia-punctata-type-1. Accessed 5/29/12.

"Rhizomelic Chondrodysplasia Punctata Type 1 (RCDP1)." 23andMe. https://www.23andme.com/health/rhizomelic-chondrodysplasia-punctata-type-1. Accessed 5/29/12.

Ring Chromosomes: A Special Topic

What Are Ring Chromosomes?

Ring chromosomes result when a chromosome breaks in two places, and the ends join together to from a round structure that looks like a ring. The ends of a chromosome are called telomeres, and sometimes, for many reasons, the end area is lost.

Normal Chromosomes

One might think chromosomes are simply holders of genetic information. However, scientists are aware that they are very dynamic structures with a highly regulated organization. The way the chromatin is organized plays a role in the regulation of what genes are expressed and how. Chromatin refers to the genetic material on the chromosome.

When DNA is replicated, the linear organization comes into full play. The duplication results in two linear sister chromatids lying side by side so that a symmetrical separation can occur at the metaphase-anaphase part. A change is in this arrangement from linear to circular could totally disrupt the sequence of events.

Two Kinds of Ring Formations

Telomere Dysfunction

The telomeres at the ends of the chromosome are lost, and the two ends fuse together. These shapes are called constitutional aberrations and are found in fetuses or newborns with developmental disorders. Several studies of in vitro and animal models have shown that when telomeres are shortened, the detachment of protective proteins that are present at the ends makes the chromosome prone to recombination. This recombination shuffles the genes, causing a genetic deformity.

Constitutional ring chromosomes occur in 1 per 50,000 human fetuses. Ring syndromes are quite a heterogeneous group. Their characteristics may differ depending on the certain chromosome that is involved and the place of the break-points within the chromosome. Many of the individuals with ring chromosomes are stillborn or do not live long after birth.

Constitutional ring syndromes include the following:

- Ring chromosome 13 syndrome: This syndrome is associated with mental retardation and distinct facial appearance.

- Ring chromosome 14 syndrome: Seizures and intellectual disabilities characterize this syndrome. Many other abnormalities include respiratory disorders, eye problems, and a distinct facial appearance.

- Ring chromosome 15 syndrome: This is related to mental retardation, dwarfism, and a very small head, a condition known as microcephaly.

- Ring chromosome 20 syndrome: This syndrome is associated with epilepsy.

- Ring formation of an X chromosome: This formation is related to a condition known as Turner syndrome.

Other complex arrangements of the rings include deletions and duplications of parts of the chromosomes.

Acquired Genetic Abnormalities

More commonly, ring chromosomes may arise as acquired abnormalities in tumors and leukemias. Radiation or other genetic damage may cause these cells to form. Ring chromosomes have been found in many cancerous growths, especially in skin cancers, gall bladder tumors, and pancreatic cancers.

Further Reading

Chromosome 18 Registry and Research Society. 2009. http://www.chromosome18.org. Accessed 2/10/12.

"Ring Chromosome 14 Syndrome." 2012. Genetics Home Reference. National Library of Medicine (U.S.). http://ghr.nlm.nih.gov/condition/ring-chromosome-14-syndrome. Accessed 2/10/12.

Ring Chromosome 20 Foundation. http://www.ring20.org. Accessed 11/11/11.

Robinow Syndrome

Prevalence   Extremely rare; 100 cases documented in literature

Other Names   acral dysostosis with facial and genital abnormalities; fetal face syndrome; mesomelic dwarfism-small genitalia syndrome; Robinow dwarfism; Robinow-Silverman-Smith syndrome

In 1969, Meinhard Robinow, a human geneticist, noted a condition of the skeletal system that had many other seemingly unrelated conditions. The syndrome was unusual in that it had both a dominant and recessive inheritance pattern. Along with physicians Frederic Silverman and Hugo Smith, he described the condition in the *American Journal of Diseases of Children*, and by 2002, over 100 cases were recognized.

What Is Robinow Syndrome?

Robinow syndrome is a disorder of the skeletal system that affects many other areas of the body. This rare genetic disease shows two inheritance patterns, one dominant and the other recessive. The two types basically differ in their severity.

Following are the characteristics of the autosomal recessive type:

- Distinct facial features—the person has a broad forehead, widely spaced eyes, a short nose, an upturned lip, and wide nasal bridge
- Very short stature
- Shortening of the long bones of the arms and legs, especially the forearms
- Very short fingers and toes, a condition called brachydactyly
- Spinal bones shaped like wedges, a condition leading to abnormal curvature of the spine
- Ribs underdeveloped or fused
- Dental problems with overgrowth of gums and crowded teeth
- Underdeveloped genitalia

- Kidney problems
- Heart defects
- Delayed development

The autosomal recessive form has been seen in families from many countries, including Turkey, where a founder gene has been traced to a town in eastern Turkey. Cases have also been identified in Oman, Pakistan, Czech Republic, Slovakia, and Brazil. Intelligence of the person is usually normal.

What Is the Genetic Cause of Robinow Syndrome?

Mutations in the *ROR2* gene, officially known as the "receptor tyrosine kinase-like orphan receptor 2" gene, cause Robinow syndrome. Normally, *ROR2* provides instructions for a protein that is part of a family called receptor protein kinases (RTKs). This group of genes is important in cell signaling, cell growth and division, cell function, survival, and cell movement.

According to researchers, *ROR2* is believed to be essential for forming the skeleton, heart, and genitalia. Here it plays an important part in embryonic development as it is involved in a chemical signaling pathway called WNT. This pathway controls the activity of genes that are needed at special times during development when cells, tissues, and organs are forming.

About 12 mutations in *ROR2* cause the autosomal recessive form of Robinow syndrome. Some mutations change only one building block or amino acid, leading to a short, nonfunctional protein. Because of the nature of this form's preventing the proper role, it is called a "loss of function" mutation. The loss of function during embryonic development is especially noted in the development of bones in the face, spine, and limbs. The mutation continues after birth when it affects other body parts. *ROR2* is inherited in both recessive and dominant patterns and is located on the long arm (q) of chromosome 9 at position 22.

What Is the Treatment for Robinow Syndrome?

Treating the manifestations of the symptoms is important to help the child have a viable life. Corrective surgery is possible for limb and spinal development, as well as facial abnormalities and orthodontal treatment. To treat genital conditions, the male may need to have scrotal surgery and hormone therapy for the other malformations.

Further Reading

Bacino, Carlos. 2011. "ROR2-Related Robinow Syndrome." *GeneReviews*. http://www.ncbi.nlm.nih.gov/books/NBK1240. Accessed 5/29/12.

"Robinow Syndrome." 2011. Genetics Home Reference. National Library of Medicine (U.S.). http://ghr.nlm.nih.gov/condition/robinow-syndrome. Accessed 5/29/12.

"*ROR2*." 2011. Genetics Home Reference. National Library of Medicine (U.S.). http://ghr.nlm.nih.gov/gene/ROR2. Accessed 5/29/12.

Rogers Syndrome

Prevalence   Rare and unknown; reported in about 30 families worldwide

Other Names   TRMA; thiamine-responsive megaloblastic anemia syndrome; thiamine-responsive myelodysplasia

Thiamine is known as vitamin B1, an essential vitamin for healthy living. However, in Rogers syndrome, which is also known as thiamine-responsive myelodysplasia, thiamine is used as a treatment. The body does not make vitamin B1. It must be obtained from many foods such as whole grains, pasta, fortified breads, cereal, lean meats, fish, and beans.

What Is Rogers Syndrome?

Rogers syndrome is a rare genetic disorder in which the person cannot use thiamine or vitamin B1. The disorder affects many seemingly unrelated body functions. The following characteristics are common in Rogers syndrome:

- Anemia: A special kind of anemia called megaloblastic anemia is present. In this condition, the person has a very low number of red blood cells, and those that present are larger than normal. The meaning of the word "megaloblast" is from the Greek: *megalo*, meaning "large," and *blast*, meaning "germ." A megaloblast is a type of very large red blood corpuscle. The person displays the typical signs of anemia such as loss of appetite and energy, headaches, pale skin, and tingling in the hands and feet.

- Hearing loss: People develop hearing loss during early childhood.

- Diabetes: The person develops diabetes during infancy or early childhood, but it is different from the type 1 diabetes that is found in children. Although it is sometimes treated with insulin, it may also be treated with doses of thiamine.

- Optic degeneration: The optic nerve, which carries images from the eye to the brain, may atrophy and cause blindness.

- Heart defects: Some individuals have both heart and blood vessel problems and may experience heart rhythm abnormalities.

- Internal organ problems: The abdominal organs may be out of position, a condition called transposition. The liver and spleen may be enlarged.

What Is the Genetic Cause of Rogers Syndrome?

Mutations in the *SLC19A2* gene, officially known as the "solute carrier family 19 (thiamine transporter), member 2" gene, cause the many symptoms of Rogers syndrome. Normally, *SLC19A2* provides instructions for the protein thiamine transporter 1. Located on the surface of the cells, thiamine transporter 1 brings thiamine or vitamin B1 into cells. Inside the cells' engines, thiamine helps convert carbohydrates into energy that is needed for the heart, muscles, and nervous system to function.

About 17 mutations in *SLC19A2* causes Rogers syndrome. The mutations may cause a very short and nonworking thiamine transporter 1. This shortened version then disrupts the internal mechanism of the cell and keeps it from reaching the cells surface so it can transport the thiamine into the cells. This disruption causes the seemingly unrelated symptoms of the syndrome. *SLC19A2* is inherited in an autosomal recessive pattern and is found on the long arm (q) of chromosome 1 at position 23.3.

What Is the Treatment for Rogers Syndrome?

Megadoses of thiamine or vitamin B1 treat Rogers syndrome. The treatment is necessary to avoid the serious symptoms of the disorder. The person begins to show symptoms between infancy and adolescence, so early diagnosis is important. The syndrome is called "thiamine responsive" or "thiamine dependent" because administering the vitamin is usually successful.

Further Reading

"Rogers Syndrome." 2011. RightDiagnosis.com. http://www.rightdiagnosis.com/medical/rogers_syndrome.htm. Accessed 5/29/12.

"*SLC19A2*." 2011. Genetics Home Reference. National Library of Medicine (U.S.). http://ghr.nlm.nih.gov/gene/SLC19A2. Accessed 5/29/12.

"Thiamine-Responsive Megaloblastic Anemia Syndrome." 2011. Genetics Home Reference. National Library of Medicine (U.S.). http://ghr.nlm.nih.gov/condition/thiamine-responsive-megaloblastic-anemia-syndrome. Accessed 5/29/12.

Romano-Ward Syndrome (RWS)

Prevalence   About 1 in 7,000 people worldwide; may be more common as some people do not experience ill effects of heart arrhythmias

Other Names autosomal dominant long QT syndrome; long QT syndrome type 1; long QT syndrome without deafness; LQTS1; RWS; Romano-Ward long QT syndrome

What Is Romano-Ward Syndrome (RWS)?

Romano-Ward syndrome is a condition that disrupts the normal heartbeat rhythm. The heartbeat is like a cycle and is measured from the beginning of one heart beat to the beginning of the next. The heartbeat is controlled by an electrical system that is basically very consistent. In Romano-Ward syndrome, this normal heart rhythm is disrupted, causing a condition known as arrhythmia.

The disorder is characterized by a heart condition that causes the heart muscle to take longer to recharge between beats. This problem with recharging is called long QT syndrome. Irregular heartbeat can lead to life-threatening conditions, such as fainting, cardiac arrest, and sudden death.

Three different types are recognized and are related to mutations in different genes. The irregular heartbeats can lead to fainting (syncope) or cardiac arrest and sudden death:

- *LQ1* is caused by mutations in *KCNQ1* or *KCNE1*, which are related to abnormal potassium channel function. This type accounts for about 60% of cases. Primary triggers include exercise and sudden emotion.

- *LQ2* is caused by mutations in *KCNH2* or *KCNE2*, leading to an abnormal potassium channel function. This type is involved in about 35% of cases. Exercise, emotion, and sleep appear to trigger this type.

- *LQT3* is caused by mutations in *SCN5A*, the gene related to the heart sodium channel. This type has fewer than 5% and is triggered by sleep.

The usual age for events to emerge is from the pre-teen years through the 20s. However, about 50% of people with the mutations may never show any symptoms.

What Are the Genetic Causes of Romano-Ward Syndrome?

Mutations in five genes—*ANK2*, *KCNE1*, *KCNE2*, *KCNH2*, *KCNQ1*, and *SCN5A*—cause Romano-Ward syndrome.

ANK2

Mutations in the *ANK2* gene, officially known as the "ankyrin 2, neuronal" gene, are suspected to cause Romano-Ward syndrome. Normally, *ANK2* instructs for making the protein ankyrin-2. Ankyrin-2 is a member of a large family of ankyrins that interact with many body cells. Their work includes making sure that the proper proteins are inserted in the correct places. They are important in all kinds of cellular functions, especially in the heart and brain. The protein maintains a complex system of ion channels that move charged atoms across the cell membranes. They are especially active in the heart to move sodium, potassium, and calcium and to maintain the proper flow of ions in and out of heart muscle cells. This protein is different from the others in that it does not form the channels but is active in moving ions across cell membranes.

About 10 mutations in *ANK2* have been associated with heart problems. Most mutations lead to abnormalities in the heart's pacemaker, called the sinoatrial node. The channels may be produced normally but the changed protein cannot send ions to their correct locations in the muscle cells, thus altering the heart's normal rhythm and causing fainting or other symptoms. *ANK2* is inherited in an autosomal dominant pattern and is located on the long arm (q) of chromosome 4 at position 25-q27.

KNCE1

Mutations in *KNCE1* gene, officially known as the "potassium voltage-gated channel, Isk-related family, member 1" gene, causes some types of Romano-Ward syndrome. Normally, *KCNE1* instructs for making a protein that regulates potassium channels. These channels are important for carrying potassium ions, which are the key to electrical signal production. *KCNE1* controls the channel of proteins produced by the *KCNQ1* gene. These channels are found in the inner ear and heart muscle, where they transport potassium ions. The KNCE1 protein is involved in the recharging of muscle after heartbeat to regulate the contractions of a regular heartbeat.

More than 30 mutations of *KCNE1* cause RWS. Usually people have only one copy of this gene on each cell, and a mutation disrupts the potential of the flow of potassium atoms, causing irregular heartbeat. *KCNE1* is inherited in an autosomal dominant pattern and is located on the long arm (q) of chromosome 21 at position 22.1-q22.2.

KCNE2

Mutations in the *KCNE2* gene, officially known as the "potassium voltage-gated channel, Isk-related family, member 2"gene, are related to RWS. Normally, *KCNE2* instructs for a protein that regulates the activity of the potassium channels, which play a big role in the electrical function of the heart. In addition, the gene protein is involved in helping the heart maintain its regular rhythm.

About 10 mutations in *KCNE2* cause Romano-Ward syndrome. The mutations involve a single protein building block and disrupt the regulating of potassium channels in heart muscle cells. The channels open more slowly and close more rapidly, causing the person to experience fainting and sudden death. *KCNE2* is inherited in an autosomal dominant pattern and is found on the long arm (q) of chromosome 21 at position 22.12.

KCNH2

Mutations in the *KCNH2* gene, officially known as the "potassium voltage-gated channel, subfamily H (eag-related), member 2" gene, cause Romano-Ward syndrome. Normally, *KCNH2* is another member of this large family of genes that encode for proteins charged with moving potassium in and out of the cells. Also these proteins are involved in the interactions with another gene, *KCNE2*, to form the proper working potassium channels.

More than 140 mutations have been found that prevent the proper assembling of channels to regulate the flow of potassium. The disruption leads to the irregular heartbeat with the risk if fainting and death. *KCNH2* is inherited in an autosomal dominant pattern and is located on the long arm (q) of chromosome at position 36.1.

KCNQ1

Mutations in the *KCNQ1* gene, officially known as the "potassium voltage-gated channel, KQT-like subfamily, member 1" gene, cause Romano-Ward syndrome. Normally, *KCNQ1* instructs for the protein that transports potassium ions and plays a key role in generating and transmitting electrical signals. All the proteins from genes in this family form subunits that create the potassium channels.

More than 140 mutations of the KCNQ1 gene cause Romano-Ward syndrome. Most are the result of a single change in an amino acid building block. All the mutations cause the disruption of the flow of the potassium channels, which can result in irregular heartbeat. *KCNQ1* is inherited in an autosomal dominant pattern and is located on the short arm (p) of chromosome 11 at position 15.5.

SCN5A

Mutations in the *SCN5A* gene, officially called the "sodium channel, voltage-gated, type V, alpha subunit" gene, cause Romano-Ward syndrome. Normally, *SCN5A* provides instruction for making sodium channels. Although it operates in a similar way to the other proteins, the SCN5A protein is different from other proteins in this group in that it transports sodium atoms rather than potassium. The channels are abundant in the heart muscle and are important in generating and transmitting electrical signals.

About 30 mutations in *SCN5A* cause a form of Romano-Ward called long QT3. The mutations cause the channels to stay open longer, disrupting the heart's normal rhythm. *SCN5A* is inherited in an autosomal dominant pattern and is located on the short arm (p) of chromosome 3 at position 21.

What Is the Treatment for Romano-Ward Syndrome (RWS)?

Management of the manifestations of the LQ1 and LQ2 types include treatment with drugs called beta-blockers and possibly the use of a pacemaker. LQ3 may need an implantable cardioverter defibrillator. All individuals with this condition should be monitored regularly for side effects of the drugs and working mechanisms.

Further Reading

"Romano-Ward Syndrome." 2011. Genetics Home Reference. National Library of Medicine (U.S.). http://ghr.nlm.nih.gov/condition/romano-ward-syndrome. Accessed 5/29/12.

"Romano-Ward Syndrome." Patiet.co.uk. http://www.patient.co.uk/doctor/Romano -Ward-Syndrome.htm. Accessed 5/29/12.

Vincent, G. Michael. 2009. "Romano-Ward Syndrome." *GeneReviews*. http:// www.ncbi.nlm.nih.gov/books/NBK1129. Accessed 5/29/12.

Rothmund-Thomson Syndrome (RTS)

Prevalence   Rare with incidence unknown; about 300 people reported world-wide

Other Names congenital poikiloderma; poikiloderma atrophicans and cataract; poikiloderma congenitale; poikiloderma congenitale of Rothmund-Thomson; RTS

In 1868, August von Rothmund noted a condition that involved the skin as well as many other body parts. However, it was not until that 1936 that Matthew Sydney Thomson published a further description of the rare disorder. The syndrome was named after the two doctors.

What Is Rothmund-Thomson Syndrome (RTS)?

Rothmund-Thomson syndrome is a condition that affects the skin as well as many other body parts. Beginning about three months of age, the child develops a very red rash on the cheeks that soon spreads to the arms and legs, causing the skin to change colors. Also, in the area of skin coloring, the tissue begins to degenerate with small clusters of enlarged blood vessels under the skin. These blood vessels are called telangiectases, which may persist for life. The collection of skin problems are known as poikiloderma.

Following are other body systems that are affected by Rothmund-Thomson syndrome:

- Hair system: Eyebrows, eyelashes, and hair are sparse.
- Abnormal teeth and nails.
- Skeletal issues: Children are short and grow slowly. Absent or malformed bones with delayed bone formation. Bone density is low, a condition known as osteopenia. Problems occur with bones in the forearms and thumbs, a condition known as radial ray malformations.
- Small stature and slow growth.
- Eye problems such as cataracts.
- Increased risk for developing bone cancer, a condition known as osteosarcoma.
- Many types of skin cancer.
- Gastrointestinal issues with chronic diarrhea.
- Genitals: Underdeveloped or missing testes in males or ovaries in females.

The condition overlaps with a few other rare conditions, including Baller-Gerold syndrome and RAPADILINO syndrome, which are caused by mutations in the same gene.

What Is the Genetic Cause of Rothmund-Thomson Syndrome (RTS)?

Mutations in the *RECQL4* gene, officially known as the "RecQ protein-like 4" gene, causes Rothmund-Thomson syndrome. Normally, *RECQL4* provides instructions for making a member of the RecQ helicase protein family. This important family of enzymes binds to DNA and unwinds the double helix. In preparation for cell division and for repairing damaged DNA, this unwinding is essential. These RecQ helicases are known as the "caretakers of the genome" because they are necessary for keeping the correct. structure of the cell.

Like other members of the RecQ family, *RECQL4* is important in normal DNA replications and repair, although it does not appear to act as a helicase within the cells. *RECQL4* is important both in embryonic development and after birth and is especially important in developing bone and skin cells.

About 25 mutations in *RECQL4* cause Rothmund-Thomson syndrome. The mutations disrupt the production of normal proteins and lead to a nonfunctional version. This version interferes with the DNA replication and repair. Over time, a great deal of damage is done to the genetic information, causing the many features of Rothmund-Thomson syndrome. *RECQL4* is inherited in an autosomal recessive pattern and is located on the long arm (q) of chromosome 8 at position 24.3.

What Is the Treatment for Rothmund-Thomson Syndrome (RTS)?

Although this is a serious genetic disorder, with good care, people with the syndrome can have a normal life expectancy. However, the symptoms of the condition cannot be reversed. Skin lesions can be treated with isotretinoin or laser therapy. Other health care professionals, such as dentists, endocrinologists, ophthalmologists, and cancer specialists may become part of the team to treat the manifestations.

Further Reading

Kugler, Mary. 2007. "Rothmund-Thomson Syndrome." About.com Rare Diseases. http://rarediseases.about.com/od/rarediseasesr/a/rothmundthomson.htm. Accessed 10/7/11.

"Rothmund-Thomson Syndrome." 2011. DermNet NZ. http://dermnetnz.org/systemic/rothmund-thomson.html. Accessed 10/7/11.

Wang, Lisa L., and Sharon E. Plon. 2009. "Rothmund-Thomson Syndrome." *GeneReviews*. http://www.ncbi.nlm.nih.gov/books/NBK1237. Accessed 10/7/11.

RP. *See* Retinitis Pigmentosa (RP)

RTT. *See* Rett Syndrome

Rubinstein-Taybi Syndrome

Prevalence   Rare; about 1 in 100,000 to 125,000 newborns
Other Names   broad thumb-hallux syndrome; RSTS; RTS

In 1957, three doctors, Michail, Matsoukas, and Theodorous, described a case in which a child had very broad great toes and large thumbs. The child had a distinct facial appearance and a mental disability. In 1963, two other doctors, Jack Herbert Rubenstein and Hooshang Taybi, studied a larger number of cases. The rare syndrome was named after these two doctors: Rubenstein-Taybi syndrome

What Is Rubinstein-Taybi Syndrome?

Rubinstein-Taybi syndrome is a disorder that affects many body parts and functions but has one distinctly different feature. The large toes and thumbs are very broad and flat, and this symptom can be noted at birth. Typical other features of this order include:

- Distinct facial appearance involving eyes, nose, and palate
- Short height, small head, and slow bone growth
- Mental disability
- Eye abnormalities
- Heart and kidney disorders
- Dental defects
- Cancerous and noncancerous tumors, including brain tumors
- Leukemia
- Cryptochidism or undescended testicles in males
- Life-threatening infections

A 2009 study added that these children are probably obese and have a short attention span and poor coordination.

What Are the Genetic Causes of Rubinstein-Taybi Syndrome?

Mutations in two genes and a chromosome are related to Rubinstein-Taybi syndrome. The genes *CREBBP* and *EP300*, as well as chromosome 16, are involved in the disorder.

CREBBP

Mutations in the *CREBBP* gene, known officially as the "CREB binding protein" gene, are related to the disorder. Normally, *CREBBP* provides instruction for making the CREB binding protein. This protein is essential for regulating many genes. It is known to control cell growth and division. Animal studies suggest that this protein may help in the formation of long-term memories. The protein is active before and after birth.

This very important protein may act as a transcriptional coactivator, which means that it connects transcription factors in the cells to help start an essential process.

About 90 mutations in *CREBBP* cause Rubinstein-Taybi syndrome. The mutations may involve deletions, insertions, or exchanges of one amino acid building block. The disruption in the process of transcription affects the development of the individual, leading to the symptoms of Rubinstein-Taybi syndrome. *CREBBP* is inherited in an autosomal dominant pattern and is located on the short arm (p) of chromosome 16 at 13.3.

EP300

Mutations in the *EP300* gene, officially known as the "E1A binding protein p300" gene, cause a small number of cases of Rubinstein-Taybi syndrome. Normally, *EP300* instructs for a protein called p300. This is an active protein that regulates many genes in the body. It is also important in controlling cell growth and division. Like *CREBBP*, it appears to activate transcription and is also a transcriptional coactivator.

Many mutations in *EP300* appear to cause only a small number of cases of Rubinstein-Taybi syndrome. Genetic changes lead to abnormal proteins that do not function and reduce the amount of p300 by about half. *EB300* is inherited in an autosomal dominant pattern and is located on the long arm (q) of chromosome 22 at position 13.2.

Chromosome 16

Chromosome 16 has between 850 and 1,200 genes that perform a variety of roles in the body. Some cases of Rubinstein-Taybi syndrome are traced to 16p.23.3 deletion. The presence of the deletion appears to cause a very serious type of the syndrome. Infants born with this form seldom survive past infancy.

What Is the Treatment for Rubinstein-Taybi Syndrome?

Because this syndrome is a serious genetic disorder, there is no cure. Treating the manifestations involves a team of doctors.

Further Reading

"*CREBBP*." 2011. Genetics Home Reference. National Library of Medicine (U.S.). http://ghr.nlm.nih.gov/gene/CREBBP. Accessed 5/29/12.

"*EP300*." 2011. Genetics Home Reference. National Library of Medicine (U.S.). http://ghr.nlm.nih.gov/gene/EP300. Accessed 5/29/12.

"Rubinstein-Taybi Syndrome." 2011. Genetics Home Reference. National Library of Medicine (U.S.). http://ghr.nlm.nih.gov/condition/rubinstein-taybi-syndrome. Accessed 5/29/12.

Russell-Silver Syndrome (RSS)

Prevalence   About 1 in 3,000 to 1 in 100,000 people; more than 400 cases reported worldwide

Other Names   RSS; Silver-Russell dwarfism; Silver-Russell syndrome; SRS.

In 1953 and 1954, Drs. H. K. Silver and A. Russell wrote about a type of dwarfism readily recognizable at birth. The children were born with the short stature but also had a distinct facial appearance and other physical problems. The condition was named after the two doctors—putting first the one whose name came first in the alphabet: Russell-Silver syndrome.

What Is Russell-Silver Syndrome (RSS)?

Russell-Silver syndrome is one of many of the types of dwarfism; however, this one is noted both before and after birth. The baby is born with an unusually low birth weight, which means that growth in the womb has been retarded. This low rate is called Intrauterine Growth Retardation. However, the head growth may be normal and appear unusually large compared to the rest of the body.

Following are the symptoms of the child with Russell-Silver syndrome:

- Distinct facial appearance: The face is triangular shaped with a pointed chin and small jaw; mouth curves downward.

- Soft spot slow to close: The fontanelle or soft spot is the opening between the bones of the skull, which enables the child to pass through the birth canal. After birth, the area closes to form the protective skull. In children with RSS, the fontanelle may be very wide and slow to close.

- Body growth: The body continues to grow slowly and does not catch up during the time of growth. Also, the body does not grow symmetrically. One side may grow faster than the other.

- Eyes: The eyes of younger children have a blue tinge to the normal white part of the eye, called the sclera.
- Skin: The color of the skin appears gray.
- Fingers: The little finger on each hand may curve inward, a condition known as clinodactyly.
- Abnormal sweating, especially at night.
- Low blood sugar: The children may experience low blood sugar, a condition known as hypoglycemia.
- Feeding problems: From birth, problems are obvious. The child is not interested in eating or may take small amounts.
- Learning disabilities: Children may be hyperactive with attention deficits and have learning disabilities.
- Puberty: Puberty may begin at an early age.

Adults with the syndrome are very short. Males may reach only about 4 feet, 11 inches; females may attain a height of about 4 feet, 7 inches.

What Are the Genetic Causes of Russell-Silver Syndrome?

Two genes, as well as two chromosomes, are involved in Russell-Silver syndrome. The causes are complex and obviously involve genes that control growth. The two genes involved are *H19* and *IGF2*, as well as areas of chromosomes 7 and 11. Most cases of RSS are sporadic, meaning they occur in people with no family history of the disorder.

H19

Mutations in the *H19* gene, officially known as the "H19, imprinted maternally expressed transcript (non-protein coding)" gene, cause RSS. Normally, *H19* provides instructions for a noncoding RNA. Although most genes encode for a protein, *H19* does not do this but functions perhaps as a tumor suppressor that keeps cells from growing and dividing too rapidly. The gene is especially active before birth to keep certain cells from growing in an uncontrolled way.

The inheritance pattern of this gene is also different. People inherit one copy of a gene from each parent, and both are active in cells. However, in *H19* only the gene inherited from the mother is active. This kind of parent-specific gene activation is called genomic imprinting. In genomic imprinting, small molecules called methyl groups are added to parts of DNA, which is a process called methylation. In genes that undergo genomic imprinting, methylation is one way in which the gene's parent of origin is marked during the formation of egg and sperm cells.

Mutations in the ICR1 regions of chromosome 11 cause Russell-Silver syndrome. Often too few methyl groups are added, leading to a loss in activity in another gene called *IGF2* and increased activity of *H19*. The improper balance of loss of *IGF2* that promotes normal growth and an increase in *H19* activity that restrains growth lead to the short stature and poor growth of people with RSS. *H19* may be inherited both a recessive and dominant pattern and is located on the short arm (p) of chromosome 11 at position 15.5.

IGF2

Mutations in the *IGF2* gene, known officially as the "insulin-like growth factor 2 (somatomedin A)" gene, cause RSS. Normally, this gene encodes for a protein called insulin-like growth factor 2. The IGF2 protein is very active before birth, controlling growth and division of cells in many tissues; it is much less active in the adult body. In this gene, genomic imprinting does occur, but this copy is inherited from the father. *IGF2* is one of several genes on chromosome 11 that experience genomic imprinting.

The mutations in this gene cause changes also in the adding of the methyl groups or methylation. Too few methyl groups are attached and lead to loss of the activity of the *IGF2* gene that promotes growth and increase in the *H19* gene activity that restrains growth. The two genes working against each other result in poor growth and the symptoms of RSS. *IGF2* is located on the short arm (p) of chromosome 11 at position15.5.

Chromosome 7

Chromosome 7 has about 1,150 genes, which represent more than 5% of the total DNA in cells. This chromosome has a region that may undergo genomic imprinting. Instead of getting one copy of a gene from each parent, the chromosome may undergo a happening called maternal uniparental disomy or UPD. In this happening, two active copies of the mother's gene are expressed rather than one active copy form mother and an inactive copy from the father. This accounts for about 7% to 10% of cases of RSS.

Chromosome 11

Chromosome 11 also has about 1,500 genes, which have many functions in the body. About 150 of the genes are responsible for the sense of smell. An area in the short arm of this chromosome has been associated with mutations in genomic imprinting that affect the regulation of genes *H19* and *IGF2* and cause RSS.

What Is the Treatment for Russell-Silver Syndrome?

Because this condition is genetic, at present there is no cure. However, many of the manifestation or symptoms can be treated. The treatments include growth hormone therapy, physical therapy, occupational therapy, and speech and language therapy. The individual qualifies for special education and an individualized education plan.

Further Reading

Child Growth Foundation. 2010. "Russell Silver Syndrome (RSS)." http://www.childgrowthfoundation.org/Default.aspx?page=ConditionsRSS. Accessed 5/29/12.

"Russell-Silver Syndrome." 2011. Genetics Home Reference. National Library of Medicine (U.S.). http://ghr.nlm.nih.gov/condition/russell-silver-syndrome. Accessed 5/29/12.

Saal, Howard M. 2011. "Russell-Silver Syndrome." *GeneReviews*. http://www.ncbi.nlm.nih.gov/books/NBK1324. Accessed 5/29/12.

S

Saethre-Chotzen Syndrome

Prevalence   About 1 in 25,000 to 50,000 people

Other Names   acrocephalosyndactyly III; acrocephalosyndactyly, type III; acrocephaly, skull asymmetry, and mild syndactyly; ACS3; ACS III; Chotzen syndrome; dysostosis craniofacialis with hypertelorism; SCS

A condition, in which the child is born with a distinct facial appearance and the bones of the skull have joined prematurely, was described several decades ago. H. Saethre, a German physician, wrote about the condition in 1931. Two years later, F. Chotzen, another German doctor, wrote about a similar condition. The condition that they described has been given many times, but the one most commonly used is named after the two doctors: Saethre-Chotzen syndrome.

What Is Saethre-Chotzen Syndrome?

Saethre-Chotzen syndrome is a condition in which the child is born with distinct facial appearances because certain bones of the skull have fused together prematurely. The coronal suture is a line in the skull that goes over the head from ear to ear. Most children with Saethre-Chotzen syndrome have the fusion of this suture line.

The following are signs of Saethre-Chotzen syndrome:

- Misshaped skull: In addition to the fusion along the coronal suture, other areas of the skull may also be malformed. The head may have a high forehead with low frontal hairline.

- Distinct facial appearance: Widely spaced eyes with drooping eyelids and a broad nasal bridge may be present. One side of the face may appear very different from the other side. Ears are small with an irregular shape.

- Hands and feet: The skin between the second and third fingers may be fused; the large toe may be broad or duplicated.
- Skeletal abnormalities: The person may be very short.
- Other systems: Less common are disorders of the bones and heart. The child may lose hearing.
- Delayed development: The person may have some learning difficulties, although most have normal intelligence.

What Are the Genetic Causes of Saethre-Chotzen Syndrome?

Mutations in the *TWIST1* gene, officially known as the "twist homolog 1 (Drosophila)" gene, cause Saethre-Chotzen syndrome. In addition, other areas of chromosome 7 are suspect. Normally, *TWIST1* instructs for a protein that is essential in embryonic development. The protein attaches to specific areas of DNA to control the activity of certain genes. This kind of action is called a transcription factor. The TWIST1 protein is a member of a large protein family called the helix-loop-helix (bHLH). This loop determines the three-dimensional shape of the protein, enabling it to bind to certain parts of the DNA and control growth of bones and muscles, especially those muscles of the hand and face. The TWIST1 protein is also important in the development of the limbs.

Other 80 mutations of *TWIST1* cause Saethre-Chotzen syndrome. Some of the mutations involve a change in just one amino acid while others delete or insert material in the gene. The defective protein then affects the development of the cells in the skull, face, and limbs, resulting in the symptoms of Saethre-Chotzen syndrome. *TWIST1* is inherited in an autosomal dominant pattern and is located on the short arm (p) of chromosome at position 21.2.

Chromosome 7

This chromosome has about 1,150 genes that are important for body functions. In addition to the *TWIST1* gene, other changes in chromosome 7 may cause Saethre-Chotzen syndrome. There may be translocations of genetic material between this and another chromosome, or there may be inversions of material within the chromosome. Ring chromosomes, the abnormal circular structure in which the ends of the chromosome are lost and parts join together, may form. All the changes however, involve the area around the *TWIST1* gene in some way and may affect nearby genes. Children with the chromosomal aberrations are more likely to have intellectual disabilities.

What Is the Treatment for Saethre-Chotzen Syndrome?

Treatment of Saethre-Chotzen syndrome includes surgery to correct the fusion of cranial structures. Reconstructive surgery may correct the webbing of the fingers and malformations of the eyelids and nose. The other symptoms such as dental problems or cleft palate may also require surgery.

Further Reading

"Saethre-Chotzen Syndrome." 2011. Genetics Home Reference. National Library of Medicine (U.S.). http://ghr.nlm.nih.gov/condition/saethre-chotzen-syndrome. Accessed 5/29/12.

"Saethre-Chotzen Syndrome." FACES—The National Craniofacial Association. http://www.faces-cranio.org/Disord/Saethre.htm. Accessed 5/29/12.

"*TWIST1*." 2011. Genetics Home Reference. National Library of Medicine (U.S.). http://ghr.nlm.nih.gov/gene/TWIST1. Accessed 5/29/12.

San Luis Valley Syndrome. *See* Recombinant 8 Syndrome

Schindler Disease

Prevalence	Very rare; only a few people with each type have been identified
Other Names	alpha-galactosidase B deficiency; alpha-galNAc deficiency, Schindler type; alpha-N-acetylgalactosaminidase deficiency; alpha-NAGA deficiency; angiokeratoma corporis diffusum–glycopeptiduria; GALB deficiency; Kanzaki disease; lysosomal glycoaminoacid storage disease–angiokeratoma corporis diffusum; NAGA deficiency; neuroaxonal dystrophy, Schindler type; neuronal axonal dystrophy, Schindler type

Since Archibald Garrod first wrote the "inborn errors of metabolism" in the early 1900s, researchers have identified many kinds of metabolic disorders. In 1988, Detlev Schindler wrote a paper detailing an inherited metabolic disease in which abnormal compounds relating to the lack of sugars and fat breakdown accumulate in the blood. This disorder is now called Schindler disease.

What Is Schindler Disease?

Schindler disease is a metabolic disease caused by problems in lysosome storage. The lysosomes act as the recycling and garbage disposal centers of the cell. Mutations in a gene produce a deficient enzyme called alpha-N-acetylgalactosaminidase (alpha-NAGA) that leads to a damaging buildup of compounds called glycosphingolipids. This accumulation over time causes serious neurological damage.

The following three types of Schindler disease are known:

Type I Infantile Schindler Disease

Infants with this type appear normal at birth and during the first months of life. But by the age of 8–15 months, developmental regression occurs. The following symptoms may occur:

- Loss of mental and physical skills that they have acquired
- Seizures
- Blindness
- Complete loss of awareness of surroundings
- Unresponsiveness
- Low muscle tone or hypotonia that later becomes muscle rigidity

These children seldom live past early childhood.

Type II Schindler Disease

This type is also called Kanzaki disease and is a milder form that appears in adulthood. Following are the symptoms of this type:

- Mild intellectual impairment
- Hearing loss caused by abnormalities in the inner ear
- Complications in the central and peripheral nervous systems
- Weakness
- Development of enlarged blood vessels that form small, dark spots on the skin called angiokeratomas

Type III Schindler Disease

Symptoms of this type range in severity between types I and II. This type usually begins in infancy. People with this type may display the following signs:

- Developmental delay
- Seizures
- Weakened and enlarged heart
- Enlarged liver
- Behavioral problems beginning in early childhood
- Autism spectrum disorders such as poor communication and socialization skills

What Is the Genetic Cause of Schindler Disease?

Mutations in the *NAGA* gene, officially known as the "N-acetylgalactosaminidase, alpha" gene, cause Schindler disease. Normally, *NAGA* provides instructions for making the enzyme alpha-N-acetylgalactosaminidase. This enzyme is active in the lysosomes, the recycler and garbage disposal of the cells, to break down and digest materials. The enzyme targets compounds called glycoproteins and glycolipids. The word "glyco" refers to sugar; lipid refers to fat, meaning the compounds are made of molecules of sugars and fats. The enzyme breaks down the glycoproteins and glycolipids, removing a molecule called alpha-N-acetylgalactosamine.

About seven mutations in *NAGA* cause Schindler disease. Most of the mutations change the shape of the alpha-N-acetylgalactosaminidase enzyme. The enzyme then cannot break down the sugars and fats. The substances gather in the lysosomes, causing the cell to eventually die. Damage to the tissues and organs lead to the neurological defects and other symptoms of Schindler disease. *NAGA* is inherited in an autosomal recessive pattern and is located on the long arm (q) of chromosome 22 at 13-qter.

What Is the Treatment for Schindler Disease?

Because no cure exists for this disease, the focus must be on the symptoms. A team of specialists will provide for the individual's care. Physical and occupational therapy may help with some of the symptoms, especially the muscle disorders of type I.

Further Reading

"Schindler Disease." 2011. Genetics Home Reference. National Library of Medicine (U.S.). http://ghr.nlm.nih.gov/condition/schindler-disease. Accessed 5/29/12.

"Schindler Disease." 2011. National Organization of Rare Diseases (NORD). http://www.rarediseases.org/rare-disease-information/rare-diseases/viewSearchResults?term=Schindler%20Disease. Accessed 5/29/12.

Schwartz-Jampel Syndrome (SJS)

Prevalence   Rare

Other Names   chondrodystrophic myotonia; myotonic myopathy, dwarfism, chondrodystrophy, ocular and facial anomalies; Schwartz-Jampel-Aberfeld syndrome; SJA syndrome; SJS

In 1962 Oscar Schwartz and Robert Jampel described two siblings, a six-year-old boy and a three-and-a-half year-old girl, who had distinct facial appearances and unusual eyelid deformities. They were small with joint disorders. The team described the first case in the *Archives of Ophthalmology* and was honored with the name Schwartz-Jampel syndrome.

What Is Schwartz-Jampel Syndrome (SJS)?

Schwartz-Jampel syndrome is characterized by very short stature and joint limitations, accompanied by muscular spasms. Distinct facial features are also part of the syndrome. Two types are recognized; these types also have subtypes.

Type IA

Type IA is the classic type that Schwartz and Jampel originally described. This type becomes apparent later in childhood and is less severe. Following are the symptoms of SJS type IA:

- Muscle stiffness like those seen in myotonic disorders, but with a unique pattern
- Stiff pattern continues through sleep or treatment with benzodiazepine
- Unusual "puckered face" appearance
- Spasms of the eyelids
- Skeletal deformities
- Risk for malignant hyperthermia
- Mental retardation—about 20% of the individuals have some intellectual disability, although some are normal or with even superior intelligence
- Lifespan not significantly shortened

Type IB

Types IA and I B are caused by mutations in the same gene. Following are the characteristics of type IB:

- Apparent immediately at birth
- Facial appearances similar to type IA
- Symptoms more severe
- Muscle stiffness similar to type IA
- Blepharospasms (eyelid movement) similar to type IA
- Effect on life span not determined

Type II

This type is definitely connected with shortened life span as most people do not survive to adulthood. A different gene causes this type. Following are the symptoms of SJS type II:

- Apparent at birth
- Patients look similar to those of Type IB
- Joint contractures

- Bone dysplasia
- Small stature
- Respiratory illnesses
- Feeding problems
- Hypotonia rather than stiffness
- Bouts with hyperthermia

Some researchers think this condition is the same as Stuve-Wiedemann syndrome and should not be considered SJS. However, because the condition is so rare and so few live to adulthood, most of the clinical symptoms pertain to SJS types IA and IB.

What Are the Genetic Causes of Schwartz-Jampel Syndrome (SJS)?

The genetic causes of the Type of SJS are related to two genes: types IA and IB to *HSPG2*, and type II to *LIFR*.

HSPG2

Mutations in the *HSPG2* gene, known officially as the "heparan sulfate proteoglycan 2" gene, cause SJS types IA and IB. Normally, *HSPG2* provides instructions for the perlecan protein. This protein is made of a large core protein with three long chains of glycosaminoglycans (heparan sulfate or chondroitin sulfate) attached. This complex protein binds and cross-links the cell-surface molecules with many components that are outside the cell. For example, perlecan interacts with molecules such as laminin, prolargin, and collagen type IV. It is active in the matrix that surrounds the blood vessels, helping to maintain balance in these cells. Perlecan also plays an important role in regulating growth factors, such as FGFBP1, FBLN2, and FGF7, and in cell maintenance..

Mutations in *HSPG2* cause SJS type I A and IB. Minor changes in the gene disrupt the function of the perlecan protein. *HSPG2* is inherited in an autosomal recessive pattern and is located on the short arm (p) of chromosome 1 at position 36-p34.

LIFR

Mutations in the *LIFR* gene, officially known as the "leukemia inhibitory factor receptor alpha" gene, causes SJS type II. Normally, *LIFR* instructs for a protein that belongs to the type I cytokine receptor family. This protein combines with another factor to form a complex that is responsible for the leukemia inhibitory factor. This factor is involved in cell proliferation in both embryo and adult.

Mutations in *LIFR* cause SJS type II. *LIFR* is inherited in an autosomal recessive pattern and is located on the short arm (p) of chromosome 5 at position 13-p12.

What Is the Treatment for Schwartz-Jampel Syndrome (SJS)?

Treatment of this syndrome involves reducing the abnormal muscle stiffness and cramping. The treatment may include Botox surgery, massage, warming, and gradual stretching. Medications such as anticonvulsants and quinine may help relieve some of the muscle spasms.

Further Reading

"Schwartz Jampel Syndrome." 2003. National Organization for Rare Diseases (NORD). http:// www.rarediseases.org/rare-disease-information/rare-diseases/byID/1058/viewAbstract. Accessed 10/15/11.

"Schwartz-Jampel Syndrome." 2007. Orphanet. http://www.orpha.net/consor/cgi-bin/OC _Exp.php?Expert=800&lng=EN. Accessed 5/29/12.

"Schwartz-Jampel Syndrome Clinical Presentation." 2011. Medscape. http:// emedicine.medscape.com/article/1172013-clinical#a0218. Accessed 5/29/12.

Shprintzen-Goldberg Syndrome (SGS)

Prevalence   Rare; 42 cases since identification

Other Names   Marfanoid-craniosynostosis syndrome; Shprintzen-Goldberg Craniosynostosis syndrome; Shprintzen-Goldberg Marfanoid syndrome

In 1981, G. Sugarman and M. V. Vogel reported on a syndrome in which the skull closed prematurely but also showed many other physical disorders. However, the rare condition was named after Drs. Robert J. Shprintzen and R. B. Goldberg, who later studied and wrote about the syndrome.

What Is Shprintzen-Goldberg Syndrome (SGS)?

Shprintzen-Goldberg syndrome (SGS) is a condition in which the sutures of the skull close prematurely, causing many problems with brain development. Because of the abnormal closing, the skull may have the appearance of a cloverleaf. Other physical problems are involved.

Following are the major characteristics of SGS:

• Closing of the sutures in the skull, a condition called craniosynostosis: In order for the baby to pass through the birth canal, the skull must be flexible. Before birth, the skull is not solid bone but becomes so after the areas grow together after birth. The areas where the closures take place are called sutures. Several sutures exist affect this condition. The coronal suture goes across the head from ear to ear; the sagittal suture is between the two bones on the side of the skull and roof of the skull; the lambdoid sutures are between the parietal bones and the upper borders of the occipital bones on the back of the skull.

- Distinct facial appearance: The child has a high forehead, small jaw, low-set ears that are rotated to the back, and wide-set eyes that slant downward.
- Skeletal problems: Scoliosis, joint hypermobility, and enlarged long limbs are apparent.
- Long, slender fingers and toes.
- Heart: The heart valves are affected, especially the mitral valve. This valve may prolapse and allow blood to flow into the wrong chamber.
- Delayed development: The individual may have impaired motor and cognitive skills, including mild-to-moderate intellectual disability.
- Brain anomalies: Certain malformations such as hydrocephalus, which is a collection of fluid on the brain may exist.
- Abdominal hernias.
- Respiratory distress.

What Is the Genetic Cause of Shprintzen-Goldberg Syndrome (SGS)?

Mutations in the *FBN1* gene, officially known as the "fibrillin 1" gene, have been reported in some individuals with Shprintzen-Goldberg syndrome. Normally, *FBN1* instructs for a very large protein called fibrillin-1 that is found in the extracellular matrix, a complex that holds cells in place. Fibrillin-1 joins with other proteins to form tiny threadlike structures called microfibrils. These tiny fibrils make up part of the elastic fibers that allow skin, ligaments, and blood vessels to expand. They also make up stronger tissues that support the lens of the eyes, nerves, and muscles. In addition, the microfibrils produce certain factors called transforming growth factor-beta or TGF-beta. When these growth factors are activated, they affect repair and growth of tissues throughout the body.

Mutations in *FBN1* cause several disorders known as fibrillopathies, which mean that fibrillin proteins are deficient. Shprintzen-Goldberg syndrome is an early-onset form of fibrillin deficiency. It is not exactly clear how the deficient proteins cause the many symptoms of the disorder. The most common type of fibrillopathy is Marfan syndrome. *FBN1* is located on the long arm (q) of chromosome 15 at position 21.1.

What Is the Treatment of Shprintzen-Goldberg Syndrome (SGS)?

Treatment of the manifestation occurs as the varied symptoms appear. First attention is to relieve pressure on the brain and to correct as much as possible the premature growth of the skull. Surgery may also be needed to correct hernia, heart, or other conditions. Physiotherapy may increase joint mobility. The child qualifies for education in a special education center.

Further Reading

"*FBN1*." 2011. Genetics Home Reference. National Library of Medicine (U.S.). http://ghr.nlm.nih.gov/gene/FBN1. Accessed 5/29/12.

Greally, Marie T. 2010. "Shprintzen-Goldberg Syndrome." *GeneReviews*. http://www.ncbi.nlm.nih.gov/books/NBK1277. Accessed 5/29/12.

Sialidosis

Prevalence   Unknown; type I more common in people with Italian ancestry

Other Names cherry red spot myoclonus syndrome; mucolipidosis I; mucolipidosis type I; myoclonus cherry red spot syndrome

This condition relates to the storage system of the lysosomes, the recycling and garbage center of the cells. Scientists have known of this disorder for several years, and about 20 years ago, J. S. Lowden and J. A. O'Brien suggested there were two types, based on symptoms and severity. The named comes from a deficiency of sialidase, which leads to buildup of sugar in the lysosomes.

What Is Sialidosis?

Sialidosis is a lysosomal storage disease. In sialidosis, a deficiency in the enzyme called sialidase or alpha-neuraminidase causes a toxic buildup of glycoproteins or sugars in the cells. The toxins cause a variety of disorders that affect many organs, especially the nervous system.

Two types generally types exist, with type II having three forms. Following are the different types and forms of sialidosis.

Sialidosis Type 1

This is the less severe type of sialidosis, characterized by the following:

- Begins between the ages of 8 and 25.
- Progressive vision problems: Cherry-red spots appear on the macular of the retina.
- Walking problems: One of the symptoms occurs as a gait problem, which is obvious when the person walks.
- Muscle twitches, a condition known as myoclonus.

- Leg tremors.
- Seizures.
- Lack of coordination, a condition known as ataxia.
- May require wheelchair.
- Normal intelligence and life expectancy.

Sialidosis Type II

This is the more severe type of the disorder. This type has three forms or subtypes.

Congenital sialidosis type II

- Symptoms before and after birth: This form begins before birth with an abnormal buildup of fluid in the stomach area of the child. Fluid accumulation causes widespread swelling before birth. This abnormal accumulation is called hydrops fetalis. The child may be stillborn or live for only a short time after birth.
- Enlarged liver and spleen: This condition is called hepatosplenomegaly.
- Distinct facial features: The term "coarse" is used to describe the features.
- Abnormal bone development: This condition is known as dysostosis multiplex. In this condition, the cartilage that is present at birth does not completely turn to bone.

Type II Infantile Form

This type begins some time during the first year of life and has some of the same symptoms as the congenital type. Following are the symptoms:

- Coarse facial features
- Enlarged liver and spleen
- Short stature
- Defective bone formation
- Cherry-red spot on the macula of the eye
- Myoclonus or muscle contractions
- Kidney problems
- Mental retardation
- Hearing loss
- Overgrowth of gums
- Widely spaced teeth
- May survive into childhood or adolescence

Sialidosis Type II Juvenile Form

This type is the least severe of all and may begin any time between the ages of 2 and 20. Following are the symptoms:

- Mildly coarse facial features
- Muscle contractions or myoclonus
- Mild bone abnormalities
- Cherry-red spots on the macula
- Angiokeratomas or dilated blood vessels under wart-like growths

What Is the Genetic Cause of Sialidosis?

Mutations in the *NEU1* gene, officially known as the "sialidase 1 (lysosomal sialidase)" gene, cause sialidosis. Normally, *NEU1* encodes for the enzyme neuraminidase 1 or sialidase, which is located in the lysosomes. The NEU1 enzyme aids in breaking down large sugar molecules called oligosaccharides that are attached to a larger group of proteins called glycoproteins. The enzyme works to remove a substance known as sialic acid.

About 42 mutations in the *NEU1* gene cause sialidosis. Most mutations involve only one change in a protein building block. When the enzyme is defective, the process of breaking down the large molecules within the lysosomes is stopped. The molecules then build up and damage the cell, and eventually tissues and organs. *NEU1* is inherited in an autosomal recessive pattern and is located on the short arm (p) of chromosome 6 at position 21.3

What Is the Treatment for Sialidosis?

At present no cure for sialidosis exists. Treatment must focus on the symptoms. Medications for myoclonus or twitching, as well as managing seizures, are essential. An ophthalmologist, neurologist, and geneticist should be part of the team for care.

Further Reading

"Sialidosis." About.com Rare Diseases. http://rarediseases.about.com/od/lysosomalstoragediseases/a/sialidosis.htm. Accessed 5/30/12.

"Sialidosis." 2009. International Advocate for Glycoprotein Storage Disorders (ISMRD). http://www.ismrd.org/the_diseases/sialidosis/history_and_overview. Accessed 5/30/12.

"Sialidosis (Mucolipidosis I)." 2011. Medscape. http://emedicine.medscape.com/article/948704-overview. Accessed 5/30/12.

Sickle Cell Anemia. *See* Sickle-Cell Disease

Sickle-Cell Disease

Prevalence   Most common inherited blood disorder in the United States; affects about 70,000 to 80,000 Americans; occurs in 1 in 500 African Americans and 1 in 1,000 to 1,400 Hispanic Americans; found in Greece, Turkey, Italy, India, Arabian Peninsula, South America, Central America, and Caribbean.

Other Names   HbS disease; hemoglobin S disease; SCD; sickle cell disorders; sickling disorder due to hemoglobin S

In 1910, James Herrick, a Chicago physician, was treating a patient from the West Indies. The patient was quite ill and showed all the signs and symptoms of anemia. When he studied a blood sample, he found unusual shaped red blood cells. He described the cells as appearing like a sickle. The name stuck and the disorder became known as sickle-cell disease.

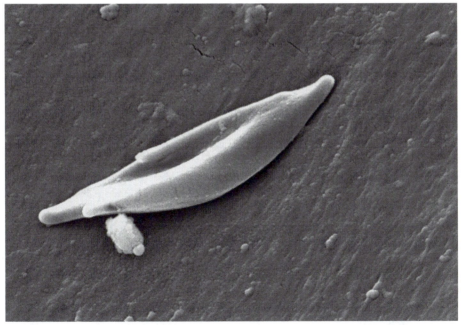

A magnified photograph of a sickle-affected red blood cell, taken with a scanning electron micrograph. Healthy red blood cells are round, and they move through small blood vessels to carry oxygen to all parts of the body. In sickle cell disease, the red blood cells become hard and sticky and look like a C-shaped farm tool called a "sickle." (CDC/ Sickle Cell Foundation of Georgia: Jackie George, Beverly Sinclair)

What Is Sickle-Cell Disease?

Sickle-cell disease, the most common of the inherited blood disorders, refers to a group of blood disorders that affect the red blood cells of the body. Normally, red blood cells under the microscope look like a red Life Saver candy. They are shaped like a disc or like a doughnut without holes. The shape helps them move smoothly through the tubelike structures of the blood vessels. The red coloring of the cell comes from an iron carrying protein called hemoglobin.

Red blood cells or erythrocytes are made in the bone marrow of large bones. They are constantly being created because the new red cells, which will live only about 120 days, must replace the old ones that will die. The red blood cells work constantly picking up oxygen in the lungs, carrying it to the body cells, and then bringing the waste product carbon dioxide back to the lungs to be exhaled.

Sickle cells are abnormal. They have a distorted shape. Instead of having the efficient disc shape, they look like a sickle or crescent. The hemoglobin is known as sickle hemoglobin or hemoglobin S and cannot carry the regular load of oxygen. In addition to the unusual shape, sickle cells are sticky and stiff. They block the flow of blood into tissues and organs and cause pain, infections, and serious damage to organs.

Sickle-cell anemia is one kind of sickle-cell disease. As a general condition, anemia is caused by a low number of red blood cells. In sickle-cell disease the cells live only about 10 to 20 days, instead of the normal 120. The bone marrow cannot

Did the Sickle Cell Evolve as a Defense from Malaria?

The dreaded disease malaria has been a deadly scourge to humankind throughout the centuries. Even today, about 400 million people have the disease, and 2–3 million die from the illness. Malaria at once was thought to be caused by "bad air"; in fact, that is what the name means. Now, scientists know that malaria is transmitted through the bite of the anopheles mosquito. The mosquito carries a one-celled organism called *Plasmodium*, which goes through stages of development in the person receiving the bite. The presence in the host reproduces and causes the ill effects of malaria.

The adaptation of the mutation against the malaria parasite is indeed intriguing. If a mutation could arise that would keep the parasite from growing, then a great mechanism could exist. The first hint came from geographers who realized that malaria did not exist in the high, dry areas of Africa; and neither did sickle-cell disease. For those living in the hot, moist tropical climates, having the sickle-cell trait provided an advantage over people with normal hemoglobin. Both traits are passed to the next generation. Unfortunately, if the offspring has both genes with the hemoglobin S mutation, he or she could have the serious complications of sickle-cell anemia. A mutation that evolved to protect has turned out to be deadly.

work fast enough to keep up with the demand of red blood cells to carry hemoglobin. In addition, when the cells are shaped like a sickle, they break down prematurely, causing anemia.

The signs of sickle cell disease begin early in childhood. Following are symptoms of the disorder:

- Fatigue
- Many infections
- Shortness of breath
- Delayed growth
- Yellowing of eyes and skin, a sign of jaundice
- Pain
- Organ damage of the lungs, spleen, kidney, and brain
- High blood pressure in the blood vessels that lead to the lungs
- Heart failure

These symptoms can vary with the individual. For some, the signs are mild; others may have to be hospitalized.

What Is the Genetic Cause of Sickle-Cell Disease?

Mutations in the *HBB* gene, known officially as the "hemoglobin, beta" gene, cause sickle cell disease. Normally, *HBB* instructs for making the protein beta-globin. Hemoglobin is made up of four protein subunits, two subunits of beta-globin and two of alpha-globin. Each of these proteins carries the iron-rich molecule called heme. It is heme that gives red blood cells the potential to pick up oxygen from the lungs, deliver it to the body cells, and then pick up the waste product carbon dioxide to carry back to the lungs. If the hemoglobin molecule is complete, each of the four heme molecules carries one molecule of oxygen, making the load total four molecules of oxygen. It is the oxygen that gives blood the bright red color.

Mutations in *HBB* cause sickle cell disease. The mutations affect one of the four protein subunits, made of two alpha-globins and two beta-globins. The *HBB* gene instructs for making beta-globin. One type of mutation in *HBB* is responsible for the abnormal version hemoglobin S (HbS). In 1956, two scientists, Vernon Ingram and J. A. Hunt, found that the amino acid valine replaced glutamic acid in the sickle cell hemoglobin molecule. Sickle cell became the first genetic disorder whose molecular basis was known. Other mutations may lead to hemoglobin C (HbC) and hemoglobin E (HbE), which cause other genetic disorders.

In people with sickle-cell disease, hemoglobin S replaces at least one of the beta-globin subunits. In people with sickle-cell anemia, hemoglobin S mutated subunits replace both beta-globin units. These versions of beta-globin distort the abnormal cells, causing them to get stuck in tiny blood vessels and developing serious complications.

People who inherit one sickle-cell gene from one parent but have a normal gene from the other parent have sickle-cell trait. They do not have the disease, but because this is a recessive gene, they have a one in four chance of passing the gene to offspring. *HBB* is inherited in an autosomal recessive pattern and is located on the short arm (p) of chromosome 11 at position 15.5.

What Is the Treatment for Sickle-Cell Disease?

Sickle-cell disease has no widely available cure. Although people have known about the condition for many years, only in the last 100 years has progress been made to treat the complications and symptoms. The symptoms, such as chronic pain and tiredness, vary from one person to another. An individual who gets proper care and treatment can still live a reasonable quality of life even into the 40s, 50s, or longer.

In 1984, a child with sickle cell received the first bone marrow transplant. The actual transplantation was done to treat acute leukemia, and the sickle-cell cure was a surprise. Now bone marrow transplants are used specifically in some of the most severe cases of sickle-cell anemia. In 1995, hydroxyurea became the first drug shown to prevent the complications of sickle-cell disease.

Because sickle cell is the most common inherited blood disease, several clinical trials are ongoing. For information about such trials, visit http://www.clinicaltrials.gov.

See also Newborn Screening: A Special Topic

Further Reading

"Sickle Cell Disease (SCD)." 2010. Centers for Disease Control and Prevention (U.S.). http://www.cdc.gov/ncbddd/sicklecell/index.html. Accessed 5/30/12.

"Sickle Cell Disease (Sickle Cell Anemia)." 2011. MedicineNet.com. http://www.medicinenet.com/sickle_cell/article.htm. Accessed 5/30/12.

Sickle Cell Disease Association of America. 2011. http://www.sicklecelldisease.org. Accessed 5/30/12.

"What Is Sickle Cell Anemia?" 2011. National Heart Lung and Blood Institute (U.S.). http://www.nhlbi.nih.gov/health/health-topics/topics/sca. Accessed 5/30/12.

Sjögren-Larsson Syndrome (SLS)

Prevalence	Incidence in United States not known; 0.4 per 100,000 people in Sweden
Other Names	FALDH deficiency; FAO deficiency; fatty alcohol: NAD+ oxido-reductase deficiency; fatty aldehyde dehydrogenase deficiency; ichthyosis, spastic neurological disorder, and oligophrenia; SLS

In 1957, Torsten Sjögren and Tage Larsson described a condition in which a cohort of Swedish patients had an unusual combination of traits that included dry, scaly skin, stiffness, muscle spasms, and mental retardation. Two decades later, the syndrome was found to be an inborn error of fat metabolism. The condition must not be confused with Sjögren's syndrome, an autoimmune syndrome named after Henrik Sjögren that attacks tears and saliva.

What Is Sjögren-Larsson Syndrome (SLS)?

Sjögren-Larsson syndrome is a metabolic disorder that is related to use of fats or lipids in the body. The condition affects many body parts, and the symptoms occur within the first two years of life. Following are the signs of SLS:

- Dry, scaly skin, a condition called ichthyosis: This clue to the diagnosis of SLS is present at birth. The dry skin develops into a severe case of pruritis, which makes the child very uncomfortable.
- Developmental delays with speech and motor skills.
- Seizures.
- Spastic diplagia or paralysis in both legs: Tetraplegia is paralysis of all four limbs.
- Mental retardation: Most patients have IQs of less than 60.
- Glistening white dots on the retina of the eye.
- Short stature.
- Preterm birth.

Fetuses with Sjögren-Larsson syndrome have shown the symptoms of ichthyosis as early as the second semester of life.

What Is the Genetic Cause of Sjögren-Larssen Syndrome (SLS)?

Mutations in the *ALDH3A2* gene, known officially as the "aldehyde dehydrogenase 3 family, member A2" gene, cause Sjögren-Larssen syndrome. Normally, *ALDH3A2* provides instructions for the aldehyde dehydrogenase enzyme, which plays a major role in breaking down aldehydes generated by the fatty alcohol phytanic acid (a long-branched fatty acid, phytol) and other substances. The breakdown results in fat storage in the membranes that cause tissue dysfunction. The breakdown is critical because of the integrity of the skin, which then responds by creating more skin cells that cause ichthyosis.

The defects in the gene disrupt the production of the proper enzyme to break down the aldehydes and lipids, causing them to build up, and in turn causing the many symptoms of Sjögren-Larssen syndrome. *ALDH3A2* is inherited in an autosomal recessive pattern and is located on the short arm (p) of chromosome 17 at position 11.2.

What Is the Treatment for Sjögren-Larssen Syndrome (SLS)?

Most of the treatments for SLS are for the symptoms. When the *ALDH3A2* gene was discovered, scientists found that a drug called zileuton reduces the itching and improves the behavior of the child. Although there is no cure, the quality of life of children with SLS can improve.

Further Reading

"*ALDH3A2.*" 2011. Genetics Home Reference. National Library of Medicine (U.S.). http://ghr.nlm.nih.gov/gene/ALDH3A2. Accessed 5/30/12.

"Sjogren-Larssen Syndrome." 2011. Genetics and Rare Diseases Information Center (GARD). National Institutes of Health (U.S.). http://rarediseases.info.nih.gov/GARD/Condition/7654/SjogrenLarsson_syndrome.aspx. Accessed 5/30/12.

"Types of Leukodystrophy." United Leukodystrophy Foundation. http://ulf.org/types-of-leukodystrophy. Accessed 5/30/12.

Smith-Fineman-Myers Syndrome (SFMS)

Prevalence   Listed as a "rare disease" by Office of Rare Diseases (ORD); exact prevalence unknown

Other Names   Carpenter-Waziri syndrome; Chudley-Lowry syndrome; Holmes-Gang syndrome; Juberg-Marsidi syndrome (JMS); mental retardation, Smith Fineman Myers type; X-linked mental retardation-hypotonic facies syndrome 1 (MRXHF1)

In 1980, Robert Fineman, Gart Myers, and Richard Smith in Sydney, Australia, reported on two brothers with distinct facial appearances, short stature, and other issues. Later, they noted five of the cases they studied were males, suggesting X-linked inheritance. Since the first descriptions, SFMS has been found in males from 11 families and one isolated case. The rare disorder was named after these three doctors—Smith-Fineman-Myers syndrome.

What Is Smith-Fineman-Myers Syndrome (SFMS)?

Smith-Fineman-Myers syndrome is a rare dysmorphy syndrome that runs in families. The word "dysmorphy" comes from two Greek roots: *dys*, meaning "with difficulty," and *morph*, meaning "form." Thus, dystrophy is another term for deformity. Following are the characteristics of Smith-Fineman-Myers syndrome:

- Very small head, a condition known as microcephaly
- Long narrow face with distinct features

- Large mouth with a small jaw but prominent incisor teeth
- Vision disorders—eyes may not focus and appear as crosses, a condition known as strabismus; the person is farsighted, and the optic nerve is underdeveloped
- Short stature
- Chest deformity
- Mental retardation
- Hypogonadism, meaning the male organs are not developed
- Obesity
- Foot deformities
- Psychomotor problems
- Serious behavior issues

What Is the Genetic Cause of Smith-Fineman-Myers Syndrome (SFMS)?

The *ATRX* gene is officially known as the "alpha thalassemia/mental retardation syndrome X-linked" gene. Normally, *ATRX* provides instructions for a protein that is essential in normal development. The ATRX protein regulates the activity of other genes through a process known as chromatin remodeling. The substance chromatin is made of the DNA and proteins that package the DNA into chromosomes. The nature of the packaging is important because it determines how the genes are expressed during embryonic development. For example, if DNA is tightly packed, gene expression or activity is lower; if it is loosely packed, the expression is higher. *ATRX* appears to control the expression of several genes.

Mutations in *ATRX* appear to cause Smith-Fineman-Myers syndrome. The disruption in the packaging procedure of DNA leads to deformities. The exact process is unknown. It is also related to the condition alpha thalassemia. *ATRX* is inherited in an X-linked recessive pattern and is located on the long arm (q) of the X chromosome at position 21.1. However, another gene located at Xq25 has been traced to a large Chinese family with SFMS.

What Is the Treatment for Smith-Fineman-Myers Syndrome (SFMS)?

Because there are so many characteristics of this syndrome, treatment for this syndrome is symptomatic. For example, eye problems usually can be corrected. The child qualifies for special education.

Further Reading

"Smith-Fineman-Myers Syndrome." 2011. Orphanet. http://www.orpha.net/consor/cgi
-bin/OC_Exp.php?lng=EN&Expert=93974. Accessed 5/30/12.

"Smith Fineman Myers Syndrome." 2011. RightDiagnosis.com. http://www.rightdiagnosis .com/medical/smith_fineman_myers_syndrome.htm. Accessed 5/30/12.

Smith-Lemli-Opitz Syndrome (SLOS)

Prevalence   Affects about 1 in 20,000 to 60,000 newborns; most common in Caucasians of Eastern European ancestry

Other Names RSH syndrome; 7-Dehydrocholesterol reductase deficiency; SLOS; SLO syndrome

In 1964, three geneticists—David Smith, Luc Lemli, and John Opitz—described a condition that they suspected as a genetic condition. The three patients had disorders in several areas, including mental disabilities and unique facial appearances. The doctors at first named the disorder RSH, using the initials of the first three patients with the condition. In 1993, Irons found that patients with the disorder had low plasma cholesterol levels, indicating an inborn error metabolism. The condition is now named after the three doctors who first described the condition—Smith-Lemli-Opitz syndrome.

What Is Smith-Lemli-Opitz Syndrome (SLOS)?

Smith-Lemli-Opitz syndrome affects many body parts and is a developmental disorder that is a classic inborn error of metabolism. The symptoms vary in severity, ranging from those who have only minor physical disorders with some behavioral problems to life- threatening conditions. Following are the general characteristic features of SLOS:

- Very small head, a condition called microcephaly.
- Distinct facial appearance: Eyelids may droop, a condition called ptosis; person may have a cleft palate.
- Feeding difficulties.
- Hypocholesterolemia: Blood cholesterol level is consistently low.
- Abnormal development of many body organs: This includes the heart, lungs, kidneys, gastrointestinal tract, and sex organs.
- Poor muscle tone.
- Very pale appearance.
- Sensory disabilities: The person may be hearing impaired. Vision may be affected with disorders such as cataracts.
- Abnormal digits: The person may have fused second and third toes or extra fingers or toes.

Cholesterol: Essential for Life

In the news, and especially in television advertisements, the word "cholesterol" appears to be something that is really bad for you. And it is true that having high cholesterol is unhealthy and can be life-threatening.

But that is not the complete story. Cholesterol is a waxy, fat-like substance that is both produced in the liver and intestines and obtained from foods that you eat. The name comes from two Greek words—*chole*, meaning "bile," and *stereos*, meaning "solid"—combined with the chemical suffix for an alcohol, which chemically it is. The substance was originally found in 1769 in the solid form in gallstones. Certain foods are high in cholesterol—for example, egg yolks, meat, poultry, fish, and dairy products.

Yet, this substance that the advertisements scream about is really essential for life. It is absolutely essential for normal embryonic development and has several roles after birth. Cholesterol is used to make hormones and is an important component of cell membranes and the fatty myelin sheath that protects nerve cells. In addition, it plays an important role in digestive acids and vitamin D. If cholesterol is low, certain serious disorders can happen to the body—one of these being Smith-Lemli-Opitz syndrome.

- Weak immune system.
- Skeletal disorders such as scoliosis and osteoporosis.
- Delayed speech.
- Learning disabilities and behavioral problems: Many children with SLOS demonstrate features of autism, such as communication disorders and the ability to bond with others.

What Is the Genetic Cause of Smith-Lemli-Opitz Syndrome (SLOS)?

Mutations in the *DHCR7* gene, officially known as the "7-dehydrocholesterol reductase" gene, cause Smith-Lemli-Opitz Syndrome (SLOS). Normally, *DHCR7* instructs for an enzyme 7-dehydrocholesterol reductase. This enzyme 7-dehydrocholesterol reductase controls the final step in the production of cholesterol. Cholesterol is essential for the development of the embryo and does important work before and after birth.

About 120 mutations in *DHCR7* cause Smith-Lemli-Opitz syndrome. The mutations involve the substitution of one amino acid for another, and this change interferes with the normal processing of the enzyme 7-dehydrocholesterol reductase. A common mutation that occurs when methionine replaces threonine affects people of Mediterranean heritage. Other mutations occur when genetic material is either added or deleted, causing a short enzyme that is dysfunctional. When the enzyme does not work properly, toxic products build up in the blood and other tissues. The combination of very low cholesterol added to the accumulation of toxic substances disrupts body growth and leads to the many characteristics of SLOS. *DHCR7* is inherited in an autosomal recessive pattern and is located on the long arm (q) of chromosome 11 at position 13.4.

What Is the Treatment for Smith-Lemli-Opitz Syndrome (SLOS)?

Because this condition is a genetic disorder, there is no cure at present. To treat the manifestations of the disease, cholesterol supplements may result in some clinical improvement. Treating the symptoms such as feeding disorders or digestive disorders can improve quality of life. The person also needs early intervention of physical/occupation/speech therapies for identified disabilities.

Further Reading

"*DHCR7*." 2011. Genetics Home Reference. National Library of Medicine (U.S.). http://ghr.nlm.nih.gov/gene/DHCR7. Accessed 5/30/12.

Irons, Mira. 2007. "Smith-Lemli-Opitz Syndrome." GeneReviews. http://www.ncbi.nlm.nih.gov/books/NBK1143. Accessed 5/30/12.

"Smith-Lemli-Opitz Syndrome." 2011. Genetics Home Reference. National Library of Medicine (U.S.). http://ghr.nlm.nih.gov/condition/smith-lemli-opitz-syndrome. Accessed 5/30/12.

Smith-Magenis Syndrome (SMS)

Prevalence	Affects about 1 in 25,000 people worldwide; some researchers believe this estimate is low because of people who are not diagnosed; they believe the estimate to be about 1 in 15,000
Other Names	chromosome 17p deletion syndrome; deletion 17p syndrome; partial monosomy 17p; 17p11.2 monosomy; 17p- syndrome; SMS

In 1986, a genetic counselor, Ann C. M. Smith at the National Institutes of Health, and Ellen Magenis, an Oregon geneticist, described a condition in which the people had a distinct facial appearance, several other common problems, and certain deletions in chromosome 17. The eponym is called Smith-Magenis syndrome after the two scientists.

What Is Smith-Magenis Syndrome (SMS)?

Smith-Magenis syndrome is a disorder that is not generally inherited but is one that is related to a deletion in chromosome 17 when the egg or sperm is forming. It is a developmental disorder that affects many parts of the body. Rarely is the gene *RAI1* involved in mutations.

Following are the common signs of Smith-Magenis syndrome:

- Distinct facial appearance: This may be subtle in early years but develops later in childhood or even adulthood. The face may be broad and square with deep-set eyes, fat cheeks, and a prominent lower jaw. The mouth turns down with a full, outward curving upper lip. Middle of the face appears flattened.
- Dental abnormalities.
- Abnormal sleep patterns: The person may be sleepy during the day but cannot sleep at night.
- Affectionate and engaging personalities.
- Mild to moderate intellectual disabilities.
- Unique behavior patterns: The individual may have frequent tantrums, aggression, impulsivity, and attention difficulties. The person may injure himself by biting, head banging, or skin picking. They may have repetitive patterns like self-hugging or licking and turning pages of a book, called lick and flip. The person may have an unusual ability to recall tiny details about people or subject-specific trivia.
- Short stature.
- Reduced sensitivity to pain and temperature.
- Curvature of the spine.
- Ear problems that lead to hearing loss.
- Eye abnormalities.
- Hoarse voice.
- Heart and kidney defects are less common.

What Is the Genetic Cause of Smith-Magenis Syndrome (SMS)?

The short arm of chromosome 17 has an area that contains a gene called *RAI1*, officially known as the "retinoic acid induced 1" gene. Normally, *RAI1* instructs for a protein that is essential for nervous system development. The gene is probably part of a complex of genes that also controls the activities of many other genes.

Mutations that result in a change in the function of the *RAI1* gene cause Smith-Magenis syndrome. Researchers suspect that one copy of the *RAI1* is lost when a region of chromosome 17 is deleted. Only a few cases are caused by mutations in the gene itself; rather, the deletions cause the disorder. In people with these mutations, many of the physical symptoms do not occur. *RA11* is located on the short arm (p) of chromosome 17 at position 11.2.

What Is the Treatment for Smith-Magenis Syndrome (SMS)?

Treatment according to symptoms is in order. For example, sleep disorders can be treated with medication. Early childhood intervention, special education, and a variety of physical and behavioral therapies may help. Families may also need social and psychological support and respite care.

Further Reading

"Chromosome 17." 2011. Genetics Home Reference. National Library of Medicine (U.S.). http://ghr.nlm.nih.gov/chromosome/17. Accessed 5/30/12.

Smith, Ann C. M., et al. 2010. "Smith-Magenis Syndrome." *GeneReviews*. http://www.ncbi.nlm.nih.gov/books/NBK1310. Accessed 5/30/12.

"Smith-Magenis Syndrome." 2011. Genetics Home Reference. National Library of Medicine (U.S.). http://ghr.nlm.nih.gov/condition/smith-magenis-syndrome. Accessed 5/30/12.

Sotos Syndrome

Prevalence   About 1 in 10,000 to 14,000 newborns; many cases not diagnosed; true incidence closer to 1 in 5,000

Other Names   cerebral gigantism; Sotos sequence; Sotos' syndrome

In 1964 in an article in the *New England Journal of Medicine*, Juan Sotos and a team of researchers described five children with unusual overgrowth, large heads, distinct facial appearances, and mental disability. The condition at the time was called cerebral giantism but was later changed to Sotos syndrome after the scientist who first wrote about the condition.

What Is Sotos Syndrome?

Sotos syndrome is a developmental disorder, similar to other syndromes such as Weaver syndrome and Fragile X. In addition to a distinct facial appearance and learning disabilities, this syndrome is characterized by overgrowth in childhood. The appearance of the face is a strong initial indicator to a trained physician of the disorder.

Following are the common characteristics of Sotos syndrome:

- Facial appearance: The appearance of the face is one of the ways a trained physician may diagnose this disorder to distinguish it from other syndromes.

The head is described as an inverted pear. Cheeks are flushed, and hair near the temporal area is spare. The individual will have a high forehead, down-slanting eyes, a long narrow face, and narrow jaw. The facial pattern or gestalt develops between the ages of one and six.

- Overgrowth: The infant body grows rapidly, and the child is much taller and larger than others of his or her age. The size of the head is also large, a condition called macrocephaly. Although the height may become normal in adulthood, the large head remains at all ages.

- Developmental delay: Early development is slow. Poor coordination and language delay is apparent. The individual may have problems with making sounds, stuttering, and a monotone voice.

- Intellectual impairment: The majority of children with Sotos syndrome have some degree of learning disability. However, the extent is variable, ranging form mild to severe, in which the person may require lifelong care.

- Behavioral issues: Frequent behavioral problems include tantrums, phobias, obsessions, impulsive behavior, and attention deficit hyperactivity disorder (ADHD).

- Abnormal curvature of the spine or scoliosis.

- Seizures.

- Heart or kidney defect.

- Hearing and vision loss.

- Jaundice with yellowing of eyes.

- Cancer most often in childhood.

What Is the Genetic Cause of Sotos Syndrome?

Mutations in the *NSD1* gene, officially called the "nuclear receptor binding SET domain protein 1" gene, cause Sotos syndrome. Normally, *NSD1* instructs for making a protein that is active in such organs and tissues as the brain, kidney, skeletal muscle, spleen, thymus, and lung. The gene may also control the genes related to normal growth and can turn these genes off or on as they are needed.

Over 100 mutations in *NSD1* cause Sotos syndrome. Incidences in the Japanese population appear to be missing the part of chromosome 5 that contains *NSD1*. However, most of the cases are changes within the gene itself that includes insertion or deletions of a small amount of DNA or changes in just one of the amino acid building blocks. Most of the mutations prevent one copy of the *NSD1* gene from making a normal protein. Scientists are still investigating how these changes lead to overgrowth and the characteristics of Sotos syndrome. About 95% of cases have no family history of Sotos, indicating new mutations have developed. The few families that do pass the mutated gene do so in an autosomal dominant pattern. *NSD1* is located on the long arm (q) of chromosome 5 at position 35.

What Is the Treatment for Sotos Syndrome?

Treatment is symptomatic. If there is a specific physical problem such as a heart, kidney, or skeletal disorder, the proper specialist should be called in. Referral to specialists is essential for management of learning disabilities, speech delays, and learning behavior.

Further Reading

"*NSD1*." 2011. Genetics Home Reference. National Library of Medicine (U.S.). http://ghr.nlm.nih.gov/gene/NSD1. Accessed 5/30/12.

Sotos Syndrome Support Association. 2011. http://www.sotossyndrome.org. Accessed 5/30/12.

Tatton-Brown, Katrina, Trevor R. P. Cole, and Nazneen Rahman. 2009. "Sotos Syndrome." *GeneReviews*. http://www.ncbi.nlm.nih.gov/books/NBK1479. Accessed 10/16/11.

Spastic Paralysis, Infantile-Onset Ascending Hereditary

Prevalence   Rare; only a small number of cases reported

Other Names   familial spastic paraparesis (or paraplegia); hereditary Charcot disease; hereditary spastic paraplegia (or paraparesis); IAHSP; spastic paralysis, infantile onset ascending; spastic paraplegia; spastic spinal paralysis; Strümpell-Lorrain disease

In the late 1800s, Adolph Strümpell, a German neurologist, observed a father with two sons who had the same gait disorders and problems with spasticity in the legs. Performing an autopsy after the brothers died, Strümpell found that the nerve fibers from the spinal cord had deteriorated. The disorder was named for Strümpell and later called Lorrain and Charcot. As one can see from the long list of names, there is no agreement at present. The National Institutes of Health uses the term infantile-onset ascending hereditary spastic paralysis, or hereditary spastic paraplegia (HSP) for people who develop the condition later in life. This article refers to the infantile-onset type.

What Is Infantile-Onset Ascending Hereditary Spastic Paralysis?

This hereditary type of spastic paralysis occurs in infancy and progressively damages motor neurons, causing stiffness and weakness in the muscles of the arms, legs, and face. The symptoms usually occur during the first two years of life.

Following are the common characteristics of spastic paraplegia:

- Paralysis during the first two years of life.
- Legs noted first: The person has noticeable weakness in the muscles of the legs, which appear as stiff and tight. Legs then become paralyzed.

- Ascending muscle weakness to upper limbs: The weakness then travels up or ascends from the legs to the arms. By late childhood, arm and hand muscles are weak and eventually paralyzed.

- Ascending muscle weakness to the neck and head: By adolescence, the person cannot move the neck, and the eye movements are slow. Problems with speech and swallowing follow.

What Is the Genetic Cause Infantile-Onset Ascending Hereditary Spastic Paralysis?

Mutations in the *ALS2* gene, officially known as the "amyotrophic lateral sclerosis 2 (juvenile)" gene, cause infantile-onset ascending hereditary spastic paralysis. Normally, *ALS2* instructs for making the protein alsin. Alsin is found in many parts of the body, but the highest amounts are in the brain. Alsin is very prevalent in the motor neurons, those specialized nerve cells located in the brain and spinal cord to control muscle movement. Alsin may play a role in the development of the parts of the neuron. The long part of neurons called axons receives messages and passes it to the cell body. The dendrites, the treelike structures, pass the message from the cell body across the synapse to the next neuron.

Several mutations in *ALS2* cause infantile-onset ascending hereditary spastic paralysis. The changes in the protein alsin occur when one or more DNA building blocks are exchanged or deleted. When this happens, alsin becomes unstable, and its loss disrupts the movement of the essential molecules in the cells. The long axons of the motor neurons may be especially vulnerable to the loss, leading to the symptoms of infantile-onset ascending hereditary spastic paralysis. *ALS2* is inherited in an autosomal recessive pattern and is located on the long arm (q) of chromosome 2 at position 33.1.

What Is the Treatment for Infantile-Onset Ascending Hereditary Spastic Paralysis?

Because this is a genetic disorder, no cure is available. Likewise, no treatments are currently available to prevent, stop, or reverse infantile-onset ascending hereditary spastic paralysis. Treatment is focused on symptom relief, such as medication to reduce spasticity. Physical therapy and exercise may help maintain flexibility, strength, and range of motion. In addition, assistive devices and communications aids may make the individual more comfortable. Life expectancy is normal, but complications or falls may shorten the person's life.

Further Reading
"About HSP (Hereditary Spastic Paraplegia)." 2011. Spastic Paraplegia Foundation. http://www.sp-foundation.org/hsp.html. Accessed 5/30/12.

Bertini, Enrico S., et al. 2011. "ALS2-Related Disorders" [including infantile-onset ascending hereditary spastic paralysis]. *GeneReviews*. http://www.ncbi.nlm.nih.gov/ books/NBK1243. Accessed 5/30/12.

"NINDS Motor Neuron Diseases Information Page." 2011. National Institutes for Neurological Disorders and Stroke (U.S.). http://www.ninds.nih.gov/disorders/motor _neuron_diseases/motor_neuron_diseases.htm. Accessed 5/30/12.

Spina Bifida

Prevalence   One of the most common neural tube defects; affects about 1 in 2,500 newborns worldwide; in the United States, occurs more frequently in non-Hispanic and Hispanic whites than other ethnic groups

Other Names   cleft spine; open spine; rachischisis; spinal dysraphism

Some scientists suggest that the condition known today as "spina bifida" occurred almost 12,000 years ago. In the seventeenth century, a Dutch professor, Nicholas Tulp, first described a situation in which the bones of the spine did not appear to

SPINA BIFIDA

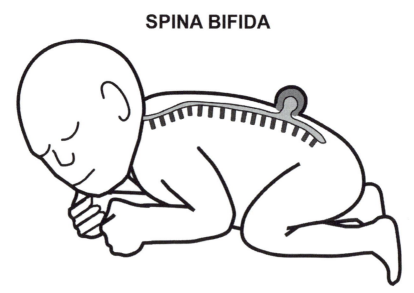

A drawing of a baby with spina bifida, a congenital condition affecting the spine. It can result in the spinal cord and meninges (the spinal cord covering tissues) protruding out of the back. (ABC-CLIO)

close; he gave it the name spina bifida. "Spina" relates to the vertebra or backbone. The word "bifida" comes from two Latin words: *bi*, meaning "two," and *fid*, meaning "divide." In spina bifida, it appears that the spine remains separated in two parts rather than joining to form the hard vertebra. Later in 1771, the Italian anatomist Giovanni Battista Morgagni linked hydrocephalus, the condition in which fluid gathers on the brain, with the lower-limb deformity of spina bifida.

What Is Spina Bifida?

Spina bifida, a condition that is apparent at birth, affects the spinal column. In this condition, the bones of the vertebra do not completely close, causing part of the unprotected spinal cord to stick out through the opening in the spine. The condition, which occurs during the early days of embryonic development, is one of several types of neural tube defects (NTD). The neural tube is a layer of embryonic cells that develops into the brain and spinal cord.

The defect can occur any place on the spine where the neural tube does not close properly. When the spinal cord protrudes through the opening, the nerves of the cord are permanently damaged. The disabilities can range from severe to mild depending on the size and location of the opening in the spine and whether part of the nerves are affected.

Three types of spina bifida are noted: myelomeningocele, meningocele, and spina bifida occulta.

Myelomeningocele

Myelomeningocele is the most serious of the three types and is the one that people usually are referring to when they mention spina bifida. Characteristics of this type are the following:

- A sac of fluid covered by skin is prominent on the baby's spine.
- Nerves and spinal cord have come through the opening into the sac; some of the nerves are damaged.
- The person may have some disabilities arising from the damage, such as a lack of bowel and bladder control and not being able to move legs.
- Some individuals may have an additional complication of fluid collecting on the brain, a condition known as hydrocephalus.
- The child may have some learning disabilities or behavioral issues, such as attention deficit hyperactivity disorder (ADHD).

Meningocele

In this type of spina bifida, the sac of fluid comes through the opening, but the spinal cord is not in the sac. Therefore, there is little or no damage to the cord. Only minor disabilities are seen with this type of defect.

Spina Bifida Occulta

The word "occult" means hidden. A small gap may be in the spine, but there is no opening or sac. The nerves and spinal cord are usually normal. This type of spina bifida is the mildest and may not be detected until later childhood or adulthood. Although it may be a source of back pain or bladder problems, no disabilities usually are present in this type.

What Is the Genetic Cause of Spina Bifida?

Mutations in the *MTHFR* gene, officially known as the "methylenetetrahydrofolate reductase (NAD(P)H)" gene, cause spina bifida. Normally, *MTHFR* instructs for the enzyme methylenetetrahydrofolate reductase. This enzyme is important in the processing of the building blocks of all proteins called amino acids. Chemical reactions are taking place constantly within the body. The B-vitamin folate is also called folic acid or vitamin B9. This vitamin combines with other chemicals to form 5,10-methylenetetrahydrofolate. The MTHFR enzyme converts this form of folate to another form, 5-methyltetrahydrofolate. Then after several other chemical steps, the amino acid homocysteine is made, which then converts to another amino acid methionine. The body is then able to use methionine to make important proteins and compounds.

Scientists are not exactly clear on how folic acid makes a difference. It is known that this vitamin is important for the synthesis of nucleic acids and forming hemoglobin. Regardless of how it works, researchers have documented that folic acid in the diet prior to the onset of pregnancy lowers the risk of spina bifida by 70% to 75%.

Mutations in *MTHFR* are related to an increase in spina bifida. Variations in the gene change the ability of methylenetetrahydrofolate reductase to process folate. Other genes may also play a minor role, but just what these genes are and how they work have not been substantiated. Most cases of spina bifida are sporadic, meaning they occur in families with no history. However, the few instances that have been seen in families do not have a consistent inheritance pattern. *MTHFR* is located on the short arm (p) of chromosome 1 at position 36.3.

What Is the Treatment for Spina Bifida?

Treatment for spina bifida may involve surgery. If hydrocephalus is present, surgery is essential to relieve the pressure created by the fluid buildup on the brain. Other symptoms, such as issues with going to the bathroom, may be treated as they occur. Generally, with good care, a person with spina bifida can have a normal life expectancy.

Further Reading

"Spina Bifida." 2011. Centers for Disease Control and Prevention (U.S.). http://www.cdc.gov/NCBDDD/spinabifida/facts.html. Accessed 5/30/12.

"Spina Bifida." 2011. *Journal of the American Medical Association.* http://jama
.ama-assn.org/content/285/23/3050.full.pdf. Accessed 5/30/12.

"Spina Bifida." 2011. Kids Health from Nemours. http://kidshealth.org/parent/medical/
brain/spina_bifida.html#cat135. Accessed 5/30/12.

Spinal Muscular Atrophy (SMA)

Prevalence   Affects 1 in 6,000 to 10,000 people

Other Names hereditary motor neuronopathy; progressive muscular atrophy;
SMA

February 28 of each year is dubbed International Rare Disease Day. The day coordinated by EURODIS encourages hundreds of patient organizations in more than
40 countries to conduct awareness activities about rare disorders. Their slogan is
"Rare but equal." Spinal muscular atrophy (SMA) is listed as one of the top 10
of the rare groups.

What Is Spinal Muscular Atrophy (SMA)?

Spinal muscular atrophy (SMA) is a disorder of the muscles that affects movement. Movement is controlled by a group of nerve cells called motor neurons,
which are located mostly in the spinal cord and the part of the brain connected to
the spinal cord called the brain stem. That is the reason for the word "spinal" in
the name. Part of the motor neuron is called the axon, a long, wirelike projection
that connects the neurons to the muscles in the limbs and trunk, causing them to
contract. The term "atrophy" in the name refers to wasting away or dying. Thus,
in SMA, motor neurons are lost, causing the weakness and wasting of the muscles.

Several forms of SMA exist, but all forms have the primary feature of muscle
weakness, followed by atrophy or loss of the muscle. Many of the features of the
disorder are secondary to the loss of muscle weakness. Following are the common
characteristics of SMA:

- Muscle weakness with poor muscle tone: The muscles needed for crawling,
  sitting up, and walking are weak. The child is late in developmental mile-
  stones that the average child may exhibit. The child may appear limp with a
  tendency to flop.

- Weak cry.

- Difficulty sucking or swallowing.

- Bell-shaped torso: This is because of abdominal breathing rather than using
  muscles around the lungs.

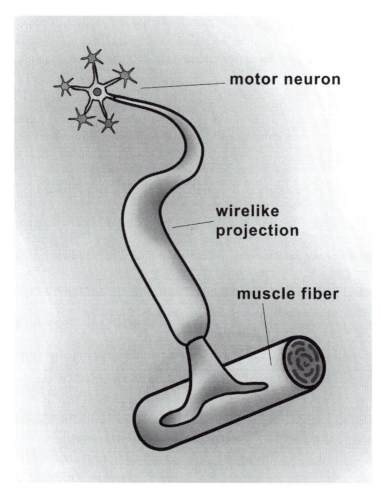

Spinal muscular atrophy affects voluntary muscles, where certain neurons, which are nerve cells in the spinal cord responsible for muscle contraction, become atrophied, fading away and dying, causing weakness and wasting of the muscles.

- Accumulation of fluid and mucus in the lungs and throat: The fluid may lead to respiratory infections.
- Unusual body posture: The head is often tilted to one side; the tongue may flicker or vibrate. The legs may be held in a froglike position.
- Clenched fists with sweaty hands.
- Bowel or bladder disorders.
- Lower-than-normal weight.

Depending on the type, features range from trivial to fatal, with every shade of impairment in between.

Four types of SMA affect children before the age of one. Some other types of SMA affect people later in life.

Type I SMA

This condition, also called Werdnig-Hoffman disease, is the severe form that has most of the issues in the characteristics list. The child will have trouble breathing, sucking, and swallowing and will not be able to sit up without support. The disorder is noted at birth or soon after. At one time, children with this disorder did not survive more than two years. Today, technological devices, such as lightweight mechanical ventilators and feeding tubes that go directly into the stomach, can extend life. Mental and emotional development is normal.

Type II SMA

This type of muscle weakness occurs between the ages of 6 months and 12 months. Children with this type can learn to sit without support but will not be able to walk or stand. The muscles closer to the body center are affected more than those that are farther from the body. For example, the muscles of the thighs, which are nearer the center, are affected before the lower legs and feet. Hands may become weak but usually remain strong enough to type on the computer. The most serious complication of type II appears to be in breathing. The children may also develop scoliosis or curvature of the spine.

Children with type II can benefit greatly from physical therapy and devices such as motorized wheelchairs. A small body of research has shown children with SMA appear usually intelligent.

Type III SMA

Also called Kugelberg-Welander disease or juvenile type, type III is a milder form of the disorder. The children can stand and walk without help but often lose this ability as they get older. The symptoms appear after 18 months of age or generally after the child starts to walk. The ability to walk continues into the 30s or 40s. People with this type can live well into adulthood and experience success in many professions.

Type IV SMA

This type appears in early infancy and appears to work differently from type I in that the weakness begins in the hands and feet and then spreads to the limbs. The diaphragm, the muscle that controls breathing, may also be affected. Symptoms appear between the ages of 6 weeks and 6 months.

X-linked SMA

The features of this type are similar to those of type I except that the infants are born with joint deformities called contractures. The children cannot move from birth. Sometimes, the children are born with broken bones and poor muscle tone.

Finkle Type SMA

This type is associated with an adult-onset type of SMA.

What Are the Genetic Causes of Spinal Muscular Atrophy (SMA)?

Several genes are indicated in SMA. These genes are the *IGHMBP2*, *SMN1*, *SMN2*, *UBA1*, and *VAPB* genes.

IGHMBP2

Mutations in the *IGHMBP2* gene, officially known as the "immunoglobulin mu binding protein 2" gene, cause type I SMA. Normally, *IGHMBP2* instructs for an enzyme called helicase. Helicase is active in the process of cell division that involves replication and duplication. In cell division, the chromosomes are duplicated so that each daughter cell has a set of identical chromosomes. When the DNA unwinds, RNA, a chemical cousin of DNA, in a process called translation, forms the template for replication. Helicase is the enzyme that attaches to special regions of DNA and causes the two spiral strands to unwind. This protein also is involved in producing proteins from RNA. Helicase is found in cells throughout the body.

About 60 mutations in *IGHMBP2* cause type I SMA. Most mutations involve a single building block, but this one change disrupts the protein's ability to unwind DNA and RNA. The motor neurons called alpha neurons, which are located in the upper part of the spinal cord or brain stem, are especially sensitive to the disruption. The loss of the neurons leads to the progressive muscle weakness and other symptoms of type I SMA. *IGHMBP2* is inherited in an autosomal recessive pattern and is located on the long end (q) of chromosome 11 at position 13.3.

SMN1

Mutations in the *SMN1* gene, officially called the "survival of motor neuron 1, telomeric" gene, cause some forms of SMA. Normally, *SMN1* instructs for making the survival motor neuron or SMN protein. This protein is the caregiver for motor neurons and is responsible for their maintenance. The SMN protein is essential in making messenger RNA or mRNA, which is the blueprint for making proteins. Before becoming mRNA, the substance goes through a process called pre-mRNA before becoming the final form. The SMN protein aids in developing the cellular machinery that processes the pre-mRNA. The protein also may have other functions, such as maintaining the dendrite and axons.

About 65 mutations in the *SMN1* gene cause a type of SMA. About 95% of people have a deletion in both copies of the gene—one from the father and one from the mother. Therefore, no SMN protein is made. The other 5% may have the mutation in one copy of the gene, and the other has a different mutation. Without SMN protein, the whole machinery needed for pre-mRNA is disrupted. Also, the shortage may impair the work if the axons and dendrites leading to their death. *SMN1* is inherited in an autosomal recessive pattern and is located on the long arm (q) of chromosome 5 at position 13.2

SMN2

Mutations in the *SMN2* gene, officially known as the "survival of motor neuron 2, centromeric" gene, cause some types of SMA. Normally, *SMN2* provides instructions for making another survival motor neuron protein. The protein is essential for maintaining motor neurons. The gene produces several versions of the proteins, only one of which is fully functional. This version called isoform d is just like the protein made by the *SMN1* gene, only in a smaller form. SMN protein of both genes is important in making the pre-mRNA and other nerve functions.

People have two copies of *SMN1* and *SMN2* in their cells. People with SMA may have altered copies of *SMN1* genes and three or more copies of the *SMN2* gene. The additional copies of *SMN2* may produce a milder course. *SMN2* is inherited in an autosomal recessive pattern and is located on the long arm (q) of chromosome 5 at position 13.2.

UBA1

Mutations in the *UBA1* gene, known officially as the "ubiquitin-like modifier activating enzyme 1" gene, cause the X-linked form of SMA. Normally, *UBA1* provides for instructions for the ubiquitin-activating enzyme E1. This enzyme is essential in the process of breaking down certain proteins in cells. Breaking down proteins is the normal process for removing damaged or extra proteins that would interfere with the functions of the cell. This enzyme is part of an important system called ubiquitin-proteasome that acts as the quality control of the cell. Ubiquitin-activating enzyme E1, the first step in this system, starts the process by turning on an important protein called ubiquitin.

About three mutations in the *UBA1* gene are linked to the X-linked form of SMA. Only one change in the DNA building block leads to the impairment of the enzyme. When the enzymes do not function, unwanted proteins build up and cause the cell to die. The motor neurons are especially vulnerable to this buildup. *UBA1* is located on the short arm (p) of the X chromosome at position 11.23.

VAPB

Mutations in the *VAPB* gene, officially called the "VAMP (vesicle-associated membrane protein)-associated protein B and C" gene, cause a type of SMA called Finkle type. Normally, *VAPB* makes a protein that is associated with the membrane around the endoplasmic reticulum, the series of mazes that transports materials from the cells to outside the cell surface. To function properly, the endoplasmic reticulum must fold the proteins properly for transport. The *VAPB* protein appears to be essential in the unfolded protein process. It is not known how the protein exactly causes late-onset SMA. *VAPB* is inherited in an autosomal dominant pattern and is located on the long arm (q) of chromosome 20 at position 13.33.

What Is the Treatment for Spinal Muscular Atrophy (SMA)?

The treatment of SMA is according to the type and symptoms. Type I may need breathing devices from the very beginning of life. Feeding tubes may also be essential.

There is an array of assistive technology products that help children to move: standers, walkers, various kinds of manual of wheeled chairs, and braces. Also, tables, chairs, and other special furniture are now available.

Further Reading

"Spinal Muscular Atrophy." 2011. Genetics Home Reference. National Library of Medicine (U.S.). http://ghr.nlm.nih.gov/condition/spinal-muscular-atrophy. Accessed 5/30/12.

SMA Foundation. 2011. http://www.smafoundation.org. Accessed 5/30/12.

Spinocerebellar Ataxia, Type 1

Prevalence   Affects 1 to 2 per 100,000 worldwide

Other Names   olivopontocerebellar atrophy I; spinocerebellar atrophy I; type 1 spinocerebellar ataxia

The spinocerebellar ataxias are a large group of disorders, each of which could be considered a separate disease. The disorders are related to the degeneration of the cerebellum and spinal cord. The cerebellum is the back and lower part of the brain that coordinates movement. With this disease, the critical control center of the brain wastes away. Spinocerebellar atrophy, type 1, or SCA1, is the most common type.

What Is Spinocerebellar Ataxia, Type 1?

Spinocerebellar ataxia, type 1, is a progressive condition that affects balance, co-ordination, and movement. The name describes the disorder: spino refers to the spinal cord; cerebellar refers to the cerebellum that cauliflower-like structure at the back of the brain that attaches to the brain stem and the spinal cord; ataxia relates to balance and coordination.

The symptoms of SCA1 can begin any time from childhood to late adulthood. Survival period is from 10 to 20 years after diagnosis. Following are the symptoms of SCA1:

• Initial problems with coordination and balance, a condition called ataxia
• Speech and swallowing difficulties

- Muscle stiffness, a condition called spasticity
- Weakness in the muscles that control the eyes; this weakness may lead to voluntary eye movements and loss of control of eye movements
- Cognitive impairment; the person may have difficulty learning and remembering information
- Numbness and tingling in legs, a condition called neuropathy
- Muscle problems, such as muscle tensing, wasting, and twitches.
- Occasional jerking, rigidity, and tremors after living many years with the condition

What Is the Genetic Cause of Spinocerebellar Ataxia, Type 1?

Mutations in the *ATXN1* gene, officially known as the "ataxin 1" gene, cause spinocerebellar ataxia. Normally, *ATXN1* provides instructions for the protein ataxin-1. This protein is located in the nucleus of cells and is possibly involved in controlling other proteins, especially those related to the first stage of transcription and processing the important chemical RNA. One region of the gene has a segment known as a CAG repeat. CAG stands for the amino acids cytosine-adenine-guanine. When these amino acids are repeated in a row, they are referred to as a trinucleotide repeat. The segment is normally repeated 4 to 39 times.

Mutations of the *ATXN1* gene that cause SCA1 may have 40 to more than 80 CAG repeats. These expanded repeats lead to a very long version of the ataxin-1 proteins, which then folds into an abnormal protein shape. These abnormal proteins clump with each other within the nucleus of the cell, disrupting the normal function and damaging the cells. The aggregates appear to be found only in the brain and spinal cord. Certain cells called Purkinje cells are located in the cerebellum, the area that controls movement. It is the aggregation in these cells that cause cell death and lead to the progressive loss of movement in SCA1. *ATXN1* is inherited in an autosomal dominant pattern and is located on the short arm (p) of chromosome 6 at position 23.

What Is the Treatment for Spinocerebellar Ataxia, Type 1?

Treatment for SCA1 is symptomatic. As the symptoms appear, individuals may receive therapy or drugs to make them comfortable. There is no cure. Various adaptive devices such as a cane, crutches, walker, or wheelchair may help the impaired gait. Other devices may assist with writing, feeding, and self-care.

Further Reading

"Spinocerebellar Ataxia." 2011. National Library of Medicine (U.S.). *Genes and Disease.* http://www.ncbi.nlm.nih.gov/books/NBK22234. Accessed 5/30/12.

"Spinocerebellar Ataxia: Making an Informed Choice about Genetic Testing." 2011. http://depts.washington.edu/neurogen/downloads/ataxia.pdf. Accessed 5/30/12.

"Spinocerebellar Ataxia Type 1." 2011. Genetics Home Reference. National Library of Medicine (U.S.). http://ghr.nlm.nih.gov/condition/spinocerebellar-ataxia-type-1. Accessed 5/30/12.

Spondyloepiphyseal Dysplasia Congenita (SED Congenita)

Prevalence   Rare; incidence unknown; over 175 cases reported

Other Names   SED, congenital type; SEDc; SED congenital; spondyloepiphyseal dysplasia, congenital type

In 1939, A. W. Jacobsen named a condition SED tarda in a report of 20 patients with short trunks. Later in 1966, Jürgen W. Spranger and Hans-Rudolf Wiedemann first described SED congenita, which led to other descriptions of the condition. The difference between the two conditions is that in SED congenita, the symptoms are noted at birth; in SED tarda, the symptoms appear later.

The long name spondyloepiphyseal dysplasia (SED) is made of several smaller Greek roots; *spondylos*, meaning "vertebra," and *epiphysis*, meaning the "ends of the long bones." The word "dysplasia" comes from *dys*, meaning "with difficulty," and *plasia*, meaning "form." A dysplasia is an abnormal growth.

What Is Spondyloepiphyseal Dysplasia Congenita (SED Congenita)?

Spondyloepiphyseal dysplasia congenita is a disorder of bone growth that results in dwarfism. Symptoms of the conditions include the following:

- Short stature from birth: The average adult height ranges from three feet to four feet.

- Skeletal abnormalities: The individual is born with very short trunk, neck, and limbs. The short bones are the result of a defect in the ends of the long bones in the arms and legs, called the epiphysis, and in the bones of the vertebrae.

- Abnormal curvature of the spine: Breathing problems are caused by the condition.

- Unstable neck bones: The instability of the bones may cause spinal cord damage.

- Upper leg bones: The upper leg bones turn inward, a condition known as coax vara. This condition may be caused by an abnormality of the hip joint.

- A broad barrel-shaped chest.

- Club foot.
- Arthritis: Joint mobility may decrease early in life
- Facial features: Cheekbones appear flat.
- Cleft palate.
- Severe nearsightedness and other eye problems.
- Hearing loss in about one-quarter of cases.

What Is the Genetic Cause of Spondyloepiphyseal Dysplasia Congenita (SED Congenita)?

Mutations in the *COL2A1* gene, officially known as the "collagen, type II, alpha 1" gene, cause spondyloepiphyseal dysplasia congenita. Normally, *COL2A1* provides instructions for making one part of type II collagen, called pro-alpha1(II) chain. Collagen has many functions including the following:

- Adds strength to connective tissue, making it strong
- Supports muscles, joints, organs, and skin
- Makes up a large part of the embryonic skeleton
- Protects the ends of the bones
- Makes up the ends of nose and ears
- Includes part of clear get that fill the eyeball, inner ear, and center portion between the vertebrae and spine

In order to make type II collagen, three pro-alpha1(II) chains twist together like a rope to make a triple-stranded procollagen molecule. Enzymes then process molecules in the cell, allowing them to then leave the cell and form long, thin fibrils that cross-link in the extracellular spaces. The links make strong type II collagen fibers.

More than 20 mutations in the *COL2A1* gene cause spondyloepimetaphyseal dysplasia congenita. In the mutations, another amino acid replaces glycine in the pro-alpha1(II) chain. The defect in the mutations makes a very short chain, and this interferes with the formation of mature and strong collagen. *COL2A1* is inherited in an autosomal recessive pattern and is located on the long arm (q) of chromosome 12 at position 13.11.

What Is the Treatment for Spondyloepiphyseal Dysplasia Congenita (SED Congenita)?

From birth, the individual with this disorder must have comprehensive care. An ophthalmologist is needed to care for pathological eye disorders. A neurologist can identify motor delay or other muscular disorder. A pulmonologist will evaluate respiratory complications. Surgery may correct some of the orthopedic disorders.

Further Reading

"Spondyloepiphyseal Dysplasia." 2011. Medscape. http://emedicine.medscape.com/article/1260836-overview. Accessed 5/30/12.

"Spondylo-Epiphyseal Dysplasia." 2011. Nemours Children's Hospital. http://www.nemours.org/service/medical/orthopedics/dysplasia/spondylo.html. Accessed 5/30/12.

"Spondyloepiphyseal Dysplasia Congenita." 2012. Genetics Home Reference. National Library of Medicine (U.S.). http://ghr.nlm.nih.gov/condition/spondyloepiphyseal-dysplasia-congenita. Accessed 5/30/12.

Stargardt Macular Degeneration

Prevalence   Affects 1 in 8,000 to 10,000 worldwide; affects over 25,000 Americans

Other Names   fundus flavimaculatus; juvenile onset macular degeneration; Stargardt disease; Stargardt macular dystrophy

In 1901, Karl Stargardt, a German ophthalmologist, reported on an early-onset form of macular degeneration. Before that time, it was thought that macular degeneration was a single disease. Now scientists know that it is a group of disorders that have the common feature—loss of central vision—and the disorders may affect people at any age. This juvenile form of macular degeneration that occurs in one in 10,000 children was named after the German doctor—Stargardt macular degeneration.

What Is Stargardt Macular Degeneration?

Stargardt macular degeneration is a progressive eye disorder that affects the retina and causes progressive vision loss. The problem arises in the retina of the eye, the special screen at the back of the eye that is sensitive to light. The macula is the area in the center of the retina that is necessary for the sharp central vision needed for reading, drawing, and recognizing faces.

Sometime between the ages of 6 and 20, the condition becomes apparent. Children begin to notice that gray, black, or hazy areas appear in the center of the page they are reading. They then may notice it takes a longer length of time to adjust between light and dark. Following is the progression of macular degeneration disorder:

- A fatty yellow-white pigment called fundus flecks or lipofuscin gathers in the cells under the macula. These cells, called retinal pigment epithelium or RPE cells, form the layer between the rod and cone cells and the choroids, the dark brown vascular structure located between the sclera and retina.

- After a period of time, the collection of the lipofuscin damages the cells that are essential for clear vision.

- Night vision and low-light vision become impaired.

- Color vision may also be affected at later stages.

- Vision loss is slow at first until the 20/40 level is reached, and then it is rapid until 20/200, the point of legal blindness, is reached.

- Impairment to both eyes begins between 6 and 20, but may not be apparent until the ages of 30 or 40.

What Are the Genetic Causes of Stargardt Macular Degeneration?

In 1997, the discovery that Stargardt macular degeneration had a strong genetic component linked the condition to a group of genes called ABC1 genes. The disorder was especially found in mutations in two specific genes: *ABCA4* and *ELOVL4*. Most cases are caused by the *ABCA4* gene; cases involving the *ELOVL4* gene are much rarer.

ABCA4

Mutations in the *ABCA4* gene, officially known as the "ATP-binding cassette, subfamily A (ABC1), member 4" gene, cause Stargardt macular degeneration. Normally, the *ABCA4* gene instructs for a protein found in the retina. This protein is made in the retina's photoreceptor cells, the specialized cells in the retina that converts sensory images into electrical signals that are then sent to the brain, where the perception of sight occurs. This process, called phototransduction, also leads to making some toxic products including one called N-retinylidene-PE. ABCA4 protein takes this substance out and keeps it from building up to a harmful level.

Over 500 mutations in *ABCA4* cause Stargardt macular degeneration. Only one change in an amino acid building block causes most of the mutations. When the *ABCA4* protein malfunctions, it cannot remove the N-retinylidene-PE from the photoreceptor cells, leading to the buildup of the fatty yellow pigment lipofuscin. The buildup over time causes the progressive loss of cells in the macula. *ABCA4* is inherited in an autosomal recessive pattern and is located on the short arm (p) of chromosome 1 at position 22.

ELOVL4

Mutations in the *ELOVL4* gene, officially known as the "elongation of very long chain fatty acids (FEN1/Elo2, SUR4/Elo3, yeast)-like 4" gene, cause Stargardt macular degeneration. Normally, *ELOVL4* instructs for a protein found in the photoreceptor cells in the retina. This protein is found in the endoplasmic reticulum of these cells. The endoplasmic reticulum is a maze inside a cell that is the channel for transport of proteins from the cell to the outside of the cell. The ELOVL4 protein adds carbon molecules to long-chain fatty acid, making them very long chains.

Three mutations in *ELOVL4* cause Stargardt macular degeneration. These mutations stop making the protein, which then disrupt the long-chain fatty acids and interfere with the cell's processing by the endoplasmic reticulum. The loss of the photoreceptor cells then causes progressive vision loss. *ELOVL4* is inherited in an autosomal recessive pattern and is located on the long arm (q) of chromosome 6 at position 14.

What Is the Treatment for Stargardt Macular Degeneration?

At present, there is no cure for this disorder and really no way to stop its progress. Wearing sunglasses may help protect from the sun and aid in comfort. People are encouraged to limit their intake of vitamin D because it may promote damage. Gene therapy is being investigated and has been given orphan status by the FDA and the European Medicines Agency.

Further Reading

"Stargardt Disease." 2011. American Macular Degeneration Foundation. http://www.macular.org/stargardts.html. Accessed 5/30/12.

"Stargardt Disease." 2011. Macular Degeneration Support. http://www.mdsupport.org/library/stargrdt.html. Accessed 5/30/12.

"Stargardt Macular Degeneration." 2011. Genetics Home Reference. National Library of Medicine (U.S.). http://ghr.nlm.nih.gov/condition/stargardt-macular-degeneration. Accessed 5/30/12.

Stickler Syndrome

Prevalence   Affects about 1 in 7,500 to 9,000 newborns

Other Names   hereditary arthro-ophthalmo-dystrophy; hereditary arthro-ophthalmopathy

In 1965, Gunnar B. Stickler reported on a condition with joint problems and distinct facial features, including both vision and hearing loss. Later, it was found that the problems were related to collagen, a connective tissue. The condition was named after the physician who first described it—Stickler syndrome.

What Is Stickler Syndrome?

Stickler syndrome is a group of disorders with distinctive facial appearances, joint problems, and other problems including those of the eyes and ears. There are four different kinds, and these vary with the signs and symptoms of the disorder. Following are some of the features of Stickler syndrome:

- Distinct facial appearance: The bones of the face, including those of the cheekbone and nose are underdeveloped.

- Robin sequence: Robin sequence is a group of physical symptoms that includes a cleft palate, large tongue, and a small lower jaw.
- Feeding problems.
- Difficulty breathing.
- Severe nearsightedness: In some types, the fluid of the eyeball is abnormal; other types have glaucoma, cataracts, and detached retina.
- Hearing loss: Hearing loss varies and may become more severe over time.
- Joints: Joints may be loose and very flexible. Arthritis may appear later in life.
- Back pain caused by vertebral abnormalities: Problems with the bones in the spine may occur, including curvature and flattened vertebra.

Four different types of Stickler syndrome have been identified. These types are related to their genetic causes and characteristic signs and symptoms. Following are the four types and the genes that cause them:

- *Type 1*: The gene *COL2A1* is related to about 75% of Stickler cases. Most of these cases have many of the "full Stickler" symptoms including eyes, joints, hearing, and cleft palates disorders.
- *Type 2*: The gene *COL11A1* is related to cases that also have the "full Stickler" symptoms, but individuals have more severe cases of deafness.
- *Type 3*: The gene *COL11A2* is related only to disorders in the joints and hearing with no eye problems. It is referred to as a non-ocular type.
- *Type 4*: The gene *COL9A1* is related to a rare recessive form of Stickler syndrome.

Another condition known as Marshall syndrome has the same facial, joint, hearing loss and other features, but it also includes short stature. Researchers have not determined if this is a separate disorder or a variation of Stickler.

What Are the Genetic Causes of Stickler Syndrome?

Four genes are related to Sticker syndrome: *COL2A1*, *COL11A1*, *COL11A2*, and *COL9A1*.

COL2A1

Mutations in the *COL2A1* gene, officially known as the "collagen, type II, alpha 1" gene, cause type 1 or classical Stickler syndrome. Normally, *COL2A1* instructs for making one part of type II collagen called the pro-alpha1(II) chain. Type II collagen is the substance that gives strength and structure to the connective tissue of the body. Type II collagen is found in the cartilage that makes up most of the skeleton during embryonic development and in connective tissue that supports muscles, joints, organs, and skin. Additionally, it is found in the clear fluid that fills the eyeball, inner ear, and the area between the discs in the vertebrae. Three pro-alpha1(II) chains twist together to make a strong triple-stranded rope-like molecule called procollagen. The procollagen is then processed to form long, thin fibrils that cross-link in spaces around cells, resulting in the strong material.

Mutations in *COL2A1* cause Stickler syndrome. The mutations either stop the production of the pro-alpha1(II) chain or make a very short, dysfunctional chain. Because of the gene mutations, only some of the collagen is made, causing the conditions associated with joints, facial features, hearing loss, and vision problems. *COL2A1* is inherited in an autosomal dominant pattern and is located on the long arm (q) of chromosome 12 at position 13.11.

COL9A1

Mutations in the *COL9A1* gene, know officially as the "collagen, type IX, alpha 1" gene, cause type 2 Stickler syndrome. Normally, *COL9A1* instructs for a molecule called type IX collagen. This family of collagens has the same function as the other collagens but is found especially in cartilage. Type IX is made up of three proteins made from three genes: one α1(IX) chain, which is produced from the *COL9A1* gene; one α2(IX) chain, which is produced from the *COL9A2* gene; and one α3 (IX) chain, which is produced from the *COL9A3* gene. This type of collagen is more flexible than other types and probably acts as a bridge between other types of collagen, such as type II.

Researchers have found one mutation in the *COL9A1* gene in a family in Morocco that stops the instruction for making the type IX collagen. The family has overly flexible joints, facial features, hearing loss, and severe nearsightedness. *COL9A1* is inherited in an autosomal dominant pattern and is located on the long arm (q) of chromosome 6 at 12-q14.

COL11A1

Mutations in the gene *COL11A1* gene, officially called the "collagen, type XI, alpha 1" gene, cause type 3 Stickler syndrome. Normally, *COL11A1* instructs for making part of type XI collagen called the pro-alpha1(XI) chain. This collagen is found is the same places as other collagens and especially in the clear gel in the eye, the inner ear, and vertebral discs. The pro-alpha1(XI) chain combines with two other collagen chains (pro-alpha2[XI] and pro-alpha1[II]) to form a procollagen molecule and then to become the long, strong fibrils that link to form strong collagen.

Mutations in *COL11A1* gene cause Stickler syndrome. Most mutations involve a change in one building block and others are the result of skipped segments of DNA when the protein is made. An abnormally short molecule causes the problems and symptoms of Stickler syndrome. *COL11A1* is inherited in an autosomal dominant pattern and is located on the short arm (p) of chromosome 1 at position 21.

COL11A2

Mutations in the *COL11A2* gene, officially known as the "collagen, type XI, alpha 2" gene, cause type 4 Stickler syndrome. Normally, COL11A2 instructs for one part of the type XI collagen called the pro-alpha2(XI) chain. Like the previous types of collagen, type XI is important in the development of the early skeleton and in the formation of muscles, joints organs, and skin. The formation is the same as the other types of collagen as it forms part of the triple-stranded chain that then makes the mature type XI collagen fibers.

Mutations in *COL11A2* lead to the abnormal production of the pro-alpha2(XI) chain, part of type XI collagen, resulting in a disruption of the proper formation. However, this mutation does not appear to affect the vision as do the other mutations in the collagen chain. In addition, *COL11A2* is inherited in a different pattern. It appears to be inherited in an autosomal recessive pattern and is located on the short arm (p) of chromosome 6 at position 21.3.

What Is the Treatment for Stickler Syndrome?

Attention to the distinctive facial features is essential. A craniofacial clinic can correct the manifestations of Robin sequence. The vision and hearing problems may be treated with standard surgery, such as retinal attachment.

Further Reading

Robin, Nathaniel H., Rocio T. Moran, Matthew Warman, and Leena Ala-Kokko. 2011.

"Stickler Syndrome." *GeneReviews*. http://www.ncbi.nlm.nih.gov/books/NBK1302. Accessed 5/30/12.

"Stickler Syndrome." 2011. Mayo Clinic. http://www.mayoclinic.com/health/stickler -syndrome/DS00831. Accessed 5/30/12.

Sudden Infant Death Syndrome. *See* MCAD Deficiency

Systemic Scleroderma

Prevalence   About 50 to 300 cases per million; women four times more likely to develop than men

Other Names   familial progressive scleroderma; progressive scleroderma; systemic sclerosis

In 1996, television actor and comic Bob Saget wrote the script for an ABC-TV movie called *For Hope*, starring Dana Delany. The story tells of a young woman who had scleroderma. The movie detailed the struggles of the woman as the disease progressively took its toll and as her body build up scar tissue in her vital organs. The film was based on the experiences of Saget's sister, Gay, who died from the disorder.

What Is Systemic Scleroderma?

Systemic scleroderma is a disorder in which the skin and internal organs become hard and fibrous. The term "scleroderma" comes from two Greek words: *sclero*, meaning "hard," and *derm*, meaning "skin." The hard tissue is called fibrous tissue, which is really a collection of the tough protein collagen. Normally, collagen strengthens and supports all the connective tissue in the body, but in scleroderma, the body's immune system turns against itself and forms extra collagen. The condition is a serious autoimmune disorder.

Following are the symptoms of scleroderma:

- Raynaud phenomenon: This condition is the first sign of the disorder and occurs when the fingers and toes turn white and blue when exposed to cold or other stresses. Blood vessels that carry blood to the extremities constrict. This happening usually occurs several years before the fibrous buildup is noted.
- Puffy and swollen hands: This second symptom of swollen or puffy hands also occurs before the hardening of the skin.
- Hair loss.
- Buildup of hard, fibrous tissue in the skin: The buildup first begins in the fingers, a condition called sclerodactyly. With the tissues in the fingers, sores or ulcers may open up, and painful bumps under the skin or enlarged blood vessels under the skin may occur. Then the symptoms move to the hands and face.
- Buildup of fibrous tissue in other organs: The scar tissue builds up in the lungs, esophagus, heart, and kidneys. The person may experience heartburn, swallowing difficulty, high blood pressure, shortness of breath, kidney problems, and diarrhea.
- Bone and muscle pain.

Scleroderma is found more often in females. Symptoms start between the ages of 30 and 50. Some people that get scleroderma have a history of being around silica dust or polyvinyl chloride.

There are three types of systemic scleroderma, defined by tissues that are affected. The three types are as follows:

- Limited cutaneous scleroderma: This type at one time was called CREST syndrome. CREST is an acronym that stands for calcinosis, Raynaud phenomenon, esophagus issues, sclerodactyly, and telangiectasia (blood vessels under the skin). The fibrous tissue development is usually limited to skin and face. This type is the least severe.

- Diffuse cutaneous systemic scleroderma: The condition progresses rapidly and widespread organ damage occurs.

- Systemic sclerosis sine scleroderma: The word *sine* means without in Latin. In this condition, organs may be affected but not the skin.

Scleroderma is an autoimmune disease, in which the body's immune system turns on itself. About 15% to 25% of the people with this disorder also have disorders that affect the connective tissue, such as lupus erythematosus, rheumatoid arthritis, and Sjögren syndrome.

What Are the Genetic Causes of Systemic Scleroderma?

Changes in several genes that are members of the human leukocyte antigen (HLA) complex are associated with scleroderma. The HLA complex is the one that helps the body recognize its own proteins from those made by outside invaders, such as bacteria or viruses. The HLA genes appear to affect the risk of developing the disorder. However, scientists have pinpointed two genes that are suspect in specific types. These two genes are especially present in people with certain types of the disorder: the *IRF9* gene and the *STAT4* gene.

IRF9

Mutations in the *IRF9* gene, officially known as the "interferon regulatory factor 5" gene, are associated with the diffuse cutaneous systemic scleroderma type. Normally, *IRF5* produces a protein called the interferon regulatory factor 5 (IRF5). This protein is a transcription factor that binds to a section of DNA and controls the activity of other genes. A virus in the cell activates the *IRF5* gene, leading it to produce the IRF5 protein. It then binds to areas of DNA to produce interferons and other cytokines to fight infection and regulate immune system activity. Specifically, interferons block the replication of viruses and activate other immune system cells called natural killer cells.

Changes in the *IRF5* gene may contribute to the risk of diffuse cutaneous systemic scleroderma. The exact process is not clearly understood, but it is thought that both genetic and environmental factors play a role in the development of the disorder. The appearance of systemic scleroderma appears to be sporadic with no family history of the disorder. *IRF5* is located on the long arm (q) of chromosome 7 at position 32.

STAT4

Variations in the *STAT4* gene, officially known as the "signal transducer and activator of transcription 4" gene, are associated with limited cutaneous systemic scleroderma. Normally, *STAT4* provides instructions for another transcription factor protein, which is activated by cytokines. The activated *STAT4* protein then increases the activity of the T cells to become specialized T cells that fight off foreign invaders.

In this type of scleroderma, *STAT4* does not work properly in response to pathogens but attacks its own body cells. Both genes and environment appear to play a role in the development of the condition. *STAT4* is located on the long arm (q) of chromosome 2 at position 32.2-q32.3.

What Is the Treatment for Systemic Scleroderma?

Although no specific treatment exists for scleroderma, certain medications may control the symptoms. Following are some examples of medications:

- Powerful anti-inflammatory medicines called corticosteroids
- Immune-suppressing medications such as methotrexate and Cytoxan
- Nonsteroidal anti-inflammatory drugs (NSAIDs)

Other treatments for specific symptoms may include:

- Medicines for heartburn or swallowing problems
- Blood pressure medications (particularly ACE inhibitors) for high blood pressure or kidney problems
- Light therapy to relieve skin thickening
- Medicines to improve breathing
- Medications to treat Raynaud's phenomenon

Treatment usually also involves physical therapy. Support groups also play a role in helping people cope. Lung problems are the most common cause of death.

Further Reading

"Scleroderma." 2011. MedicineNet.com. http://www.medicinenet.com/scleroderma/article.htm. Accessed 12/24/11.

"Scleroderma." 2011. MedlinePlus. National Institutes of Health (U.S.). http://www.nlm.nih.gov/medlineplus/ency/article/000429.htm. Accessed 12/24/11.

"Systemic Scleroderma." 2011. Genetics Home Reference. National Library of Medicine (U.S.). http://ghr.nlm.nih.gov/condition/systemic-scleroderma. Accessed 12/24/11.

T

Tangier Disease

Prevalence Very rare; only 100 cases identified worldwide; many cases may go undiagnosed

Other Names A-alphalipoprotein neuropathy; alpha high density lipoprotein deficiency disease; analphalipoproteinemia; cholesterol thesaurismosis; familial high density lipoprotein deficiency disease; familial hypoalphalipoproteinemia; HDL lipoprotein deficiency disease; lipoprotein deficiency disease, HDL, familial; Tangier disease neuropathy; Tangier hereditary neuropathy

Tangier is an island off the coast of Virginia in the Chesapeake Bay. Until recent years, its inhabitants were quite isolated from the mainland. In the 2010 census, there were 727 inhabitants. Many of these people still speak a dialect, perhaps, some believe, from the English Restoration period, when the settlement was first founded in the seventeenth century. In addition, Tangier Island is the location of individuals with a distinctive metabolic disorder.

Teddy and Elaine Laird, occupants of Tangier Island, were brought to their doctor who noted strange orange-colored tonsils, combined with enlarged liver and spleen. Researchers began to study this unusual disorder and noted the absence of high-density lipoproteins or high-density cholesterol. Because of this condition, researchers began research into statins and cholesterol. Teddy, who was five years old when he was first diagnosed, died of complications from the disorder at the age of 56 in 2011. The disorder became known as Tangier disease because of its discovery among the inhabitants of this island.

What Is Tangier Disease?

Tangier disease is a metabolic involving reduced levels of certain high-density lipoproteins called HDLs in the bloodstream. Because high levels appear to reduce the chances of developing heart disease, HDL is often called "good cholesterol." Its role is to carry cholesterol and fats called phospholipids to the liver for removal from the blood. On Tangier Island, researchers connected the occurrence of low HDLs and an increased risk of heart disease.

Following are the signs and symptoms of Tangier disease:

- Enlarged orange-colored tonsils
- Slightly elevated amount of fats or triglycerides in the blood
- Disturbances in nerve function, a condition called neuropathy
- Decreased strength
- Development of atherosclerosis, accumulated fat deposits and scarring in the arteries
- Premature cardiovascular disease
- Enlarged spleen
- Enlarged liver
- Abdominal pain and diarrhea
- Collections of yellow patches in the lining of both the intestines and the rectum
- Clouding of the cornea of the eye
- Type II diabetes

The condition is first noted in childhood and ranges from mild to severe.

What Is the Genetic Cause of Tangier Disease?

Mutations in the *ABCA1* gene, officially known as the "ATP-binding cassette, sub-family A (ABC1), member 1" gene, cause Tangier disease. Normally, this gene instructs for proteins that carry molecules from the inside of the cell and across the cell membrane. Produced in the liver and in immune system cells called macrophages, the protein moves cholesterol and certain fats across the cell membrane to outside the cell. A protein called apolipoprotein A1 (apoA1) then carries the cholesterol and fats in the bloodstream. ApoA1 is used to make the good-cholesterol high-density lipoprotein or HDL. HDL carries the substances to the liver, where they are removed from the blood. Removing this extra cholesterol from the blood is very important in having a healthy heart.

More than 30 mutations in *ABCA1* cause Tangier disease. Most of the mutations are changes in one amino acid building block. The mutations keep the cholesterol and phospholipids from being released from the cells. The toxic substances build up in certain tissues such as the tonsils, giving them the characteristic orange color. The damaged cells then die. Because of the inability to get the cholesterol and harmful substances out of the cells, people with the disorder have very low levels of HDL and increased risk of heart disease. *ABCA1* is inherited in an autosomal recessive pattern and is located on the long arm (q) of chromosome 9 at position 31.1.

What Is the Treatment for Tangier Disease?

Finding this cholesterol transport gene has led to an understanding of the relationship between the good cholesterol or HDL and heart disease. However, because the disease is so rare, detection can be a problem. Many people are misdiagnosed with other diseases such as Hansen's disease (leprosy) and given unnecessary medicines. Although no cure exists, treatment for symptoms does vary from case to case. Heart surgery and organ replacement may be necessary.

Further Reading

"Tangier Disease." 2011. *Genes and Disease.* http://www.ncbi.nlm.nih.gov/books/NBK22201. Accessed 5/30/12.

"Tangier Disease." 2011. Tangier Island Guide. http://www.tangierisland-va.com/tangierdisease. Accessed 10/20/11.

"What Is Tangier Disease?" 2011. About.com. http://cholesterol.about.com/od/hypolipidemia/a/tangierdisease.htm. Accessed 5/30/12.

TAR Syndrome (Thrombocytopenia-Absent Radius)

Prevalence   Affects fewer than 1 in 100,000 newborns

Other Names   chromosome 1q21.1 deletion syndrome, 200-KB; radial aplasia-amegakaryocytic thrombocytopenia; radial aplasia-thrombocytopenia syndrome

First identified in 1956, the condition thrombocytopenia-absent radius was simply called TAR. The word "thrombocytopenia" comes from three Greek roots: *thrombo*, meaning "clot"; *cyto*, meaning "cell"; and *penia*, meaning "lack of." The word then means lack of cells that cause blood clotting.

The "AR" part of TAR indicates the absence of the radial bone (radius) in the arm. About 13 years after the first case was identified, TAR was used to describe the condition. At that time, three families with nine newborns were noted with severe bruising and abnormally short forearms.

What Is TAR Syndrome (Thrombocytopenia-Absent Radius)?

Thrombocytopenia-absent radius (TAR) syndrome is a condition that pairs abnormal blood clotting and an absence of the radial bone in each forearm. The combination of the two conditions has the following symptoms:

- Thrombocytopenia: This condition prevents normal blood clotting. The person bruises easily and has frequent nosebleeds. The problem arises with a shortage of blood platelets, part of the clotting mechanism in the blood. The condition is life-threatening if severe bleeding occurs in the brain and other organs during the first year of life. If children survive the first year without hemorrhaging, they may expect to have a normal life with normal intellectual development.

• Skeletal abnormalities: In addition to the missing radius in the forearm, people with TAR may also have other skeletal problems. Other bones in the arms and legs may not develop properly. The face may have some unusual features including a very small jaw, a prominent forehead, and low-set ears.

• Other developmental disorders: People with TAR may also have malformation of the heart and kidney and digestive disorders. Many of the people with TAR have difficulty digesting cow's milk.

What Is the Genetic Cause of TAR Syndrome (Thrombocytopenia-Absent Radius)?

Deletions in chromosome 1 cause TAR. This chromosome is the largest human chromosome and has more than 3,000 genes, which have different roles in the body. The chromosome has about 8% of the total DNA in cells. People with TAR have a deletion in the long arm (q) of chromosome 1 at position 21.1. This section of the chromosome represents about 200,000 DNA building blocks and has about 11 genes. Loss of genes in this region is probably responsible for the symptoms of TAR. Researchers have identified the 11 genes in this area. However, other genetic components may be involved.

The inheritance pattern of TAR syndrome is complex. In some families, the deletion of the genetic material appears to be passed through the family in a dominant pattern. Some deletions may occur with no family history of the disorder, and not all people who have the deletion develop the condition. Even in a family, one member may not be affected while others have the condition.

What Is the Treatment for TAR Syndrome (Thrombocytopenia-Absent Radius)?

Treatments for people with TAR include platelet transfusions to help with the clotting mechanism in the blood. New technology has enabled transfusions to be performed before birth, which has decreased infant mortality during the important first year of life. Surgery for normalizing the appearance of bones in the arm is controversial.

Further Reading

"Chromosome 1." 2011. Genetics Home Reference. National Library of Medicine (U.S.). http://ghr.nlm.nih.gov/chromosome/1. Accessed 5/30/12.

"Thrombocytopenia-Absent Radius Syndrome." 2011. Genetics Home Reference. National Library of Medicine (U.S.). http://ghr.nlm.nih.gov/condition/thrombocytopenia-absent -radius-syndrome. Accessed 5/30/12.

"Thrombocytopenia-Absent Radius Syndrome." 2011. Medscape. http://emedicine
.medscape.com/article/959262-overview. Accessed 5/30/12.

Tay-Sachs Disease (TSD)

Prevalence   Rare in the general population; about 1 in 27 people of Ashkenazi
Jewish heritage are carriers; mutations found in some French
Canadian communities of Quebec and the Cajun population of
Louisiana; found in the Old Order Amish of Pennsylvania

Other Names   B variant GM2 gangliosidosis; GM2 gangliosidosis, type 1; HexA
deficiency; hexosaminidase A deficiency; hexosaminidase alpha-
subunit deficiency (variant B); sphingolipidosis, Tay-Sachs; TSD

In 1881, Warren Tay, a British ophthalmologist, observed a red spot on the retina
of a young patient. Three years later, he noted three cases in a single family that
were members of the Eastern European Jewish population. Later that decade,
Bernard Sachs, an American neurologist, found similar cases of the disorder, along
with intellectual disability. For a while, the disorder was thought to be only a
Jewish disease. The condition was named after the two doctors: Tay-Sachs disease.

In the twentieth century, the disorder was found to be a genetic mutation of the
HEXA gene, which also had a large number of mutations with many new ones that
were later discovered. Mutations have been reported with significant frequencies
in a population of French Canadians of Quebec, who migrated to the Acadian or
Cajun region of Louisiana.

What Is Tay-Sachs Disease (TSD)?

Tay-Sachs is a progressive disorder of the nervous system. Cells called neurons in
the brain and spinal cord slowly deteriorate and die. There are three forms of the
disorder, which appear at different stages of life.

Infantile TSD

The infant appears normal for about six months, and then certain signs begin to
appear. Following are some of the symptoms of infantile TSD:

- Lost motor skills—infants lose what skills they do have, such as rolling over,
  sitting up, and crawling
- Characteristic red spot on the eye
- Startling effect to loud noises
- Seizures

- Progressive hearing and vision loss
- Intellectual disability
- Paralysis

Children with this type seldom live past four years of age because the nerve cells are damaged.

Juvenile TSD

This type is very rare and occurs in children between the ages of 2 and 10. They may have some of the same symptoms as the infantile type with the following added symptoms:

- Loss of motor skills—because these children are older and have attained more skills, the loss is quite noticeable; they lose both cognitive and motor skills
- Speech difficulties—the children have developed normal speech patterns, then begin to slur words as the condition progresses
- Swallowing difficulties
- Unsteady gait
- Stiff joints

Children with this type usually die between 5 and 15 years of age.

Late Onset TSD or LOTS

This very rare type occurs in people in their early 20s or 30s. Frequently, this disease is mistaken for other conditions, such as Friedrich Ataxia. Following are the symptoms of LOTS:

- Unsteady gait: The person who has lived a normal life begins to notice problems with walking and motor skills.
- Neurological deterioration: The person begins to notice cognitive decline.
- Speech and swallowing difficulties.
- Psychiatric illness: The person may develop a schizophrenic-like psychosis.

This form is seldom fatal. The person may become wheelchair bound but can learn to cope. Psychiatric illness can be controlled with medication.

What Is the Genetic Cause of Tay-Sachs Disease (TSD)?

Mutations in the *HEXA* gene, officially called the "hexosaminidase A (alpha polypeptide)" gene, cause Tay-Sachs disease. Normally, *HEXA* provides instructions for making part of the enzyme beta-hexosaminidase A. One alpha unit joins with a beta subunit, which is produced by the HEXB gene to make an important working enzyme.

Beta-hexosaminidase A plays an essential role in the brain and spinal cord. It is found in the lysosomes, the recycling plant of the cells. Here in the lysosomes, the enzyme is part of a compound that breaks down a fatty substance called GM2 ganglioside. The breaking down of the fatty complex is important for health.

Over 120 mutations in *HEXA* cause TSD. The mutations either stop or reduce the activity of beta-hexosaminidase A. The fatty complex called GM2 ganglioside is not broken down, allowing toxic materials to build up, especially in the nerve cells. The gradual build up in these cells causes the progressive symptoms of TSD. Most of the mutations cause a completely nonworking form of the enzyme and hence the severe form that occurs in infancy. *HEXA* is inherited in an autosomal dominant pattern and is located on the long arm (q) of chromosome 15 at position 24.1.

What Is the Treatment for Tay-Sachs Disease (TSD)?

Because TSD is a serious genetic disorder, there is no cure for the infantile or juvenile types. The parent tries to keep the child as comfortable as possible and treat symptoms. The psychotic episodes of LOTS may be controlled with medication.

Further Reading

National Tay-Sachs and Allied Disease Foundation. 2011. http://www.ntsad.org. Accessed 12/2/11.

"Tay-Sachs Disease." 2011. Genetic and Rare Diseases Information Center (GARD). National Institutes of Health (U.S.). http://rarediseases.info.nih.gov/GARD/Disease.aspx?PageID=4&DiseaseID=7737. Accessed 12/2/11.

"Tay-Sachs Disease." 2011. Genetics Home Reference. National Library of Medicine (U.S.). http://ghr.nlm.nih.gov/condition/tay-sachs-disease. Accessed 12/2/11.

Thalassemia, Beta Type. *See* Beta Thalassemia

Thanatophoric Dysplasia

Prevalence   Occurs in 1 in 20,000 newborns; type I more common than type II.

Other Names   dwarf, thanatophoric; thanatophoric dwarfism; thanatophoric short stature

There are many different kinds of dysplasias or abnormal development of limbs. Many dysplasias are related to dwarfism as is thanatophoric dysplasia. The word "thanatophoric" comes from two Greek words: *thanatos*, meaning "death," and *phor*, meaning "to bear." Some children with this disorder are stillborn or dead at birth; however, others, with care, can live many years.

What Is Thanatophoric Dysplasia?

Thanatophoric dysplasia is a malformation of the skeletal system, combined with extra folds of skin on the arms and legs. Two major forms exist. In type I, the thigh bones are curved and the bones of the spine are flattened; type II includes the same features in addition to a large cloverleaf head and distinct facial features and chest and lung disorders.

Thanatophoric dysplasia or TD is suspected when very short bones appear in prenatal ultrasound examination. Following are the signs of TD:

• Shortening of long bones, seen about 12 to 14 weeks. During this first trimester, several short bones and a narrow chest may be observed.

• During ultrasound in the second and third trimester, the bowed femurs of type I are observed; the cloverleaf skull of type II may be seen.

• After birth, the short arms and legs.

• Folds of extra skin on the arms and legs.

• Narrow chest and underdeveloped lungs.

• Enlarged head.

• Distinct face with large forehead and large, wide-spaced eyes.

What Is the Genetic Cause of Thanatophoric Dysplasia?

Mutations in the *FGFR3* gene, officially known as the "fibroblast growth factor receptor 3" gene, cause thanatophoric dysplasia. Normally, *FGFR3* instructs for the protein fibroblast growth factor receptor 3. The protein, which is a member of a large family, is essential for several cell processes, including cell growth, division, and determination of types of cells. The protein is also related to embryonic development, blood vessel formation, and wound healing. The FGFR3 protein is a transcription factor, in which one end remains inside the cell and the other end is outside the cell to receive signals that are in the bloodstream. When these factors are transmitted into the cell, changes go on in the cell that provide for normal growth.

About 10 mutations in the *FGFR3* gene cause thanatophoric dysplasia type I. Most of the mutations substitute arginine for the amino acid cysteine. Other mutations cause the protein to be longer than normal. Only one mutation causes type II. In this mutation, glutamic acid replaces lysine in the protein. Both types of mutations cause the FGFR3 receptor to be overactivated, leading to the problems of bone growth seen in thanatophoric dysplasia. *FGFR3* is inherited in an autosomal dominant pattern and is located on the short arm (p) of chromosome 4 at position 16.3.

What Is the Treatment for Thanatophoric Dysplasia?

The child must be kept in the neonatal unit for possible treatment of respiratory distress. The infant must be kept warm, comfortable, and nourished properly.

Further Reading

"*FGFR3*." 2011. Genetics Home Reference. National Library of Medicine (U.S.). http://ghr.nlm.nih.gov/gene/FGFR3. Accessed 5/30/12.

"Thanatophoric Dysplasia." 2011. International Birth Defects Information Center. http://www.ibis-birthdefects.org/start/thanatop.htm. Accessed 5/30/12.

"Thanatophoric Dysplasia." 2011. Medscape. http://emedicine.medscape.com/article/949591-overview#a0199. Accessed 5/30/12.

Three-M (3M) Syndrome. *See* 3-M Syndrome

3-M Syndrome

Prevalence   Rare; about 50 individuals identified

Other Names   dolichospondylic dysplasia; Le Merrer syndrome; 3-MSBN; three-M slender-boned nanism; three M syndrome; Yakut short stature syndrome

Rarely is a disorder named with just letters to identify it, but this condition has been named using the initials of three people. Previously, several names were used to describe this condition, in which the child is very short at birth and has a sad-looking countenance. The gloomy face or dolichospondylic dysplasia attracted the attention of the researchers: J. D. Miller, Victor McKusick, and P. Malvaux. The condition became known as 3-M syndrome, using the initials of the last names of these three researchers.

What Is 3-M Syndrome?

3-M syndrome is a disorder of the skeletal system. The child has very short stature or dwarfism, identifiable facial features, and a host of skeletal deformities. People with 3-M grow extremely slowly before birth, and slow growth continues throughout childhood and adolescence. Their adult height seldom reaches more than 4 feet, 6 inches.

Following are other symptoms of 3-M:

- Normal-sized head that is much larger than the rest of the body. The head may be very long and narrow, a condition known as dolichocelphaly.

- Face shaped like an upside-down triangle with the forehead appearing as the base of the triangle. The chin is pointed and the middle of the face is less prominent.
- Other facial features: full eyebrows; upturned nose with a large tip; very long area between the nose and couth (called the philtrum); large mouth, full lips, and large ears.
- Short, wide neck.
- Short wide chest with square shoulders and very prominent shoulder blades.
- Abnormally curved spine with a round upper back that curves to the side or an exaggerated curve in the lower back.
- Unusual curved fingers with a very short fifth finger.
- Prominent heels.
- Loose joints.

X-rays reveal that the long bones in the arms and legs are very thin and fragile. Also, the vertebrae are narrow and weak.

Other systems may be affected in 3-M syndrome. The sex hormones do not function normally, causing the sex organs to be underdeveloped. The urethra, where the male urinates, may be on the underside of the penis, a condition known as hypospadias. In addition, the person may be at risk for aneurysms, or ballooning out of blood vessels, especially in the brain. Although there are such a large number of physical disabilities with 3-M syndrome, the person has normal intelligence and can have a normal life expectancy.

What Is the Genetic Cause of 3-M Syndrome?

Mutations in the *CUL7* gene, known officially as the "cullin 7" gene, cause 3-M syndrome. Normally, *CUL7* instructs for the protein cullin-7 that is important for the process that breaks down and destroys unwanted substances. The machinery, called the ubiquitin-proteasome system, is active in many genetic disorders. The cullin-7 protein assembles an enzyme complex known as E3 ubiquitin ligase, which identifies and tags damaged and unneeded cells. It then signals structures known as proteasomes to come and destroy or recycle the unwanted parts. This system is like a garbage disposal and recycling plant and is the quality control system of the body. The complex also regulates certain cell growth activities.

About 25 mutations have been seen in people with 3-M syndrome. Most of the mutations are created with the exchange of only one building block. The gene mutations create a protein that is short and does not work properly. The tagging of the unwanted proteins is disrupted and damaged proteins build up in the cells, causing the unique symptoms of 3-M syndrome. *CUL7* is inherited in an autosomal recessive pattern and is located on the short arm (p) of chromosome 6 at position 21.1.

What Is the Treatment for 3-M Syndrome?

Specific symptoms of 3-M determine the best course of treatment and supportive devices. The disorder, which is identified usually before or at birth, requires the coordinated efforts of a team of specialists—pediatricians, orthopedists, dental specialists, and others who will determine the best course. The individual with 3-M may have corrective surgery for various symptoms and malformations. Most of the treatment is supportive and symptomatic.

Further Reading

"3-M Syndrome." 2011. Genetics Home Reference. National Library of Medicine (U.S.). http://ghr.nlm.nih.gov/condition/3-m-syndrome. Accessed 5/1/12.

Espinasse, Muriel Holder. 2012. "3-M Syndrome." *GeneReviews*. http://www.ncbi.nlm.nih.gov/books/NBK1481. Accessed 2/6/12.

Madisons Foundation. http://www.madisonsfoundation.org/index.php/component/option,com_mpower/Itemid,49/diseaseID,402. Accessed 5/1/12.

Tourette Syndrome

Prevalence At one time considered rare; affects between 1 and 10 children per 1,000; many cases undiagnosed; more prevalent in males; seen in all ethnic groups; affects about 200,000 Americans

Other Names chronic motor and vocal tic disorder; Gilles de la Tourette's syndrome; Gilles de la Tourette syndrome; GTS; TD; Tourette disorder; Tourette's disease; TS

Georges Albert Edouard Brutus Gilles de la Tourette, a French physician and neuropsychiatrist, was known as an expert on hypnotism, hysteria, and epilepsy. He first noted the condition in an 85-year-old French noblewoman. In 1884, he wrote a paper describing nine patients who had unusual involuntary movements and outbursts. Some of the outbursts were "notorious cursing." He believed the disorder to be hereditary but not progressive and degenerative. He also noted that the tics could come without warning, but stress or going without sleep would make the attacks more frequent. His friend and mentor Jean-Martin Charcot named the disorder after Tourette, who died at the age of 45. The disorder that Tourette described and studied is now called Tourette syndrome.

What Is Tourette Syndrome?

Tourette syndrome (TS) is a disorder of the nervous system characterized by motor and vocal tics. A tic is a habitual spasm involving repetitive movements.

These tics usually come and go. The signs and symptoms of Tourette syndrome may include the following:

- First noticed in childhood, with onset between ages of 7 and 10.

- Symptoms most severe in adolescents and may become milder as person ages; the severity appears to vary over time.

- Simple motor tics: Simple motor tics are brief, sudden movements, which involve a small number of muscles. Simple tics include rapid eye blinking, facial grimaces, shoulder shrugging, head or shoulder jerking, or nose twitching.

- Complex motor tics: Complex motor tics are those that involve many muscle groups and are coordinated patterns of movements. Complex tics include jumping, kicking, hopping, bending, twisting, spinning, or sniffing or touching objects. Disturbing tics such as punching oneself in the face or other forms of self-harm are possible but rare.

- Vocal tics: Sudden involuntary noises include grunting, sniffing, throat-clearing, repeating others' words, or repeating own words. Use of obscene language is possible but uncommon.

- Behavioral disorders: Tourette syndrome is occasionally related to attention deficit hyperactivity disorder (ADHD), obsessive-compulsive disorder (OD), depression, anxiety, and problems with sleep.

- Environment: Tics appear less frequently during calm and focused activities. They may worsen with excitement. Also, certain things like tight collars or hearing another person sniff or clear the throat may evoke the tic. Tics may also be present during sleep but are less severe.

- Prognosis: People with TS have normal intelligence and have normal life spans.

Most people with TS improve in late teen and early adulthood; however, about 10% have a progressive course that lasts into adulthood. Some people can learn to suppress, camouflage, or manage the tics, although they are involuntary.

What Is the Genetic Cause of Tourette Syndrome?

Mutations in the *SLITRK1* gene, officially known as the "SLIT and NTRK-like family, member 1" gene, may cause Tourette syndrome. Normally, *SLITRK1* provides instructions for making a protein that is a member of the SLITRK family found in the brain. The protein appears to be essential for growth of nerve cells and is especially essential to help the axons and dendrites that allow nerve cells to communicate.

Mutations in the *SLITRK1* gene have been found in some people with TS. One mutation involves only a single exchange of an amino acid building block but leads to a very short version of the protein. Another mutation interferes with the production of the protein. *SLITRK1* is inherited in an autosomal dominant pattern and is located on the long arm (q) of chromosome 13 at position 31.1.

Environment may also play an important part with TS. People with TS may also have genetic risks for other behavioral disorders such as depression or substance abuse.

What Is the Treatment for Tourette Syndrome?

Many scientists and neuroscientists are studying Tourette syndrome. Although the exact cause is not known, current research does indicate that in certain areas of the brain, including the basal ganglia, frontal lobes, and cerebral cortex, the neurotransmitters may not function properly. Neurotransmitters such as dopamine, serotonin, and epinephrine are responsible for communication among cells. Because of the complex nature of the symptoms of TS, finding the cause is also likely to be very complex.

Students with TS can function in a normal classroom, but the tics combined with other disorders, such as ADHD or learning disabilities, may interfere with their education. The ideal setting is the least restrictive environment; but sometimes for the child, a special setting or even a special school may be in the best interest.

Most people with TS do not need medication for their tics. If the tic is interfering with their functioning, certain neuroleptic drugs, such as haloperiodol and pimozide, may help with some people. It is important to remember that all drugs can have side effects, which may be more undesirable than the tic. Although TS is not a psychological problem, psychotherapy may help the person understand the disorder and deal with secondary issues that may be present.

The National Institute of Neurological Disorders and Stroke (NINDS), a branch of the National Institutes of Health, is supporting and conducting research on TS and other neurological disorders.

Further Reading

"Gilles de la Tourette Syndrome (GTS)." 2011. http://www.tourette-syndrome.info/gts.pdf. Accessed 5/30/12.

National Tourette Syndrome Association. 2011. http://www.tsa-usa.org. Accessed 5/30/12.

"Tourette Syndrome Fact Sheet." National Institute for Neurological Disorders and Stroke (U.S.). http://www.ninds.nih.gov/disorders/tourette/detail_tourette.htm. Accessed 5/30/12.

Treacher Collins Syndrome

Prevalence   Affects about 1 in 50,000 people

Other Names Franceschetti-Zwahlen-Klein syndrome; mandibulofacial dysostosis (MFD1); Treacher Collins–Franceschetti syndrome; zygoauromandibular dysplasia

In 1900, Edward Treacher Collins, an English surgeon and ophthalmologist, noted some patients with abnormal facial conditions. Later in 1949, Franceschetti and Klein described the same condition but called it mandibulofacial dysotosis. The later name is now used to describe the clinical symptoms, but Treacher Collins is used to identify the condition. Note the name does not have a hyphen as many genetic disorders do when two people are involved, because Treacher Collins is the name of one person.

What Is Treacher Collins Syndrome?

Treacher Collins syndrome is a condition affecting the face. Bones and tissues of the face may be underdeveloped, causing the person to have a distinct facial appearance. The syndrome is caused by a defective protein called treacle and is passed down from one generation to the next. The symptoms may vary in severity and many include the following:

- Underdeveloped facial bones, especially the cheek bones, which may appear flat
- A very small jaw and chin
- Very large mouth
- Outer part of ears missing or are abnormal
- Hearing loss
- Scalp hair that reaches to the cheeks
- Abnormal eye shape
- Defect in lower eyelid; lashes may not be on lower eyelid
- Cleft palate in roof of mouth
- Restricted airway, which may be life-threatening

People with Treacher Collins can have a normal life span and have normal intelligence.

What Is the Genetic Cause of Treacher Collins Syndrome?

Mutations in the *TCOF1* gene, officially known as the "Treacher Collins-Franceschetti syndrome 1" gene, cause Treacher Collins syndrome. Normally, *TCOF1* instructs for a protein called treacle. This protein is essential in the embryonic development of bones and other tissues of the face. Researchers have found that treacle is part of the production of a molecule called ribosomal RNA or rRNA within the nucleolus of the cells. The chemical helps make the protein building blocks or amino acids into working proteins.

About 150 mutations in *TCOF1* cause Treacher Collins syndrome. Most of the mutations either add or reduce a small number of building blocks, leading to a reduction in the production of treacle. When the protein is not present, signal cells allow the facial bones to self-destruct. The abnormal cell death leads to the symptoms that affect the face and adjoining tissues seen in Treacher Collins. *TCOF1* is inherited in an autosomal dominant pattern and is located on the long arm (q) of chromosome 5 at position 32-q33.1.

What Is the Treatment for Treacher Collins Syndrome?

Depending upon the severity of the symptoms, some of the manifestations of the disorder may be treated. A plastic surgeon may perform a series of corrective surgeries to correct such facial defects as the receding chin. Hearing loss may be treated with specific hearing aids. The cleft palate repair is done at ages one to two years.

Further Reading

Katsanis, Sara Huston, and Ethylin Wang Jabs. 2011. "Treacher Collins Syndrome." *GeneReviews*. http://www.ncbi.nlm.nih.gov/books/NBK1532. Accessed 5/30/12.

"Mandibulofacial Dysostosis (Treacher Collins Syndrome)." 2011. Medscape. http://emedicine.medscape.com/article/946143-overview. Accessed 5/30/12.

"Treacher Collins Syndrome." 2011. Genetics Home Reference. National Library of Medicine (U.S.). http://www.nlm.nih.gov/medlineplus/ency/article/001659.htm. Accessed 5/30/12.

Triosephosphate Isomerase Deficiency (TPI Deficiency)

Prevalence   Very rare; fewer than 100 patients worldwide

Other Names   autosomal recessive inheritance hemolytic anemia, neurological involvement; triose phosphate isomerase deficiency

In 1965, Schneider described a metabolic disorder with severe anemia, heart disorders, and brain dysfunction. Many of the children died in early infancy. Although only a few cases have been reported, the frequency of the recessive gene is the population appears to be between 3% and 8%. Scientists think that TPI deficiency is a more common condition than it appears and that individuals with TPI are stillborn, or the condition is poorly recognized. A unique thing about this condition is that all carriers of the mutation have a common ancestor that lived in France or England more than 1,000 years ago.

What Is Triosephosphate Isomerase Deficiency (TPI Deficiency)?

Triosephosphate isomerase deficiency (TPI deficiency) is one of the inborn errors of metabolism, a metabolic disease that involves many systems. Characteristics of TPI are the following:

- Anemia: This chronic type of anemia is called hemolytic anemia, meaning the red blood cells are broken down. The anemia is early onset and is always present. Fifty percent of the cases appear at birth.

- Jaundice at birth.
- Neurological impairment: This impairment is progressive and starts between 6 months and 30 months of age. Motor defects start in the legs and then progress to other areas.
- Heart disorders: The child develops cardiomyopathy, which is a disease of the muscles of the heart.
- Diaphragm paralysis: The child will need assisted ventilation.
- Many infections: The white blood cells appear abnormal.

Most children with TPI die early; it is rare, but people with less severe cases can live into adulthood.

What Is the Genetic Cause of Triosephosphate Isomerase Deficiency (TPI Deficiency)?

Mutations in the *TPI1* gene, officially called the "triosephosphate isomerase 1" gene, cause TPI deficiency. Normally, *TPI1* encodes for an important housekeeping enzyme called triosephosphate isomerase (TPI). This enzyme is responsible for glycolysis or breaking down sugars and also acts as a catalyst for the interconversion of dihydroxyacetone phosphate and glyceraldehyde-3-phosphate. Interconversion is a process in which these two things are converted into each other. The housekeeping gene affects many processes, especially in embryonic development.

Thirteen mutations in *TPI1* have been found to cause the deficiency. If dihydroxyacetone phosphate accumulates, it damages the red blood cells, causing anemia. Additional reduction of TPI is believed to cause the other symptoms, although the process has not been completely identified. *TPI1* is inherited in an autosomal recessive pattern and is located on the short arm (p) of chromosome 12 at position 13.

What Is the Treatment for Triosephosphate Isomerase Deficiency (TPI Deficiency)?

Managing the symptoms of anemia may require transfusions during the time when the breakdown of red cells is acute. Assisted ventilation may be used for diaphragm paralysis. Treating the progressive neuromuscular disorders is the same as for any progressive disorder of this type.

Further Reading

"*TPI1*." 2011. Genetics Home Reference. National Library of Medicine (U.S.). http://ghr.nlm.nih.gov/gene/TPI1. Accessed 5/30/12.

"Triose-phosphate Isomerase Deficiency." 2011. Orphanet. http://www.orpha.net/data/patho/GB/uk-TPI.pdf. Accessed 10/22/11

"Triosephosphate Isomerase Deficiency Causes, Symptoms and Treatment and Related Disorders." 2008. HealthWise Everyday Health.

http://www.everydayhealth.com/health-center/triosephosphate-isomerase-deficiency.aspx. Accessed 5/30/12.

Triple X Syndrome

Prevalence   About 1 in 1,000 newborn girls; in United States, 5 to 7 girls with XXX are born each day

Other Names   47,XXX; Triplo X syndrome; Trisomy X; XXX syndrome

In 1959, Dr. Patricia Jacobs and a team at the Western General Hospital in Edinburgh, Scotland, published a report in the British journal *Lancet* called "Evidence for the Existence of the Human 'Super Female.' " At the time, the sensational title stirred quite a bit of interest. The team then described how they found an extra X chromosome in the karyotype of a 35-year-old woman, who was 5 feet, 9 inches, weighed 128 pounds, and had premature ovarian failure at the age of 19. Her mother was 41 and her father 40 when she was conceived. Since that time, several other scientists, especially in Denmark, have studied the condition that is known as trisomy X or 47,XXX.

What Is Triple X Syndrome?

Triple X, also known as trisomy X, is characterized by an extra X or female chromosome in each one of the cells. Sometimes, it is designated at 47,XX because each cell has a total of 47 chromosomes, rather than just the normal 46. The extra chromosome is associated with the following characteristics:

- Tall stature
- Learning disabilities
- Developmental delays in sitting, standing, and walking
- Weak muscle tone
- Behavioral difficulties with emotional disorders
- Kidney abnormalities
- Seizures
- No unusual facial features

All the symptoms may vary with individual girls. Some girls have none of the symptoms, except for the tall stature. They may have normal sexual development and are able to conceive children. Some girls have the extra X chromosome in only some of the cells, a condition called mosaicism, or 46,XX/47,XXX mosaicism. These girls are less likely to display symptoms. In extremely rare instances, females may have 48,XXXX or 49,XXXXX. The increase in the number of chromosomes poses more of the risk for learning disabilities, intellectual disorders, birth defects, and other health problems.

What Are the Genetic Causes of Triple X Syndrome?

An extra X chromosome causes Triple X syndrome. Normally, human beings have 46 or 23 pairs of chromosomes, which include 22 pairs of body or somatic chromosomes and two sex chromosomes. The sex chromosomes are known as X and Y. Females have two X chromosomes (46,XX), and males have one X and one Y chromosome (46,XY).

In triple X syndrome, the female has an extra X chromosome, which is designated as 47,XX. The condition is not inherited but is a random event that occurs during the formation of egg or sperm. During the process of meiosis, precursors to the egg and sperm undergo divisions where only half of the 46 chromosomes are packaged in each cell. However, an error in cell division of the reproductive cells can cause an abnormal number. This process is called nondisjunction. As a result, the egg or sperm can have an extra copy of the X chromosome. If an atypical egg or sperm is fertilized, the child will have the extra X chromosome in each cell. If the abnormal division occurs late in the process, the girl may have the extra X only in some cells, causing the mosaic condition.

The X chromosome is a unique sex chromosome in that it is large and has about 5% of the total DNA in cells. Lots of genetic material is carried on the X chromosome, including many rare genetic disorders, such as color blindness. In contrast, the Y chromosome is quite small and determines little except the sex of the child.

What Is the Treatment for Triple X Syndrome?

Because Triple X is a chromosomal aberration, no cure exists. The syndrome is so individual and thus treatment is symptomatic. Behavioral and learning disabilities qualify the girl for special education in school.

Further Reading

"Triple X Syndrome." http://ghr.nlm.nih.gov/condition/triple-x-syndrome. Accessed 11/26/11.

"Trisomy X." 2011. Orphanet. http://www.orpha.net//consor/cgi-bin/OC_Exp.php?Lng=EN&Expert=3375. Accessed 11/26/11.

"X Chromosome." 2011. Genetics Home Reference. National Library of Medicine (U.S.). http://ghr.nlm.nih.gov/chromosome/X. Accessed 11/26/11.

Trisomy 13 (Patau Syndrome)

Prevalence   About 1 in 16,000 newborns; increased risk of having a child with trisomy 13 as woman gets older

Other Names   Bartholin-Patau syndrome; complete trisomy 13 syndrome; Patau's syndrome; Patau syndrome; Trisomy 13 syndrome

In 1657, Thomas Bartholin observed a birth defect in which several anomalies were present. Many of the children died, but those who did live had multiple defects and intellectual disabilities. However, it was not until 1960 that Dr. Klaus Patau found the disorder was caused by an extra chromosome 13.

Questions about the disorder arose when native tribes living in the Pacific Islands where the atomic bomb had been tested developed symptoms of chromosome 13. Radiation from the test was suspect, and the government moved the people to an area away from the site. Later, they moved the people back to the area, not realizing that radiation can linger long after a nuclear explosion. In England and Wales in 2008–2009, about 172 cases were diagnosed, with about 91% diagnosed prenatally. Trisomy 13 is one of the conditions that are tested in newborn screening in most states.

What Is Trisomy 13 (Patau Syndrome)?

Trisomy 13 or Patau syndrome is the least common and most severe of the chromosomal trisomies. Several physical and mental defects are present. Following are the characteristics of this syndrome:

- Body core problems: Heart defects, brain disorders, and spinal cord defects are part of this syndrome.
- Face and head: The child has very small and undeveloped eyes, a condition known as microphthalmia. A cleft lip and/or cleft palate are common.
- Abnormal feet: Extra fingers and toes are usual. Fingers may be held in a flexed position. Misshapened feet resemble a rocker bottom.
- Poor muscle tone.
- Many neural tube defects.
- Heart defects.
- Mental disability.

Infants normally die within the first few days; only 5% to 10% live past the age of a year.

What Is the Genetic Cause of Trisomy 13 (Patau Syndrome)?

Trisomy 13 or Patau syndrome is caused by an extra chromosome 13. Normally, humans have 23 pairs of chromosomes, with two copies of each chromosome including chromosome 13. Chromosome 13 has about 114 DNA building blocks or base pairs, accounting for about 3.5% to 4% of the total DNA. Chromosome 13 has about 600 to 700 genes, compared to a total of 20,000 to 25,000 in the human genome.

Children with trisomy 13 have three copies of the chromosome in their body cells. If the child has the extra copy in only a few cells, the condition is called mosaicism. Sometimes a part of chromosome 13 attaches to another chromosome with part of the material replicating. This process is called translocation.

The extra copies of the chromosome or extra genetic material occur during the early development of the sperm or egg. Scientists believe that the extra genetic material affects the normal course of development and the unusual medical symptoms of the disorders.

What Is the Treatment for Trisomy 13 (Patau Syndrome)?

This most severe of all trisomies has no prevention or treatment. Keeping the child as comfortable as possible is really the only treatment. Several support groups exist for parents of children with trisomy 13.

Further Reading

Living with Trisomy 13. 2010. http://www.livingwithtrisomy13.org. Accessed 5/30/12.

"Patau Syndrome." 2011. Medscape. http://emedicine.medscape.com/article/947706 -overview. Accessed 5/30/12.

"Patau Syndrome (Trisomy 13)." http://medgen.genetics.utah.edu/photographs/pages/ trisomy_13.htm. Accessed 5/30/12.

Trisomy 18 (Edwards Syndrome)

Prevalence   One in 6,000 to 8,000 births; at the time of first-trimester screening, one in 400, but spontaneous loss results in the very low birth incidence; more common in females

Other Names   complete trisomy 18 syndrome; Edwards syndrome; trisomy 18 syndrome

John Hilton Edwards was a pediatrician and geneticist in both England and at the Children's Hospital in Philadelphia. While at Oxford, he noted a severe and complex syndrome that was associated with more than 130 abnormalities. In 1960, Edwards and colleagues published a paper in the British journal *Lancet* called "A New Trisomic Syndrome." That syndrome found an extra chromosome 18 in the children with the abnormalities. The condition was named Edwards syndrome for his findings but is generally called trisomy 18.

What Is Trisomy 18 (Edwards Syndrome)?

Trisomy 18 (47XX+18), or Edwards syndrome, is a disorder caused by the presence or all or part of an extra 18th chromosome. The additional chromosome happens before conception, during meiosis or the formation of sperm or egg. The effects of the extra chromosome vary and depend on the extent of the extra copy, genetic history, or just plain chance. Like that of Down syndrome (a repeat of chromosome 21), chances of chromosome aberration increase with maternal age. The condition is three times more common in girls than in boys.

The symptoms may vary in number and severity. Following are some of the manifestations of the disorder:

- At birth, a characteristic clinched hand with index finger overriding the middle finger and fifth finger overriding the fourth finger
- Crossed legs
- Feet with rounded bottom, a condition called rocker-bottom feet
- Distinct strawberry shaped head; small head, low-set, malformed ears, very small jaw, cleft lip and or palate, widely spaced eyes, epicanthic folds; drooping of upper eyelids
- Underdeveloped fingernails
- Undescended testicles in male
- Unusual shaped chest
- Heart defects
- Kidney problems
- Part of the stomach outside the intestinal tract
- Esophagus not connected to stomach
- Severe developmental delays and possible mental retardation

With so many serious conditions, 5% to 10% do not live past their first birthday.

What Are the Genetic Causes of Trisomy 18 (Edwards Syndrome)?

Trisomy 18 is caused by having three copies or parts of copies of chromosome 18. Normally, the egg and sperm have individual chromosomes that contribute the 23 pairs needed to form a normal individual with 46 chromosomes. Meiosis is the process by which the number of chromosome is reduced to form the chromosomes that are packaged in egg or sperm cells. The process, called oogenesis or spermatogenesis, occurs in two phases—meiosis I and meiosis II. Before meiosis begins, as in mitosis or normal cell division, the centromeres of the cell hold the double-stranded chromosomes together. During meiosis I, the parts line up and then divide to form half the chromosomes still holding the double-strand DNA. With meiosis II, the centromeres split and the strands separate, moving to

Trisomy 18: You're Not Alone

Little Caleb Adamyk had just celebrated his second birthday. He had a group of friends, not from the neighborhood but from around the nation. What was so unusual? Caleb had trisomy 18, or Edwards syndrome, and had so far beaten the odds when most of these children die before their first birthday. Jeannette, the mother of Caleb, has been instrumental in establishing online sources for people with children with trisomy 18. Present also were a child with trisomy 18 who was nearly three, and another child who was "born sleeping" on Christmas Day. The support system that these parents created has taught them how to live with children with the syndrome.

To celebrate, Caleb's parents, Jeannette and Steven Adamyk, hosted a party in Ocala, Florida, and invited several other special-needs children. One mom, Susan Budd, drove from her home in South Carolina to Florida with her daughter Rebekah, who also has Edwards syndrome and was soon expecting her second birthday.

The two mothers met online. Although Edwards syndrome is not common, Budd said she has connected with moms all over the world, and at least four families with trisomy 18 children live in her area. They rejoice with each other over small victories such as a new smile or a new laugh. The Adamyks tell how they kept their artificial Christmas tree in the living room because Caleb loved to look at it. They changed ornaments for each season.

But doctors discovered stones in one of Caleb's kidneys and he needed to have his adenoids removed. An online blog (buddzoo.blogspot.com) encourages parents of others whose child has been born with Edwards syndrome and so many multiple disabilities. The purpose of the site is to help them realize they are not alone. Bella Santorum, the three-year-old youngest daughter of Rick Santorum, onetime 2012 Republican candidate for president, has trisomy 18.

Postnote: Caleb died on August 30, 2011.

Source: Ocala Star Banner, Ocala, FL. http://www.ocala.com/article/20110830/articles/ 110839970. Accessed 6/1/12.

each end of the cell. Then four cells are produced, each with only half the number of chromosomes of the original pair—23 chromosomes. Then when fertilization occurs, nature's arithmetic will be true: 23 chromosomes from the mother and 23 from the father.

With trisomy 18, numerical errors can occur at either of the two meiotic conditions and cause errors in the divisions of the daughter cells. In trisomy 18, a gamete from either sperm of egg is formed with 24 copies of chromosome of chromosome, with three copies of chromosome 18. This results in an extra chromosome, which will make the number 24 rather than 23.

There are three types of trisomy 18:

- Full trisomy 18: This type occurs in about 95% of all cases, and the extra chromosome is found in basically all cells of the body. It is neither hereditary nor due to anything that the parents do during pregnancy.
- Partial trisomy 18: This very rare condition occurs when only part of an extra chromosome is present. A piece of chromosome 18 attaches to another chromosome before or after conception. In this possible hereditary condition, people have two copies of chromosome 18 plus a small piece from chromosome 18.
- Mosaic trisomy 18: This condition is not hereditary and is also very rare. The extra chromosome is not present in all the cells. This type is attributed to a random occurrence taking place during cell division.

What Is the Treatment for Trisomy 18 (Edwards Syndrome)?

Treatment is on a case-by-case basis and addresses the individual's personal health concerns. Complications depend on the specific defects and symptoms. Fifty percent of children with this condition do not survive the first week; some have lived to their teenage years, but usually with serious mental and physical problems.

Further Reading

"Trisomy 18." 2011. Genetics Home Reference. National Library of Medicine (U.S.). http://ghr.nlm.nih.gov/condition/trisomy-18. Accessed 5/30/12.

"Trisomy 18." 2011. MedlinePlus. National Institutes of Health (U.S.). http://www.nlm.nih.gov/medlineplus/ency/article/001661.htm. Accessed 5/30/12.

"Trisomy 18." 2011. Medscape. http://emedicine.medscape.com/article/943463-overview. Accessed 5/30/12.

Trisomy 18 Foundation. 2010. http://www.trisomy18.org/site/PageServer. Accessed 5/30/12.

Trisomy 21. *See* Down Syndrome (DS)

Trisomy X. *See* Triple X Syndrome

Tuberous Sclerosis Complex

Prevalence　　Affects about 1 in 6,000 people

Other Names　Bourneville disease; Bourneville phakomatosis; cerebral sclerosis; epiloia; sclerosis tuberose; tuberose sclerosis

In 1880, a French doctor, Désiré-Magloire Bourneville, first described a condition in which hard tumors grow all over the body and also in many organs, including the brain. Although the condition is sometimes named after Dr. Bourneville, mostly it is known by the descriptive name. The Latin word *tuber* means "growth or swelling," and the Greek word *skleros* means "hard."

What Is Tuberous Sclerosis Complex?

Tuberous sclerosis complex is a disorder characterized by many benign tumors on the outside of the body and in inside organs. The hard tumors occur in the skin, brain, kidneys, and other organs. The growths cause numerous health problems including developmental disorders.

Following are the symptoms of the complex:

- Patches of raised bumps on the skin that are thick and light-colored
- Growths under the nails
- Tumors on face called angiofibromas; these growths begin in childhood.
- Growth in the brain, causing seizures
- Behavioral problems such as hyperactivity or aggression
- Learning disabilities
- Possible autistic features
- Fatal brain tumors
- Kidney tumors; these can be life-threatening.
- Tumors in the lungs and heart
- Tumors in the retina of the eye

What Are the Genetic Causes of Tuberous Sclerosis Complex?

Two gene mutations, *TSC1* and *TSC2*, cause tuberous sclerosis complex. Both of these genes work together to accomplish the function of the proteins hamartin and tuberin. Both of these genes are tumor growth suppressors.

TSC1

Mutations in the *TSC1* gene, known also as the "tuberous sclerosis 1" gene, cause tuberous sclerosis complex. Normally, *TSC1* instructs for the protein hamartin. Scientists believe that hamartin interacts with another protein called tuberin.

Tuberin is a product of a sister gene, *TSC2*. The two proteins together act to control cell growth and size. Hamartin and tuberin are both tumor suppressors working in concert to regulate a number of proteins and prevent the overgrowth of cells that lead to the growth of tumors in the skin and other organs.

Over 400 mutations in *TSC1* cause tuberous sclerosis complex. Most of the mutations involve either a deletion or addition of a small piece of DNA. The mutation produces a stop signal for the protein hamartin. The loss of the protein allows the cell to divide wildly. *TSC1* is inherited in an autosomal dominant pattern and is located on the long arm (q) of chromosome 9 at position 34.

TSC2

Mutations in the second gene, *TSC2*, officially known as the "tuberous sclerosis 2" gene, cause the disorder. Normally, *TSC2* instructs for the protein tuberin. Tuberin interacts with hamartin, which comes from the *TSC1* gene. The two work together to keep the cells from growing wildly.

More than 1,100 mutations in the *TSC2* gene cause tuberous sclerosis. The mutation is the result of an addition or deletion of a small amount of DNA. The mutations disrupt the function of the protein, keeping the protein from interacting with hamartin and causing the symptoms of this disorder. *TSC2* is inherited in an autosomal recessive pattern and is located on the short arm (p) on chromosome 16 at position 13.3.

What Is the Treatment for Tuberous Sclerosis Complex?

Treatment is symptomatic. Medication, such as vigabatrin and other antiepileptic drugs, may treat seizures. Surgery may be essential. Removing the enlarged giant cell astrocytomas before they develop and become locally invasive is essential. Other symptoms may be treated as they occur. These children qualify for special education and may need to be in special schools.

Further Reading

Northrup, Hope, Mary Kay Koenig, and Kit-Sing Au. 2011. "Tuberous Sclerosis Complex." *GeneReviews*. http://www.ncbi.nlm.nih.gov/books/NBK1220. Accessed 5/30/12.

"Tuberous Sclerosis Complex." 2011. Genetics Home Reference. National Library of Medicine (U.S.). http://ghr.nlm.nih.gov/condition/tuberous-sclerosis-complex. Accessed 5/30/12.

Tuberous Sclerosis Complex Alliance. 2011. http://www.tsalliance.org. Accessed 5/30/12.

Turner Syndrome (TS)

Prevalence   About 1 in 2,500 newborn girls worldwide; may be more prevalent because many are stillborn or miscarriages; in the United States, over 71,000 girls and women

Other Names   45,X; monosomy X; TS; Turner's syndrome; Ullrich-Turner syndrome

In 1938, Dr. Henry Turner, an endocrinologist in Oklahoma City, noted some patients that almost always were very short, infertile, and with a distinct appearance. Turner studied and wrote often about this condition and was honored with the name Turner syndrome. It was not until 1964 that it was discovered that the condition was caused by a chromosomal abnormality.

What Is Turner Syndrome (TS)?

Turner syndrome is a developmental condition caused by a missing or defective X chromosome that affects only females. The common feature of TS is that the girls are short in stature, beginning at about age five. The average adult height of a person with Turner syndrome is about 4 feet, 8 inches.

Several other characteristic features may be noted:

- Early loss of function of the ovaries: At first the ovaries develop normally, but then the eggs in the ovaries die prematurely. Some girls have only streaks of ovarian tissue that may become fibrous.
- Absence of a menstrual period, a condition called amenorrhea.
- Webbed neck: Many girls with Turner syndrome have extra folds of skin on the back of the neck.
- Low hairline at the back of the neck.
- Puffiness or swelling of hands and feet, a condition called lymphedema.
- Short, up-turned nails.
- Shortness of the fourth metacarpal in the hand.
- Small fingernails.
- Obesity: Later in life, obesity may cause diabetes.
- Distinctive facial appearance with small lower jaw.
- Low-set ears.
- Broad chest, described as a shield chest, with widely spaced nipples.
- Skeletal abnormalities: Osteoporosis may develop.
- Horseshoe-shaped kidney: These abnormalities may cause urinary infections.
- Eye deformities: Impairment may be to the sclera or cornea. The individual may develop glaucoma.

- Ear infections and hearing loss.
- Heart disorders: One half of the girls have a narrowing of the large artery of the heart or of the aortic valve that connects the aorta and heart.
- Risk of autoimmune diseases, such as Hashimoto disease.
- Hips are not much bigger than waist.
- Attention problems with concentration and memory, hyperactivity in child-hood and adolescence.
- Nonverbal learning disability, especially with math and spatial concepts: Most girls have normal intelligence, and many have higher-than-average educational achievements.

Turner syndrome appears differently in each person; no two individuals have the same symptoms.

What Is the Genetic Cause of Turner Syndrome (TS)?

Turner syndrome occurs in girls. One of the X chromosomes or sex chromosome is missing or is incomplete. The missing genes are present at conception or following the first cell division. In the girls with 45,X karyotype, about two-thirds are miss-ing the X chromosome from the father. The missing genetic material affects both the development before birth and after birth. Turner can also occur if the second X chromosome is partially missing or rearranged. Some women have the chromo-somal change only in some cells, a condition known as Turner mosaicism.

Although researchers have not determined all genes responsible for most of the features of TS, one gene called the *SHOX* gene, officially known as the "short stature homeobox" gene, is especially related to the short stature and skeletal abnormalities. Normally, *SHOX* instructs for a protein that regulates other genes. It is called a transcription factor. *SHOX* is part of a large family of homeobox genes that are essential in embryonic development. The protein is especially important for many body structures including the development of the skeleton and maturation of bones in arms and legs. *SHOX* is found on the short arm (p) of the X chromosome at position 22.33; it is also found on the short arm (p) of the Y chromosome at position 11.3.

Researchers believe that most cases of TS are not inherited. If there is only one X chromosome, the abnormality occurs as a random event during the formation of egg and sperm. A cell division error called nondisjunction resulted in the loss of an X chromosome in the egg or sperm cell. Likewise, mosaic TS is not inherited but occurs as a random error in cell development in early fetal development.

What Is the Treatment for Turner Syndrome (TS)?

Because it is a chromosomal condition, no cure for TS exists. However, many interventions can minimize symptoms. Early diagnosis may lead to administering growth hormone therapy to increase adult height. Estrogen therapy can be used to induce puberty. Because every case is individual with different symptoms, the other disorders such as hearing and vision loss, heart, bladder, or other infection can be dealt with as they occur.

Further Reading

"Turner Syndrome." 2011. Genetics Home Reference. National Library of Medicine (U.S.). http://ghr.nlm.nih.gov/condition/turner-syndrome. Accessed 5/30/12.

"Turner Syndrome." 2011. Medscape. http://emedicine.medscape.com/article/949681 -overview#showall. Accessed 5/30/12.

Turner Syndrome Foundation. 2011. http://www.turnersyndrome.org. Accessed 5/30/12.

U

Ullrich Congenital Muscular Dystrophy (UCMD)

Prevalence   About 1 in 1 million individuals

Other Names   UCMD; Ullrich disease; Ullrich scleroatonic muscular dystrophy

In 1930, O. Ullrich, a German doctor, wrote about a severe muscle condition that children were born with. He called the disorder "congenital atonic sclerotic muscular dystrophy." Later, scientists added the name of the German doctor and dropped part of the long description. Today, the condition is known as Ullrich congenital muscular dystrophy, or UCMD.

What Is Ullrich Congenital Muscular Dystrophy (UCMD)?

UCMD is a disorder of the skeletal muscles that appears soon after birth. The meaning of the term "congenital muscular dystrophy" can be explained in the following way. The word "congenital" comes from two Greek words: *con*, meaning "with," and *gen*, meaning "birth." A congenital condition is one that the person is born with. The word "dystrophy" comes from the Greek words *dys*, meaning "with difficulty," and *troph*, meaning "nourish." A dystrophy is a disorder in which the body parts do not get the nourishment they need to function. In this case, it is the muscles.

Following are the symptoms of Ullrich congenital muscular dystrophy:

- Muscle weakness that is evident soon after birth
- Joint stiffness in knees and elbows
- Large range of movement called hypermobility in wrists and ankles
- Weak respiratory muscles, requiring mechanical ventilation at night
- Skin bumps, which develop around elbows and knees
- Soft, velvety palms and soles of feet

• Wounds that split open and have little bleeding; these form scars over the years

The individual will never learn to walk without assistance.

What Are the Genetic Causes of Ullrich Congenital Muscular Dystrophy (UCMD)?

Mutations in the *COL6A1*, *COL6A2*, and *COL6A3* genes cause Ullrich congenital muscular dystrophy. Each of these genes is related to making a part of one type of collagen called type VI collagen.

COL6A1

Mutations in the *COL6A1* gene, officially known as the "collagen, type VI, alpha 1" gene, cause UCMD. Normally, *COL6A1* instructs for making one part of the type VI collagen, a protein that surrounds muscle cells in an area called the extracellular matrix. This protein is called the alpha(α)1(VI) chain of type VI collagen. When this part of the chain combines with two other types of α(VI) chains (the α2 and α3 chains), it makes a molecule called the collagen VI monomer. Two of these monomers combine to form collagen VI dimers (meaning 2). Two of the dimers combine to form a tetramer (meaning 4), which is the complete collagen VI molecule. This large collagen protein provides support to the muscle cells, allowing their stability and growth.

At least 12 mutations in *COL6A1* cause UCMD. Most mutations change one amino acid in the α1(VI) chain. Because of the mutation, the chain cannot function properly and the formation of the extracellular support system is impaired. This leads to the muscle weakness and other severe symptoms of the disorder. *COL6A1* is inherited in an autosomal recessive pattern and is located on the long arm (q) of chromosome 21 at position 22.3.

COL6A2

Mutations in the *COL6A2* gene, officially known as the "collagen, type VI, alpha 2" gene, cause UCMD. Normally, this gene instructs for a second component of type VI collagen called the alpha(α)2(VI) chain. The chain combines with the two other type α(VI) chains (the α1 and α3 chains), which will eventually become the type VI collagen molecule.

About 23 mutations in the *COL6A2* molecule are related to UCMD. These mutations disrupt the formation of the α2(VI) chain, leading to the deficiency of the type VI collagen and the muscle weakness of UCMD. *COL6A2* is inherited in an autosomal recessive pattern and is found on the long arm (q) of chromosome 21 at position 22.3.

COL6A3

Mutations in the *COL6A3* gene, officially known as the "collagen, type VI, alpha 3" gene, cause UCMD. Normally, this gene is similar to the others in this family and encodes for one part of the type VI chain called the alpha(α)3(VI) chain. This chain combines with the two others to form the collagen molecule.

Mutations cause one change in an amino acid in the amino acid. This change disrupts the formation of the α3(VI) chain, leading to the interruptions in type VI collagen. This gene is inherited in an autosomal recessive pattern and is located on the long arm (q) of chromosome 2 at position 37.

What Is the Treatment for Ullrich Congenital Muscular Dystrophy (UCMD)?

Treating the serious conditions such as respiratory failure must be considered first, as well as aggressive treatment for lung infections. Physical therapy can assist muscle disorders. In addition, attention to nutrition is important to manage any feeding difficulties.

See also Duchenne/Becker Muscular Dystrophy; Emery-Dreifuss Muscular Dystrophy (EDMD); Fascioscapulohumeral Muscular Dystrophy (FSHMD)

Further Reading

"Congenital Muscular Dystrophy." 2011. Medscape. http://emedicine.medscape.com/article/1180214-overview. Accessed 5/13/12.

Lampe, Anne Katrin, Kevin M. Flanigan, and Katharine Mary Bushby. 2007. "Collagen Type VI–Related Disorders." *GeneReviews*. http://www.ncbi.nlm.nih.gov/books/NBK1503. Accessed 5/13/12.

"Ullrich Congenital Muscular Dystrophy." 2011. Neurology MedLink. http://www.medlink.com/cip.asp?UID=mlt002hj&src=Search&ref=33956121. Accessed 5/13/12.

Unverricht-Lundborg Disease (ULD)

Prevalence	Rare; occurs in Finland in 4 out of 100,000 people; worldwide prevalence unknown
Other Names	Baltic myoclonic epilepsy; Baltic myoclonus; Baltic myoclonus epilepsy; EPM1; Lundborg-Unverricht syndrome; Mediterranean myoclonic epilepsy; myoclonic epilepsy of Unverricht and Lundborg; PME; progressive myoclonic epilepsy; progressive myoclonus epilepsy 1; ULD; Unverricht-Lundborg syndrome

This disorder was known to be one of the first inherited epilepsies but was considered to be two different conditions. The reason for the differences came from the location of the person who had it. If the person was from the area around the Baltic Sea, it was known as Baltic myoclonus; if the person was from around the

Mediterranean area, it was known as Mediterranean myoclonus. Eventually, Dr. Heinrich Unverricht of Germany, who described the disease in 1891, and Dr. Herman Bernhard Lundborg of Sweden, who described the disease in 1903, realized the two were the same disease. The condition was named for the two doctors: Unverricht-Lundborg disease or ULD.

What Is Unverricht-Lundborg Disease (ULD)?

Unverricht-Lundborg disease is a very rare form of progressive myoclonus epilepsy. A myoclonus epilepsy is the type in which the muscles begin to jerk all over. Following are some of the features of the disorder:

- Involuntary muscle jerking and twitching
- May lose consciousness
- May experience muscle rigidity and convulsions like grand mal seizures
- Begins between the ages of 6 and 15
- Brought on by physical exertions, stress, light, or other stimuli
- Progressive problems with balance
- Difficulty speaking
- Within 5 to 10 years, may interfere with walking
- Everyday activities also disrupted
- Depression
- Mild intellectual decline

People with ULD may live into adulthood and, depending on the severity and treatment, may have normal life expectancy.

What Is the Genetic Cause of Unverricht-Lundborg Disease (ULD)?

Mutations in the *CSTB* gene, officially known as the cystatin B (stefin B) gene, causes Unverricht-Lundborg disease. Normally, *CSTB* instructs for the protein cystatin B, which inhibits the enzymes called cathepsins. These enzymes are active in the lysosomes, working to break down and recycle cell materials. A region of the *CSTB* gene has a repeating sequence of 12 DNA building blocks, which as written as CCCCG-CCCCG-CG and called a dodecamer. This series is repeated two or three times in the gene to regulate the cystatin B protein production.

Mutations in *CSTB* cause the dodecamer to have more than 30 repeats in both copies of the *CSTB* gene. The repeats disrupt the function of the cystatin B protein, allowing cathepsin levels to be increased and causing the symptoms of Unverricht-Lundborg disease. CTSB is inherited in an autosomal recessive pattern and is located on the long arm (q) of chromosome 21 at position 22.3.

What Is the Treatment for Unverricht-Lundborg Disease (ULD)?

When symptoms begin, the person should have the first drug of choice called valproic acid; clonazepam is an add-on therapy. Other drugs may be needed. The person may then be given psychosocial support and rehabilitation.

Further Reading

Lehesjoki, Anna-Elina, and Reetta Kälviäinen. 2009. "Unverricht-Lundborg Disease." *GeneReviews*. http://www.ncbi.nlm.nih.gov/books/NBK1142. Accessed 12/3/11.

"Unverricht-Lundborg Disease." 2011. Epilepsy Therapy Project. http://www.epilepsy.com/epilepsy/epilepsy_unverrichtlundborg. Accessed 12/3/11.

"Unverricht-Lundborg Disease." 2011. Orphanet. http://www.orpha.net/consor/cgi-bin/OC_Exp.php?Lng=EN&Expert=308. Accessed 12/3/11.

Urea Cycle Disorders (UCD)

Prevalence About 1 in 30,000 births; partial defects much higher
Other Names None

What Are Urea Cycle Disorders?

Nitrogen and Its Role in the Body

Nitrogen is an important colorless, odorless, tasteless gas that makes up about 78% of air, along with oxygen and other elements. It is also found in the body as part of normal body metabolism. The nitrogen cycle, an important biological cycle, moves the element from the air into various organic compounds and then back into the atmosphere. The element nitrogen is an essential part of the building blocks of the body, the amino acids (NH_2), and the proteins and nucleic acids such as DNA and RNA. Neurotransmitters, the chemicals that help move nerve impulses from one neuron to another, have nitrogen as a component. In fact, about 3% of the human body is nitrogen by weight.

Waste products from the body are made of nitrogen. In the absence of disease, the waste product urine is a sterile liquid secreted by the kidneys. When cells undergo metabolism, waste products rich in nitrogen are produced and are then carried in the bloodstream to the kidneys, where they are processed and eventually eliminated in the urine. Analyzing urine is a traditional part of any physical exam. Unusual products in the urine are quite telling about the health of the individual. The best-known test is for the presence of sugar or glucose in the urine, an indicator of diabetes.

Urine and the Piss Prophets

Observing or even tasting urine at one time was the way to diagnose disease and predict the future. The art of uroscopy was known from ancient times, reached its height of popularity in the Middle Ages, and continued well into the nineteenth century. Hippocrates, the Greek physician and father of medicine, related fever in a person to the smell and color of urine, and he believed that excessive urination was a sign of illness. In addition to diagnosing illness, the Romans added the idea that swirling urine could predict the future. By the Middle Ages, the church forbade physicians to touch body parts; urine could be discreetly passed behind a screen for examination. Even during the Renaissance and renewed interest in scientific discovery, uroscopy blossomed as physicians observed and tasted the gold-colored substance.

Color was very important. Court physicians were concerned about the purple color of King George's urine; he is believed to have had a metabolic condition called porphyria. In 1674, Thomas Willis, an English physician, noted how sweet the urine of a sick patient tasted; the patient had what we know today as diabetes. A group of physicians, known as the "piss prophets," emerged and used urine to tell the fortunes of people, for a fee of course. For example, urine with lots of bubbles meant one would soon be rich.

When Alexander Garrod noted the black urine of a person with alkaptonuria, he coined the term "inborn error of metabolism." Today, testing urine is recognized as an indicator of many conditions, including diabetes and pregnancy.

Sometimes the body's processes do work properly, and an inborn error may produce defects in the metabolism of protein and other nitrogen-containing molecules. This defect causes ammonia (NH_3) or other harmful products to build up in the cells. Urea cycle disorders (UCD) are the result of disruptions in this breakdown.

What Is the Genetic Cause of Urea System Disorders (UCD)?

Urea system disorders are the result of deficiencies in the enzymes produced by mutations in certain genes. Infants with a severe USD may appear normal at birth but within a few days begin to show signs of lethargy, seizures, anorexia, slow or rapid breathing, hypothermia, or coma.

Following are five catalytic enzymes, a cofactor producing enzyme, and two transporters that are related to disorders in the urea cycle pathway:

- *Carbamoylphosphate synthetase I (CPS1)*: This deficiency is the most severe UCD. Very soon after birth, the ammonia begins to build up, causing a crisis. The ones who do survive will probably have continued bouts of hyperammonemia, or build up of excessive ammonia in the bloodstream. This disorder is inherited in an autosomal recessive manner. There is no newborn screening test for it at present.

- *Ornithine transcarbamylase (OTC)*: This condition was involved in a well-publicized case involving gene therapy. Jesse Gelsinger, whose OTC was controlled with medication, was tapped for an experimental process called gene therapy. He died after one injection. This condition also causes ammonia to be built up and can be as serious as CPS1. OTC is acquired in an X-linked manner or acquired as a new mutation while in the uterus. Female carriers may develop symptoms of the disorder later in life.

- *Argininosuccinic acid synthetase (ASS1)*: Also called citrullinemia type I, this disorder is also quite severe. Individuals are unable to incorporate nitrogen wastes into cycle intermediates. However, it is somewhat easier to treat than other types. The disorder is inherited in an autosomal recessive manner.

- *Argininosuccinic acid lyase (ASL)*: This deficiency presents itself rapidly during the newborn period. The liver is seriously affected. Another symptom may be the presence of fragile hair that appears like a node on the head. People who survive may have serious developmental disabilities. ASL is inherited in an autosomal recessive pattern.

- *Arginase (ARG)*: The onset of ammonia is not rapid and early, but may develop progressively along with tremors, ataxia, and affected growth. This condition is also an autosomal recessive condition.

- *A cofactor producing enzyme N-acetyl glutamate synthetase (NAGS)*: This cofactor produces an enzyme whose symptoms mimic CPSI deficiency. NAGS is inherited in an autosomal recessive pattern.

- *Two transporters*: Ornithine translocase (ORNTI) and citrin are transporters of the elements of the urea cycle that may also be affected.

What Does Ammonia Buildup Do to the Body?

Several things can happen to the body, and all of them are bad. Following are some of the damaging aspects of these UCDs:

- Evokes brain swelling; this condition called cerebral edema affects the water and potassium balance in the brain.
- Causes stroke
- Disrupts neurotransmitters
- Affects mitochondrial function
- Develops seizures

What Is the Treatment for Urea System Disorders (UCD)?

Efforts to reduce plasma ammonia are the goal and may require dialysis or blood filtration. Restriction of protein may be replaced with the diet consisting of carbohydrates and fat. Other symptoms may be treated as they appear. In 2003 the National Institutes of Health created a Urea Cycle Disorders Consortium to explore and research the urea disorders.

Further Reading

Lanpher, Brendan C., et al. 2011. "Urea Cycle Disorders Overview." *GeneReviews*. http://www.ncbi.nlm.nih.gov/books/NBK1217. Accessed 5/30/12.

National Urea Cycle Disorders Foundation. 2011. http://www.nucdf.org. Accessed 5/30/12.

Urea Cycles Disorders Consortium. 2011. http://rarediseasesnetwork.epi.usf.edu/ucdc. Accessed 5/30/12.

Uromodulin-Associated Kidney Disease

Prevalence　　Unknown; less than 1% of all kidney disease cases

Other Names　familial gout-kidney disease; familial gouty nephropathy; familial juvenile hyperuricemic nephropathy; FJHN; MCKD2; medullary cystic kidney disease type 2; UMAK; UMOD-related kidney disease; uromodulin storage disease

This disorder is named from the protein uromodulin that is produced by the gene *UMOD*.

What Is Uromodulin-Associated Kidney Disease?

Uromodulin-associated kidney disease is a condition that affects the kidneys with symptoms that may vary even in families. Following are the signs of the disorder:

- High levels of the waste product uric acid: Normally, the kidneys remove the uric acid from the blood so it can be excreted in the urine.
- Gout: Buildup of uric acid causes gout, a form of arthritis in which uric acid crystals are deposited in the joints. These gout symptoms usually begin during the teenage years.
- Progressive kidney disease.
- Small kidney cysts called medullary cysts.

Eventually, the person will experience kidney failure and must be on dialysis or have a kidney transplant, sometime between the ages of 30 and 70.

What Is the Genetic Cause of Uromodulin-Associated Kidney Disease?

Mutations in the *UMOD* gene, officially known as the "uromodulin" gene, cause uromodulin-associated kidney disease. Normally, *UMOD* instructs for the protein uromodulin, which is produced in the kidneys and excreted in urine. Uromodulin is the most abundant protein found in urine. It may perform the dual function of protecting against urinary tract infections and controlling the amount of water in the urine.

About 40 mutations in *UMOD* are connected to uromodulin-associated kidney disease. With a change in only one amino acid, the protein structure is changed, disrupting the release from the kidney cells and causing the abnormal buildup of uromodulin, which then damages the kidneys. *UMOD* is inherited in an autosomal dominant pattern and is located on the short arm (p) of chromosome 16 at position 12.3.

What Is the Treatment for Uromodulin-Associated Kidney Disease?

A nephrologist is a physician who monitors and treats kidney function. The specialist may prescribe allopurinol or probenecid for gout. When the disease progresses, dialysis may replace renal function, but the only cure is a kidney transplant.

Further Reading

"About UAKD." Wake Forest Baptist Health. http://www.wakehealth.edu/nephrology/gout. Accessed 12/3/11.

Bleyer, Anthony J., and P. Suzanne Hart. 2011. "*UMOD*-Associated Kidney Disease." *GeneReviews*. http://www.ncbi.nlm.nih.gov/books/NBK1356. Accessed 12/3/11.

"Uromodulin-Associated Kidney Disease." 2011. Genetics Home Reference. National Library of Medicine (U.S.). http://www.ghr.nlm.nih.gov/condition=uromodulinassociatedkidneydisease. Accessed 12/3/11.

Usher Syndrome

Prevalence	Accounts for about 3% to 6% of all childhood deafness and about 50% of deaf-blindness in adults
Other Names	deafness-retinitis pigmentosa syndrome; dystrophia retinae pigmentosa-dysostosis syndrome; Graefe-Usher syndrome; Hallgren syndrome; retinitis pigmentosa-deafness syndrome; Usher's syndrome

In 1858, Albrecht von Gräfe, a pioneer in the field of ophthalmology, reported on a patient who had both hearing and eye disorders. He noted that two brothers also had the same condition. Later, one of his students, Richard Liebreich, studied a population in Berlin looking for those with a pattern of both hearing disorders and a condition that had been called retinitis pigmentosa. Liebreich found several families and noted that it was inherited in a recessive pattern. In 1914, Charles Usher, a British ophthalmologist, wrote about 69 cases of a disorder in which patients were both blind and deaf. Because of his extensive study of the condition, the disorder as named after him—Usher syndrome.

What Is Usher Syndrome?

Usher syndrome is a disorder characterized by both progressive hearing loss and vision loss. An eye disease called retinitis pigmentosa (RP) causes the vision loss. RP occurs when the light-sensing cells of the retina begin to deteriorate. The first thing to go is night vision, and then blind spots occur in the peripheral or side vision. Later, the blind spots merge to produce a limited narrow field of vision, called tunnel vision. Some people may develop cataracts. However, some people with Usher syndrome may retain central vision throughout life.

Three major types are characterized by the age when the symptoms occur and the severity of the symptoms. Each of these types may have subtypes. Following are the types and subtypes:

- *Type I*: In this most severe type, infants are born profoundly deaf or lose their hearing in the first year of life. During the early years of life, retinitis pigmentosa begins to cause progressive blindness. In addition, the inner ear of the child is affected, causing problems with balance and learning to sit and walk. Type I is divided into seven subtypes, designated as IA through IE; each subtype has its subtle characteristics. Usher type I is more common among the descendants of Ashkenazi Jews and the Acadian people of Louisiana. In this population, it may occur in about 4 people per 100,000. Mutations in the genes *CDH23*, *MYO7A*, *PCDH15*, *USH1C*, or *USH1G* may cause type I. In addition, two other genes may be involved.

- *Type II*: In type II, the most common form of Usher syndrome, children are born with hearing loss that ranges from mild to severe and affects the high tones, which are the soft speech sounds that have the letters d and t. Vision loss begins in their adolescence or early adulthood. People with this type do not have the balance disorders associated with the inner ear. Three subtypes have been characterized, designated as subtypes IIA, IIB, and IIC. Mutations in at least two known genes—*USH2A* and *GPR98*—and two unknown genes cause type II.

- *Type III*: Progressive hearing loss and progressive vision loss in the first decades of life characterize type III. Infants have normal hearing, but the loss begins in late childhood or adolescence, fortunately after the children have learned to speak. The deafness progresses so that by middle age, they are completely deaf. Vision loss begins in late childhood or adolescence. The people may have balance problems. In this type, the problems vary greatly among individuals. Usher type III is the rarest form of the total number of cases, except in Finland, where about 40% of all cases are found. Mutations in the *CLRN1* gene and one other unidentified gene cause this type.

What Are the Genetic Causes of Usher Syndrome?

Several genes have been related to Usher syndrome. Mutations in the *CDH23*, *CLRN1*, *GPR98*, *MYO7A*, *PCDH15*, *USH1C*, *USH1G*, and *USH2A* genes appear to cause the syndrome.

CDH23

Mutations in the *CDH23* gene, whose official name is "cadherin-related 23," cause Usher type ID. Normally, *CDH23* encodes a protein called cadherin 23. This protein is essential in helping cells adhere to one another. Depending on the type of cell, specific versions of the protein are created. For example, a short version is made in the retina of the eyes and a longer version in the inner ear. The proteins work with other proteins in the cell membrane to create an attachment complex. The cadherin 23 complex creates the hairlike structures called stereocilia that respond to sound waves. It is the response of these hairlike structures that aids in converting sound waves into the nerve impulses that go to the brain where one perceives sound. Although the role is not exactly understood, cadherin 23 protein also plays a critical role in the light-sensitive cells of the retina.

More than 30 mutations in the *CDH23* gene cause Usher type ID. Most of the changes involve an exchange of one building block or other minor additions or deletions that lead to a small dysfunctional version of cadherin23. The changes may result in improper developing of the inner ear and the retina, which results in hearing and vision loss. *CDH23* is inherited in an autosomal recessive pattern and is located on the long arm (q) of chromosome 10 at position 22.1.

CLRN1

Mutations in *CLRN1*, whose official name is "clarin 1," causes Usher syndrome. Normally, this gene encodes a protein called clarin 1. Clarin 1 has been found in many body tissues, especially in the hair cells in the inner ear. The protein appears to have a strong role in the nerve communication that results in both hearing and vision. It may function at the synapse, the junction between two cells.

About 10 mutations in the *CLRN1* gene cause Usher syndrome. Two mutations have been found especially in the Finnish population and are related to type III. The changes in the gene produce very short versions of the protein, leading to the signs and symptoms of Usher syndrome. *CLNR1* is inherited in an autosomal recessive pattern and is located on the long arm (q) of chromosome 3 at position 25.

GPR98

Mutations in the *GPR98* gene, officially called "G protein-coupled receptor 98," cause Usher type IIC. Normally, *GPR98* instructs for the protein G protein-coupled receptor 98. This protein, which is very large, is found in the inner ear and in the retina. It appears to have a role in the development of the cochlear, an important inner-ear structure. The cochlea is where the sound waves are converted into nerve impulses. *GPR98* also plays a role in the maintenance of the light-sensitive cells of the retina.

About four mutations in *GPR98* are related to type IIC. The mutations either stop the production of the protein or cause an abnormally small version. The protein affects mostly hearing, with vision problems appearing relatively mild. *GPR98* is inherited in an autosomal recessive pattern and is located on the long arm (q) of chromosome 5 at position 13.

MYO7A

Mutations in *MYO7A*, whose official name is "myosin VIIA," causes Usher IB. Normally, *MYO7A* encodes for a protein called myosin VIIA. This protein appears to interact with actin to transport other molecules. The protein is made in the inner ear, where it is important in the development of cilia. Found also in the cells of the retina, its role is to carry small sacs of pigment to the retinal pigment epithelium.

About 120 mutations in *MYO7A* are related to type 1B. The changes either stop or cause a small version of the protein to be made. The results are a nonworking protein that affects development of cells of the inner ear and retina. *MYO7A* is inherited in an autosomal recessive pattern and is located on the long arm (q) of chromosome 11 at position 13.5.

PCDH15

Mutations in the gene *PCDH15*, whose official name is "protocadherin-related 15," cause Usher Type IF. Normally, *PCDH14* provides instructions for the protein protocadherin 15. This protein, which is made in the inner ear and retina, helps cells stick together. This protein is a member of an important superfamily of gene called the CDH or cadherin. Its role is similar to the protein cadherin 23.

About nine mutations in *PCDH15* cause Usher IF. In the Ashkenazi Jewish population, the mutation is the result of the replacement of the amino acid arginine. An abnormally small version of the protein is made that causes the characteristic symptoms of Usher type IF. *PCDH15* is inherited in an autosomal recessive pattern and is located on the long arm (q) of chromosome 10 at position 21.1.

USH1C

Mutations in *USH1C*, known officially as "Usher syndrome 1C (autosomal recessive, severe)" cause Usher IC. Normally, *USH1C* encodes the protein harmonin. This protein binds many other proteins in the cell membrane and provides for the support of the internal framework of the cell. It also is important in the development of the cilia in the inner ear. Harmonin also is part of the light-sensitive structure of the retina.

At least 10 mutations in USH1C cause Usher I. This type, which results when adenosine replaces guanine, is found mostly in the Acadian population of Louisiana.

Changes in the gene disrupt the blueprint for harmonin and causes the signs and symptoms of Usher syndrome. USH1C is inherited in an autosomal recessive pattern and is located on the short arm (p) of chromosome 11 at position 14.3.

USH1G

Mutations in *USH1G* or "Usher syndrome 1G (autosomal recessive)" cause Usher type IG. Normally, this gene provides instructions for a protein called SANS. This protein is found in both the retina and inner ear and is part of a large complex that develops the important structures of the ear and eye.

About five mutations of USH1G cause Usher syndrome IG. These mutations cause very short nonfunctional versions of the protein, leading to hearing and vision loss. *USH1G* is inherited in an autosomal recessive pattern and is located on the long arm (q) of chromosome 17 at position 25.1.

USH2A

Mutations in the *USH2A* gene or the "Usher syndrome 2A (autosomal recessive, mild)" gene causes Usher type 2A. Normally, *USH2A* encodes for the protein usherin, an essential part of the basement membrane or thin sheets that line cells and tissues. The protein appears to play a part in the development of the inner ear and the retina.

Over 200 mutations have been identified in Usher type IIA. The mutations cause an abnormally short version of the usherin protein and lead to hearing and vision loss. *USH2A* is inherited in an autosomal recessive pattern and is located on the long arm (q) of chromosome 1 at position 41.

What Is the Treatment for Usher Syndrome?

Currently, no cure exists for Usher syndrome. The treatment is symptomatic. It is imperative that the condition is identified early so educational programs can begin. Treatment typically includes hearing aids, assistive learning aids, cochlear implants, sign language, Braille instruction, and other training. Some eye doctors believe that a high dose of vitamin A palmitate may slow the progress of retinitis pigmentosa.

Further Reading

Guest, Mary. "Usher Syndrome: A Condition Which Affects Hearing and Sight." A–Z to Deafblindness. http://www.deafblind.com/usher.html. Accessed 12/6/11.

"Usher Syndrome." 2011. Foundation Fighting Blindness. http://www.blindness.org/index.php?option=com_content&view=article&id=56&Itemid=81. Accessed 5/30/12.

"Usher Syndrome." 2011. MedicineNet. http://www.medicinenet.com/usher_syndrome/article.htm. Accessed 12/6/11.

V

VACTERL Association

Prevalence Relatively common multiple malformation; incidence 1.6 in 10,000
Other Names VATER/VACTERL

What Is VACTERL Association?

VACTERL is an acronym for a group of conditions that occur together.

- *V—vertebrae*: Defects in the spinal column appear, as small vertebrae or only one half of the bone is formed. About 70% of people with the syndrome have some kind of vertebral issue. The problems rarely cause difficulty in early life but later appear as scoliosis or curvature of the spine.

- *A—Deformed anus*: The anus, which is the opening end of the digestive tract, may not open to the outside. This condition is referred to as imperforate anus or anal atresia. Anal atresia is seen in about 55% of patients and requires surgery the first days of life to reconstruct the intestine and anal canal.

- *C—Cardiac abnormalities*: About 75% of infants will have some type of congenital heart disease as ventricular septal defect, atrial septal defects, and tetralogy of Fallot. These defects may or may not require surgery.

- *TE—Tracheoesophageal fistula*: The trachea is the windpipe leading to the lungs, and the esophagus is the tube from the mouth to the stomach, through which food passes. A fistula is an abnormal opening or connection between the two. The abnormality occurs in about 70% of patients.

- *R—Renal abnormalities*: Renal abnormalities refer to the kidneys. Seen in about 50% of people with the association, the defect may be the result of one umbilical artery instead of two, with obstruction of urine or backup of urine into the kidneys. These problems can lead to kidney failure and may require a transplant early in life.

And sometimes:

- *L—Limb abnormalities*: Those with these abnormalities may have absent or displaced thumbs, extra digits, fusion of digits, or forearm defects. Many of the disorders can be corrected surgically before any damage occurs.

In order for a child to be diagnosed with probable VATER/VACTERL association, he or she must have at least three of the symptoms.

What Is the Genetic Cause of VACTERL Association?

No specific genes have been identified with VACTERL association. It has been seen more often in infants with Trisomy 18 and children of diabetic mothers.

What Is the Treatment for VACTERL Association?

This condition is very complex, with treatment depending upon the symptoms. With early diagnosis and attention, the prognosis for a normal life is good.

Further Reading

"Handbook of Genetic Counseling." 2008. VATER Association. 2008. http://en.wikibooks .org/wiki/Handbook_of_Genetic_Counseling/VATER_Association. Accessed 5/31/12.

"VACTERL Association." 2012. Genetics Home Reference. National Library of Medicine (U.S.). http://ghr.nlm.nih.gov/condition/vacterl-association. Accessed 2/29/12.

"VACTERL or VACTERAL Association." 2011. Cincinnati Children's Hospital. http:// www.cincinnatichildrens.org/health/v/vacterl. Accessed 5/31/12.

Van der Woude Syndrome

Prevalence	Occurs in 1 in 35,000 to 1 in 100,000 based on data from Europe and Asia
Other Names	cleft lip and/or palate with mucous cysts of lower lip; lip-pit syndrome; VDWS; VWS

In 1954, Dr. Anne Van der Woude wrote about a condition in which infants were born with a cleft and lip palate with wet-appearing pits on the inside of the lower lip. In writing in the *American Journal of Human Genetics*, she noted that over 70% of individuals with the disorder had the two conditions. The disorder was named after her: van der Woude syndrome.

What Is Van der Woude Syndrome?

Van der Woude syndrome affects the development of the face. The condition appears to vary in severity from individual to individual. Both males and females are affected, as well as people from all ethnic groups. Following are the symptoms of the disorder:

- A split in the roof of the mouth, known as a cleft palate. The split may also occur in the upper lip and on one or both sides of the mouth.
- Small pits or depressions in the inside center of the lower lip. This pit may appear moist because salivary glands and mucous glands are present in the area.
- Small mounds of tissue on the lower lip.
- Missing lower teeth.
- Narrow, high arched palate.
- Tongue deformity.
- Rarely, anomalies of heart, limbs, or other systems.
- Delayed language and learning disabilities.

The average IQ of people with van der Woude syndrome is about the same as the general population.

What Is the Genetic Cause of Van der Woude Syndrome?

Mutations in the *IRF6* gene, or the "interferon regulatory factor 6" gene, causes van der Woude syndrome. Normally, *IRF6* instructs for a protein that is essential in embryonic development. *IRF6* attaches to other regions of DNA to control the activity of specific genes. This makes it a transcription factor. It is especially present in the tissues that form the face and head.

Several mutations in the *IRF6* gene cause a shortage in the function of the protein and affect the development of certain tissues in the skull and face. This disruption causes the signs and symptoms of the syndrome, including the clefts in palate and lip and pits in the lower lip. *IRF6* is inherited in an autosomal recessive pattern and is located on the long arm (q) of chromosome 1 at position 32.2-q41.

What Is the Treatment for Van der Woude Syndrome?

Beginning a plastic surgery regimen as soon as possible can repair the lip and palate and remove the lip pits. Dental surgeons, working with plastic surgeons, can repair other problems with the teeth and jaw.

Further Reading

"Van der Woude Syndrome." 2011. Genetics Home Reference. National Library of Medicine (U.S.). http://ghr.nlm.nih.gov/condition/van-der-woude-syndrome. Accessed 12/7/11.

"Van der Woude Syndrome." 2011. Seattle Children's Hospital Research Foundation. http://www.seattlechildrens.org/medical-conditions/cleft-lip-palate/van-der-woude. Accessed 12/7/11.

"Van der Woude Syndrome Treatment and Management." 2011. Medscape. http://emedicine.medscape.com/article/950823-treatment. Accessed 12/7/11.

VATER Association. *See* VACTERL Association

Vitelliform Macular Dystrophy

Prevalence   Very rare; unknown

Other Names   Best disease; macular degeneration, polymorphic vitelline; vitelliform macular dystrophy, adult onset; vitelliform macular dystrophy, early onset; vitelliform macular dystrophy, juvenile onset

In 1905, Dr. Franz Best, a German ophthalmologist, noted a condition usually appearing in childhood, in which the macular part of the retina had a yellow or orange yolk-like appearance. Vitelliform macular dystrophy has a unique name. The vitelline is another name for the yolk of an egg; vitelliform means shaped like the yolk of an egg. This is what Dr. Best observed in the eyes of his patients. Watching the condition over a number of years, he found the disease progressed rapidly. The disorder was named after Dr. Best but is commonly called vitelliform macular dystrophy.

What Is Vitelliform Macular Dystrophy?

Vitelliform macular dystrophy is a disorder of the eye, which causes progressive vision loss. The macula is an area on the retina where sensory receptors receive light and then carry the image through the optic nerve to the brain. When a doctor looks at the eye of a person with this condition, he or she sees an orange-looking structure resembling an egg yolk on the area of the retina called the macula. A fatty yellow pigment called lipofuscin builds up under the cells of the macula and causes the orange color. This condition disrupts the central vision, making it difficult for the person to read, drive, or recognize faces. It does not affect peripheral vision or night vision.

Two following two forms may exist, which are recognized as an early-onset form and an adult-onset form:

- Early-onset form, known also as Best disease: This form appears in childhood and has varying symptoms and severity
- Adult- onset form: This type begins in mid-adulthood and progresses over time.

Both forms have the characteristic appearances to the macula that is noted in eye examinations.

What Are the Genetic Causes of Vitelliform Macular Dystrophy?

Two genes are involved—*BEST1* and *PRPH2*.

BEST1

Mutations in the *BEST1* gene, officially known as the "bestrophin 1" gene, cause vitelliform macular dystrophy. Normally, *BEST1* instructs for making the protein bestrophin, which appears to play a key role in vision. Found in the thin layer at the back of the eye called the retinal pigment epithelium or RPE, this protein assists the RPE in maintaining the growth and development of the eye and normal function of the light-sensitive cells. Specifically, bestrophin regulates the entry of calcium ions into the cells of the RPE.

Over 100 mutations in the *BEST1* gene are related to vitelliform macular dystrophy. The mutations can cause both types of the disorder. Both types have the fatty yellow deposits in the RPE. Over time, the buildup causes major damage to the eyes. The mutations in *BEST1* change or delete one or more proteins. It was designed to regulate chlorine in and out of cells, but the abnormal bestrophin disrupts the flow and leads to the progressive loss of central vision. *BEST1* is inherited in an autosomal recessive pattern and is located on the long arm (q) of chromosome 11 at position 13.

PRPH2

Mutations in the *PRPH2* gene, officially known as the "peripherin 2 (retinal degeneration, slow)" gene, cause vitelliform macular dystrophy. Normally, *PRPH2* instructs for a protein called peripherin 2, which is essential for normal vision. Peripherin 2 is also found in the light-sensitive structures of the retina and is necessary for detecting light and color.

Several mutations in *PRPH2* are related to the adult-onset form. Most mutations involve changes in only one building block, leading to a very short or nonworking version of the protein. When the protein is changed, the light-sensing cells degenerate, causing the progressive nature of vitelliform macular dystrophy. *PRPH2* is inherited in an autosomal recessive pattern and is found on the short arm (p) of chromosome 6 at position 21.1-p12.3.

What Is the Treatment for Vitelliform Macular Dystrophy?

Treating the symptoms of low vision with aids as needed. If bleeding occurs on the retina, laser surgery may be used.

Further Reading

"Best Disease." 2011. Medscape. http://emedicine.medscape.com/article/1227128-overview. Accessed 12/4/11.

MacDonald, Ian M., and Thomas Lee. 2009. "Best Vitelliform Macular Dystrophy." *GeneReviews*. http://www.ncbi.nlm.nih.gov/books/NBK1167. Accessed 12/4/11.

"Vitelliform Macular Dystrophy." 2011. Genetics Home Reference. National Library of Medicine (U.S.). http://ghr.nlm.nih.gov/condition=vitelliformmaculardystrophy. Accessed 12/4/11.

Vitiligo

Prevalence   Common disorder affecting between 0.5% and 1% of the population; more noticeable in dark-skinned people

Other Names None known

After Michael Jackson, the famous pop star, died in June 2009, some news reports said that the autopsy may have shown that Jackson had vitiligo. Jackson had often said in interviews that he did not bleach his skin. Vitiligo is more noticeable in dark-skinned people but may cause anyone great cosmetic concern.

What Is Vitiligo?

Vitiligo is a condition in which the skin loses pigment and appears as patchy coloring. Pigment is related to the substance melanin, which gives skin and hair their color. The condition appears to progress over time as large areas may lose the pigment. Although it can appear at any age, the onset age is generally in the mid-20s.

Several forms of vitiligo exist.

Generalized Vitiligo

This is the most common form of vitiligo. Symptoms include:

- Loss of pigment all over the body
- Loss of pigment usually on face, hair, neck, and body openings such as the mouth and genitals
- Loss in areas of frequent rubbing such as the hands and arms
- Loss in areas where bones are close to the skin, such as elbows and shins
- Loss in areas of trauma

Segmented Vitiligo

- Patches on underside of the body
- Patches all over the body referred to as universal vitiligo

Focal Vitiligo

- One spot on the body
- Few spots on one specific area of the body, usually on the face, neck, or trunk
- Localized spots may progress to generalized or segmented form

Acrofacial Vitiligo

With acrofacial vitiligo, pigment loss is on the fingers and face.

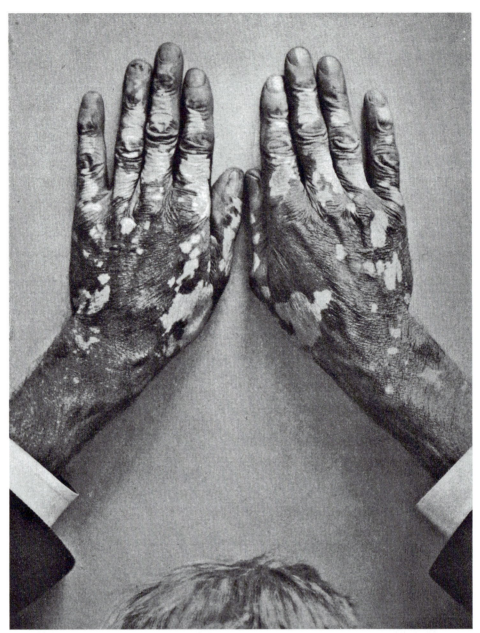

An example of vitiligo, a splotchy discoloring of the skin. (VintageMedStock/ Archive Photos/Getty Images)

Mucosal Vitiligo

With mucosal vitiligo, pigment loss is in the mucous membranes, such as the lips.

Trichrome Vitiligo

Trichrome vitiligo has three conditions associated with it:

- Patches of normal skin pigment
- Large areas of no skin pigment
- Areas with skin a color in between

What causes vitiligo? It basically is considered an autoimmune disorder, in which the body's immune cells turns against the body's own tissues or organs. In vitiligo, the body's immune cells attack the melanin in the skin. Another feature is that about 25% of people with the disorder have another autoimmune disease, such as rheumatoid arthritis, psoriasis, or lupus. However, if the other diseases are not present, the person can be healthy, although they may be concerned for his or her appearance.

What Are the Genetic Causes of Vitiligo?

The condition appears to run in families and is possibly connected with two genes: *NLRP1* and *PTPN22*.

NLRP1

Mutations in the *NLRP1*, officially known as the "NLR family, pyrin domain containing 1" gene, cause the person to be at risk for vitiligo. Normally, *NLRP1* instructs for a family of proteins called the nucleotide-binding domain and leucine-rich repeat containing (NLR) proteins. The proteins are active in the immune system, helping to control inflammation. When an injury occurs, the white blood cells race to the site and sends signaling molecules to fight the outside invaders and repair tissue. When the process is complete, information is sent to stop so that the body's own cells are not damaged. The protein produced by *NLRP1* assembles the molecular complex called an imflammasome, which tells the immune system that a foreign body is present.

At least two variations in *NLRP1* appear as a risk of vitiligo. One type occurs when the amino acid building block leucine is changed to histidine. A second type involves an area in the gene, called the promoter region, where the protein is produced. The variations change in the way the *NLRP1* protein operates, and the control mechanism that prevents the body from attacking its own tissues is impaired. What causes the system to turn on itself is not clear and is probably the result of both environment and genetics. *NLRP1* is inherited in an autosomal recessive pattern and is located on the short arm (p) of chromosome 17 at position 13.2.

PTPN22

Mutations in the *PTPN22* gene, officially known as the "protein tyrosine phosphatase, non-receptor type 22 (lymphoid)" gene, appear to play a role in vitiligo. Normally, *PTPN22* instructs for a protein that is part of the protein tyrosine phosphatases family. PTPs are essential in signal transduction, the process that relays information from the outside to the cell nucleus and tells the cell when to act. The specialized *PTPN22* protein regulates the T cells, part of the immune system. Variations in the *PTPN22* gene appear to increase the risk of vitiligo. With changes, the protein cannot function to keep the immune system from attacking its own cells. The exact trigger is unknown and probably a combination of both genetics and environment. *PTPN22* is inherited in an autosomal recessive pattern and is located on short arm (p) of chromosome 1 at position 13.2.

What Is the Treatment for Vitiligo?

Although popular myths abound about the treatment of vitiligo, with proper care, it is very treatable. Four options are available: use of sunscreens, covering up with makeup, restoration of normal skin color, and bleaching of normal skin with topical creams to make the skin an even color. Spot treatment with a topical steroid is the first option for medical treatment. If that is not effective, treatment with topical oxsoralen may be used. Other options include mini-skin grafting and whole body treatment with oral psoralen and ultraviolet light, known as PUVA treatment.

Further Reading

"Vitiligo." 2011. Genetics and Rare Disease Information Center (GARD). National Institutes of Health (U.S.). http://rarediseases.info.nih.gov/GARD/Disease.aspx?PageID=4&DiseaseID=10751. Accessed 12/2/11.

"Vitiligo." 2011. Genetics Home Reference. National Library of Medicine (U.S.). http://ghr.nlm.nih.gov/condition/vitiligo. Accessed 12/2/11.

"Vitiligo Treatments." 2011. American Vitiligo Foundation. http://www.avrf.org/treatments/treatments.htm. Accessed 12/2/11.

VLDLR-Associated Cerebellar Hypoplasia

Prevalence	Rare; prevalence unknown; first described in Canada and the United States; reported in families from Iran and Turkey
Other Names	autosomal recessive cerebellar ataxia with mental retardation; autosomal recessive cerebellar hypoplasia with cerebral gyral simplification; cerebellar disorder, nonprogressive, with mental retardation; cerebellar hypoplasia, VLDLR-associated; cerebellar hypoplasia and mental retardation with or without quadrupedal locomotion; CHMRQ1; DES-VLDLR; dysequilibrium syndrome-VLDLR; VLDLRCH; VLDLR-CH

As can been seen by the number of different names, there is a lot of different ideas about this disorder. VLDLR is an acronym meaning "very low density lipoprotein receptor," the protein that is associated with this disorder. In 2008, Ozcelik noted a large family in Turkey that walked on all fours, a condition called quadrupedal locomotion. Some researchers proposed that the name VLDLR include "quadrupedal locomotion"; however, not all people have this part of the disorder. It is referred to as *VLDLR*-associated cerebellar hypoplasia.

What Is VLDLR-Associated Cerebellar Hypoplasia?

VLDLR-associated cerebellar hypoplasia or very low density lipoprotein receptor-associated cerebellar hypoplasia is a disorder that affects the cerebellum, the back part of the brain. The cerebellum is the part of the brain that controls balance and coordination. The term "hypoplasia" comes from two Greek words: *hypo*, meaning "under," and *plas*, meaning "form." In cerebellar hypoplasia, the cerebellum is small and underdeveloped. The children who have little stability move or crawl around on all fours. Although some never learn to walk independently, others may learn to walk in later childhood, perhaps after the age of six. A family was found in rural Turkey that walked on their hands and feet.

The following additional symptoms may be associated with this condition:

- Intellectual disability: The disability may range from moderate to profound.
- Impaired speech: The person may not be able to speak or may have severely impaired speech.
- Eye problems: The person may have eyes that do not look in the same direction, a condition known as strabismus.
- Short stature.
- Flat feet.
- Seizures.

Even with all the disabilities, the person may have a normal life expectancy.

What Is the Genetic Cause of *VLDLR*-Associated Cerebellar Hypoplasia?

Mutations in the *VLDLR* gene, officially known as the "very low density lipoprotein receptor," cause *VLDLR*-associated cerebellar hypoplasia. Normally, *VLDLR* instructs for making the protein very low density lipoprotein (VLDL) receptor. This protein is essential to many organs, especially the heart, muscles, kidneys, fatty tissue, and the developing brain. The VLDL receptor works like a key in a lock with the protein reelin, which starts reactions within the cells. This signaling pathway is especially active in early embryonic development to guide immature nerve cells to the proper location in the brain.

At least six mutations cause VLDLR-associated cerebellar hypoplasia. The mutations disrupt the cells from producing a working protein. In the absence of the protein, the young immature nerve cells do not reach the part of the brain where they should function, primarily in the cerebellum. This leads to the problems with balance, impaired speech, and other symptoms. Other region of the brain may also be affected causing the intellectual disability. *VLDLR* is inherited in an autosomal recessive pattern and is located on the short arm (p) of chromosome 9 at position 24.

What Is the Treatment for *VLDLR*-Associated Cerebellar Hypoplasia?

Physical therapy may help the person maintain balance and learn to walk. Occupational therapy may be need for fine motor skill development. Other symptoms may be treated as needed. Educational support is essential.

Further Reading

Boycott, Kym M., and Jillian S Parboosingh. 2008. "*VLDLR*-Associated Cerebellar Hypoplasia." *GeneReviews*. http://www.ncbi.nlm.nih.gov/books/NBK1874. Accessed 12/4/11.

"*VLDLR*-Associated Cerebellar Hypoplasia." 2011. Genetics Home Reference. National Library of Medicine (U.S.). http://ghr.nlm.nih.gov/condition/vldlr-associated-cerebellar-hypoplasia. Accessed 12/4/11.

Von Hippel–Lindau Syndrome

Prevalence   About 1 in 36,000 people

Other Names angiomatosis retinae; cerebelloretinal angiomatosis, familial; Hippel-Lindau disease; VHL syndrome; von Hippel–Lindau disease

In 1904, Eugen von Hippel, a German ophthalmologist, observed tumors in the eyes that were filled with vascular tissue or unusual colored blood vessels. Later, in 1927, Arvid Lindau described similar blood vessels in the cerebellum and spine. The condition was called Von Hippel–Lindau syndrome.

A researcher from Vanderbilt University wrote an article that made a unique historical connection between Von Hippel–Lindau syndrome and the Hatfield-McCoy feud of the southern U.S. mountains. The scientist speculated that the hostility between the two families was the consequences of the growths called pheochromocytomas, which caused the hostile rage attacks. The McCoy family appeared to be predisposed to the disease because many of them were shown to have pheochromocytomas, which produced excess adrenaline and explosive tempers.

What Is Von Hippel–Lindau Syndrome?

Von Hippel–Lindau syndrome is a disorder in which many fluid-filled tumors or pheochromocytomas occur in many different parts of the body. These tumors can be both cancerous and noncancerous and usually appear in young adulthood, although they may be seen in people of all ages.

Following are the characteristics of Von Hippel–Lindau syndrome:

- Blood-filled tumors in the abdominal area: These tumors, called hemangio-blastomas, are made of collections of newly formed blood vessels. The cysts may develop into masses in the abdomen, kidneys, pancreas, adrenal glands, and male genital tract. Sometimes, they produce no symptoms but may lead to the increase in hormones that may cause high blood pressure. The tumors may also cause bursts of adrenaline, which may lead to aggressive behavior.
- Blood-filled tumors in the retina: These tumors may lead to vision loss.
- Blood-filled tumors or hemangioblastomas in the brain and central nervous system: These cause headaches and vomiting.
- Loss of muscle coordination.
- Noncancerous tumors in the inner ear.

The average age for diagnosis is about 26 years; however the symptoms can occur anytime from infancy to 70 years.

What Is the Genetic Cause of Von Hippel–Lindau Syndrome?

Mutations in the *VHL* gene or "von Hippel–Lindau tumor suppressor" cause Von Hippel-Lindau syndrome. Normally, *VHL* instructs for a protein that works with a group of other proteins called the VCB-CUL2 complex. This complex acts to break other proteins down when they are no longer needed. In this sense, *VRL* is a tumor suppressor gene, preventing cells from growing in an uncontrolled way. The complex targets large proteins, especially on called hypoxia-inducible factor or HIF. HIF is especially related to cell division and the formation of new blood vessels.

Over 370 mutations in the *VHL* gene cause von Hippel–Lindau syndrome. The mutations lead to an abnormal protein. Mutations are autosomal dominant but differ from the regular dominant pattern. In about 20% of cases, the altered gene is the result of a new mutation that happened when the egg or sperm was developing. Unlike most dominant conditions, just one mutated copy is not enough to cause the disorder. The second mutation must occur during the person's lifetime within specific cells such as the brain, retina, or kidneys. When the cells have two altered copies, they cannot make the VHL protein, allowing the cysts and tumors to develop. *VHL* is located on the short arm (p) of chromosome 3 at position 25.3.

What Is the Treatment for Von Hippel–Lindau Syndrome?

Treatment of such a multifaceted condition varies depending on the location and size of the cyst. Most of the growing cysts require surgery before they become harmful. Sometimes radiation may be used to treat the tumor. All conditions here require a medical team familiar with the disorder.

Further Reading

"NINDS Von Hippel–Lindau Disease (VHL) Information Page." 2011. National Institute for Neurological Disorders and Stroke (U.S.). http://www.ninds.nih.gov/disorders/von_hippel_lindau/von_hippel_lindau.htm. Accessed 5/31/12.

"Von Hippel–Lindau Syndrome." 2011. Genetics Home Reference. National Library of Medicine (U.S.). http://ghr.nlm.nih.gov/condition/von-hippel-lindau-syndrome. Accessed 5/31/12.

"Von Hippel–Lindau Syndrome." 2011. Medscape. http://emedicine.medscape.com/article/950063-overview#showall. Accessed 5/31/12.

Von Recklinghausen Disease. *See* Neurofibromatosis Type 1 (NF1)

Von Willebrand Disease

Prevalence   The most common genetic bleeding disorder; scientists differ on the number of people affected; some estimate 1 in 100, others report 1 in 10,000.

Other Names   angiohemophilia; Von Willebrand's factor deficiency

In 1924, Dr. Erik von Willebrand, a physician in Helsinki, Finland, tended to a five-year-old girl, who was brought to him with a serious bleeding disorder. The doctor was aware of hemophilia, which is seen mostly in males and determined to be carried by the mother. Interested in the girl who lived in the isolated community of the Aland Islands, he assessed 66 members of her family and determined that here was an undescribed bleeding disorder. This bleeding disorder appeared to be different from hemophilia both in its inheritance patterns and in presentation of symptoms. In 1950, researchers found that there was a special plasma factor and named the factor "von Willebrand factor" or VWF. The disorder became known as Von Willebrand disease.

What Is Von Willebrand Disease?

Von Willebrand disease is a bleeding disorder caused by a defect in the blood-clotting process. People with this disease do not have a blood factor, called the Von Willebrand factor, which allows for normal clotting and are susceptible to bruising, nosebleeds, or prolonged bleeding after even minor surgery. For example, having a tooth pulled or minor trauma may cause excessive bleeding. Women with this condition may experience heavy menstrual flow or unusual bleeding during childbirth. Symptoms may vary according to type and may change over time. Increased age, pregnancy, exercise, and stress may cause the Von Willebrand factor to rise, causing less frequent symptoms.

The Von Willebrand factor is a glue-like protein that is essential to the beginning stages of blood clotting. Cells that line the blood vessel walls produce the protein, which then interacts with the blood cells called platelets to from a clot. People without enough of the factor or with an abnormal factor are not able to make this clot.

Von Willebrand disease (VWD) is divided into the following three types:

- *Type I*: Type I is the mildest and most common form of VWD. People with this type have lower levels of the Von Willebrand factor, with levels of factor VIII (FVIII), another clotting factor also reduced. This type is inherited in a dominant pattern.

- *Type II*: In this type, the Von Willebrand factor itself has an abnormality. There are subtypes of the disorder that are intermediate in severity. The four subtypes vary primarily in the reduction of the factor and the ability of the platelets to clump together. This type is inherited in a recessive pattern.

- *Type III*: This form is the most severe. People have a total absence of the Von Willebrand factor and less than 10% of factor VIII. This type appears to be inherited in a dominant pattern.

- *Acquired VWD*: Mutations do not cause this type, but it is seen in conjunction with other orders, such as bone marrow or immune cell dysfunction. Usually appearing in adulthood, this type is characterized by abnormal bleeding into the skin or other soft tissues.

What Is the Genetic Cause of Von Willebrand Disease?

Mutations in the *VWF* gene or the "Von Willebrand factor" gene cause Von Willebrand Disease. Normally, *VWF* provides instructions for making the clotting protein Von Willebrand factor. This protein binds to specific cells and proteins to form a blood clot. The role of the clot is to keep the body from losing blood and to seal off damaged blood vessels.

The clotting process is quite complex. The Von Willebrand factor is made in the cells that line the blood vessels and the bone marrow cells and is made up of subunits. An enzyme called ADAMTS13 cuts the subunits into smaller parts.

These parts help the platelets to stick to each other and to hold on to the walls of the blood vessels at the site to plug the wound. The Von Willebrand factor also carries the clotting factor VIII to the area of the clot.

Over 300 mutations in *VWF* cause the various types of Von Willebrand disease. People with type I have the factor, but with reduced amounts. Mutations that cause the subtypes of type II usually have a change in only one of the protein building blocks. These changes disrupt the ability of the cells to stick together to form clots. People with type III, the most severe type, usually have an abnormally short, nonworking form of the factor. Typical of all three types is some malfunction of the factor, which then leads to the bleeding episodes. According to the type, *VWF* is inherited in both recessive and dominant patterns and is located on the short arm (p) of chromosome 12 at position 13.3.

What Is the Treatment for Von Willebrand Disease?

Knowing the type of the disorder is important. Working with a medical team, the individual must be constantly aware to prevent severe bleeding episodes. A blood transfusion may be necessary. At one time the blood supply was tainted with unwanted viruses. The supply is safe from HIV today, but people are still advised to get hepatitis vaccinations. Mild bleeding episodes may be treated with desmopressin, a type of vasopressin, which helps constrict the blood vessels.

Further Reading

Goodeve, Anne, and Paula James. 2011. "Von Willebrand Disease." *GeneReviews*. http://www.ncbi.nlm.nih.gov/books/NBK7014. Accessed 5/31/12.

"Von Willebrand Disease." 2011. Genetics Home Reference. National Library of Medicine (U.S.). http://ghr.nlm.nih.gov/condition/von-willebrand-disease. Accessed 5/31/12.

"Von Willebrand Disease." 2011. National Hemophilia Foundation. http://www.hemophilia.org/NHFWeb/MainPgs/MainNHF.aspx?menuid=182&contentid=47&rptname=bleeding. Accessed 5/31/12.

W

Waardenburg Syndrome (WS)

Prevalence   Affects about 1 in 10,000 to 20,000 people; types I and II most common; types III and IV rare; 2% to 3% of students in deaf schools have this condition

Other Names   Waardenburg-Klein syndrome Mende's syndrome II; Van der Hoeve–Halbertsma–Waardenburg syndrome; Ptosis-Epicanthus syndrome; Van der Hoeve–Halbertsma–Gualdi syndrome; Waardenburg–Shah Syndrome; Waardenburg's syndrome; Waardenburg's syndrome II and Vogt's syndrome; Waardenburg type Pierpont, Van der Hoeve–Waardenburg–Klein syndrome

In 1947, Petrus Johannes Waardenburg a Dutch ophthalmologist, noted a condition in which the patient had hearing loss, an unusual eye pattern, and different eye pigments. He studied the condition in other patients with similar symptoms and wrote about it in 1951. He described the most common type, known now as type I. Later, three other types were identified and were named after the Dutch eye doctor—Waardenburg syndrome.

What Is Waardenburg Syndrome (WS)?

Waardenburg syndrome is a rare disorder characterized by varying degrees of hearing loss, defects arising from the embryonic structure of the neural crest, and changes in the color of hair, skin, and eyes.

Following are the symptoms of Waardenburg syndrome:

- Hearing loss: Hearing loss ranges from profound hearing loss in one or both ears to no hearing loss.

- Unusual colored part of the eye: The colored part of the eye is called the iris. People with this condition have very pale-blue eyes, iris bicolor, or two eyes each with a different color.
- White hair in patches: The individual may have a white forelock or white hair on other parts of the body.
- White patches of skin on the body.
- Broad nose.
- Eyebrows flare.
- Premature graying of hair before age 30.
- Lateral displacement of the canthi or the eyes.

Four types of Waardenburg syndrome are known:

- *Type I*: In addition to some of the above symptoms, people with this type always have widely spaced eyes.
- *Type II*: Hearing loss occurs more frequently in people with this type.
- *Type III*: This type is also known as Klein-Waardenburg syndrome and is characterized by abnormalities of the upper limb, in addition to hearing loss and changes in skin and hair pigment.
- *Type IV*: This type is also called Waardenburg-Shah syndrome and has many of the symptoms of Waardenburg syndrome in addition to blockage of the intestine and severe constipation.

What Are the Genetic Causes of Waardenburg Syndrome (WS)?

Mutations in the *EDN3*, *EDNRB*, *MITF*, *PAX3*, *SNAI2*, and *SOX10* genes cause Waardenburg syndrome.

EDN3

The *EDN3* gene, officially called the "endothelin 3" gene, normally encodes for the protein endothelin 3, a member of the endothelin family found in many cells and tissues. Endothelin 3 interacts with a receptor on the surface of cells called endothelin receptor B. These two proteins are essential in embryonic development in an area called the neural crest. The cells move from the developing spinal cord to other parts of the embryo to give rise to the nerves of the large intestine and also pigment-forming cells called melanocytes. These cells will eventually give color to skin, hair, and eyes. Melanin cells are also active in the development of the inner ear.

Some mutations in *EDN3* are related to Waardenburg syndrome type IV. This type not only has changes in the pigment but also a severe intestinal condition called Hirschsprung disease. The mutations result in a change in one amino acid building block in the gene, which produces a protein that disrupts the function of endothelin 3. The embryonic cell types do not form properly, causing the many symptoms of Waardenburg type IV. *EDN3* is inherited in an autosomal dominant pattern and is located on the long arm (q) of chromosome 20 at 13.2-q13.3.

EDNRB

Mutations in the *EDNRB* gene, officially called the "endothelin receptor type Burg" gene, are also related to Waardenburg type IV. Normally, *EDNRB* encodes for the endothelin receptor, which is located on the surface of the cells. This protein interacts with the protein made by the *EDN3* gene to initiate many essential actions in the embryo. The two proteins are important for the action of the neural crest cells to form melanin and the nerves of the large intestine.

About three mutations in *EDNRB* are related to type IV Waardenburg syndrome. The mutations disrupt the normal production of endothelin receptor type B, leading to an abnormally, small nonworking version. The protein is necessary for the formation of the pigment and the intestinal nerves. Its lack of function leads to the symptoms of type IV Waardenburg syndrome. *EDNRB* is inherited in an autosomal dominant pattern and is located on the long arm (q) of chromosome 13 at position 22.

MITF

Mutations in the *MITF* gene, officially called the "microphthalmia-associated tran-scription factor" gene, cause Waardenburg syndrome type II. Normally, *MITF* pro-vides instructions for the protein microphthalmia-associated transcription factor. This factor is essential for the working of certain kinds of cells. It attaches to areas of DNA to control genes in the particular location of the pigment-producing cell called melanocytes. These cells will produce melanin that gives skin, eye, and hair a specific color. These cells are also found in the inner ear and provide nourish-ment to the retina.

Several mutations in *MITF* are related to type II. The mutations cause a change in the amino acids used to make microphthalmia-associated transcription factor. As a result, a short form of the factor is created, disrupting the development and leading to a shortage of melanocytes. This shortage causes the loss of pigment characteristic of Waardenburg syndrome. *MITF* is inherited in an autosomal dominant pattern and is located on the short arm (p) of chromosome 3 at 14.2-14.1.

PAX3

Officially called "paired box 3," the *PAX3* gene provides instructions for a protein that attaches to DNA and controls other genes especially during embryonic development. The PAX3 protein is essential in cells called the neural crest, a stage in the development of the spinal cord before birth. The PAX3 protein directs the migration of cells to form specialized tissues, such as bones in face and skull, muscles, nerves, and the pigment melanin.

Several mutations in the *PAX3* gene causes most cases of Waardenburg type I, and some cases of type III. Some mutations are changes in one of the amino acid building blocks, and others lead to very small version of the protein. All the mutations have the same effect; they disrupt the ability of the protein to bind to regions of the DNA to regulate it. As a result, none of the activity necessary for development of melanocytes is present, leading to the characteristics of Waardenburg syndrome. *PAX3* is inherited in an autosomal dominant pattern and is found on the long arm (q) of chromosome 2 at position 35-q37.

SNAI2

Mutations in *SNAI2*, officially called "snail homolog 2 (Drosophila)," is another gene related to Waardenburg syndrome. Like the *PAX3* gene, this gene is a transcription factor that controls other genes. Normally, this gene protein is essential for the development for the neural crest during embryonic development. The protein directs the development of tissues and melanocytes. This protein appears to have a greater role in the survival of melanocytes.

Mutations in *SNAI2* have caused copies of the protein to go missing. When no copies are present, the development of melanocytes is disrupted, causing the characteristic features of type II. *SNAI2* is inherited in an autosomal dominant pattern and is located on the long arm (q) of chromosome 8 at position 11.

SOX10

Mutations in the *SOX10* gene, officially known as the "SRY (sex determining region Y)-box 10" gene, cause Waardenburg syndrome, type IV. Normally, *SOX10* is also active in embryonic neural crest cells. It directs other genes such as *MITF* to signal the cells to become other cells. The SOX10 protein is essential in forming nerves in the large intestine and melanocytes.

About 15 mutations in *SOX10* are related to Waardenburg, type IV. Most mutations cause a very small form of the protein that cannot perform its normal functions in the neural crest. As a result, the nerves that are part of the intestine do not form and melanocytes are missing. *SOX10* is inherited in an autosomal dominant pattern and is located on the long arm (q) of chromosome 22 at position 13.1.

What Is the Treatment for Waardenburg Syndrome (WS)?

Treatment for this condition depends on the type and severity. Some of the malformations of the bones can be corrected with surgery. A medical team is essential to access the condition.

Further Reading

"Genetics of Waardenburg Syndrome." 2011. Medscape. http://emedicine.medscape.com/article/950277-overview#showall. Accessed 5/31/12.

"Waardenburg Syndrome." 2008. *New York Times*. http://health.nytimes.com/health/guides/disease/waardenburg-syndrome/overview.html. Accessed 5/31/12.

"Waardenburg Syndrome." 2011. Genetics Home Reference. National Library of Medicine (U.S.). http://ghr.nlm.nih.gov/condition/waardenburg-syndrome. Accessed 11/14/11.

Wagner Syndrome

Prevalence   Rare; exact prevalence unknown; affects 50 families worldwide

Other Names   hyaloideoretinal degeneration of Wagner; VCAN-related vitreoretinopathy; Wagner disease; Wagner vitreoretinal degeneration

In 1938, Hans Wagner of Zurich, Switzerland, noted 13 members of a family that had a suspicious growth in the vitreous humor and retina of the eye. Later in 1960, Boehringer observed 10 more affected members of the same family. Jansen in Holland in 1962 described two families that had 39 members affected by the same condition. At one time, the syndrome was thought to be the same as Stickler syndrome; gene analysis solved this problem. The one condition is now known as Wagner syndrome

What Is Wagner Syndrome?

Wagner syndrome is a disorder of the eyes that leads to vision loss. The loss begins usually in adulthood and progressively worsens with age. In this syndrome, the following things occur:

- The light-sensitive cells that line the back of the eye (retina) become very thin.
- The retina may become detached.
- Abnormal blood vessels are also at the back of the eye.
- The fluid that fills the ball of the eye becomes very thin and watery.
- The fovea, the part of the retina where sharpened vision takes place, is not located in the proper place.
- The person may be very nearsighted.
- The person may have a clouding of the lens of the eye, called a cataract.

What Is the Genetic Cause of Wagner Syndrome?

Mutations in the *VCAN* gene, officially known as the "versican" gene, cause Wagner Syndrome. *VCAN* instructs for the protein versican, which is located in the areas surrounding many tissues and organs. That substance called the extracellular matrix forms a lattice of proteins and other molecules to fill the spaces around the areas. Versican is essential in making this matrix by helping cells attach to one another. Versican also is active in forming blood vessels, wound healing, and preventing growth of tumors. *VCAN* produces four versions of the protein, which can vary in size depending on their location in the body.

About six mutations in *VCAN* cause Wagner syndrome. All the mutations disrupt the function of the versican protein, either by decreasing it or increasing it when it is not needed. The imbalance of versican appears cause the problems in the eye. *VCAN* is inherited in an autosomal dominant pattern and is located on the long arm (q) of chromosome 5 at position 14.3.

What Is the Treatment for Wagner Syndrome?

A qualified ophthalmologist can treat the symptoms of Wagner syndrome. Near-sightedness can be corrected by glasses. Retinal problems can be corrected with laser surgery. If necessary, cataract surgery can be performed. Early diagnosis is essential.

Further Reading

Kloeckener-Gruissem, Barbara, and Christoph Amstutz. 2009. "*VCAN*-Related Vitreore-tinopathy (Wagner Syndrome)." *GeneReviews*. http://www.ncbi.nlm.nih.gov/books/NBK3821. Accessed 12/5/11.

"Wagner Syndrome." Genetics Home Reference. National Library of Medicine (U.S.). http://ghr.nlm.nih.gov/condition/wagner-syndrome. Accessed 12/5/11.

"Wagner Syndrome." http://www.wagnersyndrome.eu. Accessed 12/5/11.

WAGR Complex. *See* Wilms Tumor and WAGR Syndrome

Walker-Warburg Syndrome (WWS)

Prevalence	Incidence unknown; survey in northeastern Italy reported rate of 1.2 per 100,000 births; 1 in 149 in the Ashkenazi Jewish population
Other Names	cerebroocular dysgenesis (COD) or cerebroocular dysplasia–muscular dystrophy syndrome (COD-MD); Chemke syndrome; HARD syndrome (Hydrocephalus, Agyria and Retinal Dysplasia); Pagon syndrome; Warburg syndrome

In 1942, Arthur Earl Walker noted a type of weakness of the voluntary muscles, as well as brain and eye abnormalities. Later in 1971, Mette Warburg wrote about a similar syndrome of mirophthalmia in the mentally retarded. The syndrome was named after these two doctors—Walker-Warburg syndrome.

What Is Walker-Warburg Syndrome (WWS)?

Walker-Warburg syndrome (WWS) is the most severe form of congenital muscular dystrophy. Those who are affected usually die before the age of three. Following are the characteristics of the disorder:

- Malformation of the head and brain at birth
- Weakness and atrophy of voluntary muscles

- Seizures
- Eye abnormalities
- Developmental delays
- Mental retardation

What Is the Genetic Cause of Walker-Warburg Syndrome (WWS)?

Enzymes produced by two genes *POMT1* and *POMT2* must be present for normal processes to be carried out. If there is a mutation in one of the genes and it is dysfunctional, the person may display the symptoms of Walker-Warburg syndrome.

POMT1

Mutations in the *POMT1*, or the "protein-O-mannosyltransferase 1" gene, are related to WWS. Normally, *POMT1* encodes for O-mannosyltransferase, an enzyme that requires another enzyme produced by the *POMT2* gene. The protein is found in the cell membrane of the endoplasmic reticulum, a maze-like structure that carries materials from the cell to outside the cell.

Mutations in *POMT1*, in addition to the enzyme produced by *POMT2*, cause Walker-Warburg syndrome. The mutations disrupt the function of the enzyme O-mannosyltransferase. *POMT1* is inherited in an autosomal recessive pattern and is located on the long arm (q) of chromosome 9 at position 34.1.

POMT2

Along with the enzyme produced by *POMT1*, the gene *POMT2*, officially known as the "protein-O-mannosyltransferase 2" gene, causes Walker-Warburg syndrome. Normally, this gene also encodes for O-mannosyltransferase, which is found in the endoplasmic reticulum. Coexpression of both *POMT1* and *POMT2* are necessary for enzyme activity. If one is missing, the other does not work. *POMT2* is inherited in an autosomal recessive pattern and is found on the long arm (q) of chromosome 14 at position 24.

What Is the Treatment for Walker-Warburg Syndrome?

No cure and no specific treatment exists for Walker-Warburg syndrome. Management of the infant is supportive or preventive. If the child has seizures, he may be given anticonvulsants. A shunt may remove fluid collected on the brain. Physical therapy may help muscle action.

Further Reading

"Walker Warburg Syndrome." 2009. National Organization for Rare Disorders (NORD). http://www.rarediseases.org/rare-disease-information/rare-diseases/byID/970/viewAbstract. Accessed 5/31/12.

"Walker-Warburg Syndrome." 2011. Chicago Center for Jewish Genetic Disorders. http://www.jewishgenetics.org/?q=content/walker-warburg-syndrome. Accessed 5/31/12.

"Walker-Warburg Syndrome." 2011. National Center for Biotechnology Information. National Library of Medicine (U.S.). http://www.ncbi.nlm.nih.gov/pmc/articles/PMC1553431. Accessed 5/31/12.

Weill-Marchesani Syndrome

Prevalence   Rare; estimated 1 in 100,000 people

Other Names   congenital mesodermal dysmorphodystrophy; Marchesani syndrome; spherophakia-brachymorphia syndrome; WMS

In 1932, Georges Weill in Alsace-Lorraine (now part of France) described eight patients with dislocations of the lens of the eye, short stature, and other common features. In 1939, Oswald Marchesani in Germany reported on an eight-year-old body and three siblings that had a similar lens disorder, short fingers, and short stature. In 1966, the condition was described in the Old Order Amish community in the United States. Later researchers combined the name to form Weill-Marchesani syndrome.

What Is Weill-Marchesani Syndrome?

Weill-Marchesani syndrome is a disorder of the connective tissue. The seemingly unimportant connective tissue is what holds the parts of the body together. It provides support for the bones and joints of the body framework and adds strength to muscles. Connective tissue forms the body's supportive framework, provides structure and strength to the muscles, joints, and organs, and firms the skin. The following are characteristic symptoms of Weill-Marchesani syndrome:

- Short stature: Adults' height for men ranges from 4 feet, 8 inches to 5 feet, 6 inches; women range from 4 feet, 3 inches to 5 feet, 2 inches.
- Very short fingers and toes, a condition called brachydactyly.
- Eye abnormalities: An eye condition called microspherophakia causes the eye to be small with the lens shaped like a sphere and positioned abnormally.
- Eye conditions: The unusual shape of the eye causes nearsightedness. Some people develop glaucoma, which can lead to blindness.
- Heart disorders: Some people may develop an abnormal heart rhythm.

What Are the Genetic Causes of Weill-Marchesani Syndrome?

Two genes are involved in Weill-Marchesani syndrome: *ADAMTS10* and *FBN1*.

ADAMTS10

Mutations in the gene *ADAMTS10*, officially known as the "ADAM metallopepti-dase with thrombospondin type 1 motif, 10" gene, cause Weill-Marchesani syn-drome. Normally, *ADAMTS10* instructs for an enzyme that is part of the metalloproteases family. These proteases cut apart enzymes containing the metal zinc and other metals. The enzyme, which is found in many cells and tissues, is very important in the development of structures both before and after birth.

Five mutations in *ADAMTS10* cause Weill-Marchesani syndrome. Each of the enzymes prevents the production of a working ADAMTS10 enzyme, disrupting development of the skeleton and the lens of the eye and impairing vision. It may also cause defects in the heart. *ADAMTS10* is inherited in an autosomal recessive pattern and is found on the short arm (p) of chromosome 19 at position 13.2.

FBN1

Mutations in the *FBN1* gene, officially known as the "fibrillin 1" gene, cause Weill-Marchesani syndrome. Normally, *FBN1* instructs for the large protein fibril-lin-1, which moves in and out of cells into the extracellular matrix that holds cells in place. In the matrix, fibrillin-1 attaches to other molecules to form small threads called microfibrils. These microfibrils become part of the elastic bodies that let skin, ligaments, and blood vessels stretch. They also support the lens of the eye, nerves, and muscles. In addition, microfibrils control important growth factors called TGF-beta growth factors. When the TGF-beta proteins are released from cells, they determine the growth and repair of tissues throughout the body.

Only one mutation in *FBN1* has been found in one family with Weill-Marchesani syndrome. This mutation deletes part of the gene, causing an unstable fibrillin-1 protein. This protein then disrupts the microfibrils, causing weakened connective tissue and the abnormalities associated with the disorder. *FBN1* is inherited in an autosomal dominant pattern and is located on the long arm (q) of chromosome 15 at position 21.1.

What Is the Treatment for Weill-Marchesani Syndrome?

Early detection of the condition and removal of the lens can help vision. The other symptoms can be dealt with on a case-to-case basis.

Further Reading

Tsilou, Ekaterini, and Ian M. MacDonald. 2007. "Weill-Marchesani Syndrome." *GeneReviews.* http://www.ncbi.nlm.nih.gov/books/NBK1114. Accessed 12/5/11.

"Weill-Marchesani Syndrome 2." 2011. University of Arizona Hereditary Ocular Disease Center. http://disorders.eyes.arizona.edu/disorders/weill-marchesani-syndrome-2. Accessed 12/5/11.

Weissenbacher-Zweymüller Syndrome

Prevalence   Very rare; only a few families known worldwide

Other Names heterozygous OSMED; heterozygous otospondylomegaepiphyseal dysplasia; Pierre Robin syndrome with fetal chondrodysplasia; WZS

At one time this group of symptoms was considered simply the newborn
expression of Wagner syndrome. It also has some things in common with Pierre
Robin syndrome and Stickler syndrome. However, in 1964, G. Weissenbacher
and Ernst Zweymüller characterized the group of conditions as a separate disorder.
The condition was called Weissenbacher-Zweymüller syndrome.

What Is Weissenbacher-Zweymüller Syndrome?

Weissenbacher-Zweymüller syndrome is a condition present at birth that affects
bone growth and several other systems. Following are the characteristics of
Weissenbacher-Zweymüller syndrome:

- Short limbs
- Abnormal thickness at joints, especially the knees and elbows
- Sensorineural hearing loss, especially with high tones
- Bones of the spine abnormally shaped
- Distinctive facial features, which include wide-set protruding eyes, a small
  upturned nose with flat bridge, and small lower jaw
- Cleft palate

The features are similar to those of otospondylomegaepiphyseal dysplasia
(OSMED). However, as the child grows, the skeletal conditions tend to vanish.
Adults are not unusually short but may still retain other features of the syndrome.

What Is the Genetic Cause of Weissenbacher-Zweymüller Syndrome?

Mutations in the *COL11A2* gene, officially known as the "collagen, type XI, alpha 2"
gene, cause Weissenbacher-Zweymüller syndrome. Normally, *COL11A2* instructs for
making one part of a type XI collagen called the pro-alpha2(XI) chain. Type XI
collagen has many functions, including the following:

- Found in cartilage, the tough, flexible material that makes up most of the
  skeleton during early development
- Strengthens the connective tissues that support muscles, skin, joints, and organs
- Found in the clear gel that fills the eyeball, called the vitreous humor
- Found in the inner ear
- Located in the center of the vertebrae and the ends of bones

Type XI collagen is made of several chains of ropelike molecules that enzymes in the cells can process. To make the collagen, pro-alpha2(XI) chain combines with two other collagen chains (pro-alpha1[XI] and pro-alpha1[II]) to form a procollagen molecule that then cross-links into long thin fibrils and fills spaces around cells. Type XI collagen also aids in the spacing of type II collagen, an important part of the eye and cartilage tissue.

At least one mutation in *COL11A2* causes Weissenbacher-Zweymüller syndrome. In this mutation, glutamic acid replaces glycine in the pro-alpha2(XI) chain. This disruption keeps the strong ropelike molecule of type XI collagen from forming properly. The disruption results in the symptoms of Weissenbacher-Zweymüller syndrome. *COL11A2* is inherited in an autosomal dominant pattern and is located on the short arm (p) of chromosome 6 at position 21.3.

What Is the Treatment for Weissenbacher-Zweymüller Syndrome?

The treatment for this condition is symptomatic. Conditions such as the cleft palate can be treated with plastic surgery. Many of the skeletal conditions become less problematic with age. Other issues such as hearing or eye problems are treated on an individual basis.

Further Reading

"Weissenbacher-Zweymüller Syndrome." 2005. *Gale Encyclopedia of Public Health*. http://www.healthline.com/galecontent/weissenbacher-zweymuller-syndrome-1. Accessed 5/31/12.

"Weissenbacher-Zweymüller Syndrome." 2011. Genetics Home Reference. National Library of Medicine (U.S.). http://ghr.nlm.nih.gov/condition/weissenbacher -zweymuller-syndrome. Accessed 5/31/12.

"Weissenbacher-Zweymüller Syndrome." 2011. RightDiagnosis.com. http://www .rightdiagnosis.com/medical/weissenbacher_zweymuller_syndrome.htm. Accessed 5/31/12.

Werner Syndrome

Prevalence	Affects about 1 in 200,000 in the United States; seen more frequently in Japan, with 1 in 20,000 to 1 in 40,000 affected
Other Names	adult premature aging syndrome; adult progeria; Werners syndrome; Werner's syndrome; WS

In 1904, Otto Werner noted a condition in which a patient appeared to age before his or her time. In their early teens or young adulthood, the person would develop many of the symptoms associated with people much older, such as wrinkled thin skin, cataracts, or other conditions. Later the disorder was named for this doctor—Werner syndrome. Like Hutchinson-Gilford syndrome, which occurs in young children, this condition is one of the progerias. The word comes from two Greek words: *pro*, meaning "before," and *ger*, meaning old. (This is the same root word as seen in geriatrics, the science of aging.) Thus, the meaning of the word progeria is getting old before it is time to do so.

The images of this disease evoke not only the imagination, but scientific interest. Studying the process of progeria sheds light on understanding the processes of normal aging. The film community has also used characters in their series to show people with progeria. In the television series *Bones*, the victim in an episode called "Stargazer in a Puddle" had Werner syndrome. In the film *Jack*, Robin Williams played a character who aged four times faster than normal.

What Is Werner Syndrome?

Werner syndrome is a condition in which the person ages prematurely. For example, a person in their early 40s may look like a person in the late 70s. The person with the disorder may grow and develop normally until puberty. Teenagers with the condition may not have a growth spurt and remain short. Most of the cases begin in the 20s with the following symptoms:

- Hair: The person may have premature gray hair. He or she may lose the hair prematurely.
- Voice: The voice becomes hoarse and shaky.
- Skin: The skin becomes thin, inflexible, and hard.
- Distinct appearance: The face becomes "bird-like."
- Thick trunk due to abnormal fat deposits.
- Thin arms and legs.
- Disorders associated with aging: Person may develop cataracts in both eyes, hardening of the arteries, thinning of the bones or osteoporosis, type 2 diabetes, and loss of fertility. The individual may develop several rare types of cancer during life.
- No mental retardation or deterioration is observed.

The person may die in the late 40s or 50s, usually from cancer or hardening of the arteries.

Scientists who study aging know that the process involves errors as the cells divide over and over again. In this process called mitosis that occurs after the postreproductive stage of life, the instability of the genome activates stress kinases that cause more errors in cell division.

What Is the Genetic Cause of Werner Syndrome?

Mutations in the *WRN* gene, officially known as the "Werner syndrome, RecQ helicase-like" gene, cause Werner syndrome. Normally, *WRN* instructs for the Werner protein, which functions as an enzyme called a helicase. Helicases are enzymes that unwind the double strands of the DNA so division can take place. The Werner protein also plays the role of an exonuclease, an enzyme that cuts off damaged broken ends of the DNA by removing those unneeded parts. First, the protein unwinds the DNA and then cuts the abnormal structures that have been generated accidentally. This protein may have the role of being the watchdog for the telomeres (the ends of the chromosomes) and for maintaining the health of the genome.

Over 60 mutations in *WRN* cause Werner syndrome. Most of the mutations cause an abnormally shortened gene that cannot be transported into the cell's nucleus to unwind the DNA. The DNA replication and repair is disrupted, leading to the signs and symptoms of Werner syndrome. *WRN* is inherited in an autosomal recessive pattern and is located on the short arm (p) of chromosome 8 at position 12.

What Is the Treatment for Werner Syndrome?

Treating the life-threatening conditions that develop, such as cancers and hardening of the arteries, is first priority. Other symptoms can be treated on an individual basis. In 2010, researchers found that supplementing vitamin C reversed damage in mouse models of Werner syndrome. Vitamin C appears to increase tissue repair in these mice.

See also Aging and Genetics: A Special Topic

Further Reading

"Werner Syndrome." 2011. About.com Rare Diseases. http://rarediseases.about.com/od/rarediseasesw/a/100304.htm. Accessed 12/5/11.

"Werner Syndrome." 2011. Genetics Home Reference. National Library of Medicine (U.S.). http://ghr.nlm.nih.gov/condition/werner-syndrome. Accessed 12/5/11.

Werner Syndrome Web Page. 2011. University of Washington. http://www.wernersyndrome.org. Accessed 12/5/11.

Williams Syndrome

Prevalence   About 1 in 7,500 to 20,000 people

Other Names   Beuren syndrome; elfin facies syndrome; elfin facies with hypercalcemia; hypercalcemia-supravalvar aortic stenosis; infantile hypercalcemia; supravalvar aortic stenosis syndrome; WBS; Williams-Beuren syndrome; WMS; WS

In 1961, Dr. J. C. P. Williams, a New Zealand physician, noted a developmental disorder in which the child had a distinct facial appearance and conditions that affected many body systems. Although other physicians had studied the condition and the disorder has had many names, it is most commonly called Williams syndrome.

What Is Williams Syndrome?

Williams syndrome affects many areas of the body. Following are the many symptoms of the disorder:

- Heart and blood vessel disorders: The blood vessels of the aorta, the larger artery that carries blood from the heart to the body, are narrow. In addition, the artery leading to the lungs may be narrow. The term used to describe the narrowing of such structures is "stenosis."
- High blood pressure.
- High blood calcium levels: These high levels may cause seizures and muscle rigidity.
- Feeding disorders: The infant may not eat or vomit what has been eaten. She may also have colic and reflux.
- Inward bend of little finger, a condition called clinodactyly.
- Distinct facial appearance: The individual has a small upturned nose, flattened nasal bridge, ridges that run from nose to upper lip, large lips, open mouth, and skin covering inner part of the eye.
- Partially missing or widely spaced teeth with defective enamel.
- Sunken chest.
- Developmental delay, especially speech.
- Spatial difficulties: Children with Williams syndrome may have difficulty with tasks such as drawing and assembling puzzles.
- Attention deficit disorder (ADD).
- Phobias.
- Mild to moderate intellectual disability.
- Farsightedness.
- Short stature, compared to rest of the family.
- Complications later in life: Calcium deposits may lead to kidney disorders. Heart failure may occur due to narrowed blood vessels.

In spite of the many physical difficulties, children with Williams syndrome are friendly, outgoing, and interested in other people. They may do very well in music and learning by rote memory. However, most individuals do not live as long as normal because of possible complications. They may require full-time caregivers and often live in supervised group homes.

What Are the Genetic Causes of Williams Syndrome?

Deletions in a certain area of chromosome 7 cause Williams syndrome. In this area are about 25 genes, including the following five genes that play a part in the many abnormalities: *CLIP2*, *ELN*, *GTF2I*, *GTF2IRD1*, and *LIMK1*. Although these genes appear as autosomal dominant, the loss of the region of the chromosome is not inherited but thought to be a random mistake occurring during the formation of sperm or egg.

CLIP2

Missing segments of chromosome 7 that include *CLIP2*, officially called the "CAP-GLY domain containing linker protein 2" gene, cause some of the symptoms of Williams syndrome. Normally, *CLIP2* instructs for making a protein called CAP-GLY domain containing linker protein 2, which is found primarily in the brain. The protein regulates the structural framework of the cytoskeleton and is especially associated with the rigid, hollow microtubules that make help determine size, shape, and movement of the cell.

Loss of the *CLIP2* gene disrupts the normal regulation of the cytoskeleton and affects the structure of the nerve cells in the brain, leading to the distinct behavior patterns of people with this disorder. *CLIP2* is found in cells as an autosomal dominant pattern and is located on the long arm (q) of chromosome 7 at position 11.23.

ELN

Missing segments of chromosome 7, which include the *ELN* gene or the "elastin" gene, cause some of the symptoms of Williams syndrome. Normally, *ELN* makes a protein called elastin. This protein makes up the slender bundles of elastic fibers that are located between spaces in the cells. This area is called the extracellular matrix. Elastin gives support to all the major tissues and organs of the skin, heart, blood vessels, and lungs.

As a result of the deletion, people have a missing copy of *ELN* in the cells, reducing elastin by half. This disruption of the normal structure of elastin makes blood vessels thicker and less resilient, leading to constricted flow and serious complications such as heart failure, shortness of breath, and chest pain. *ELN* is seen in the cells as a dominant pattern and is located on the long arm (q) of chromosome 7 at position 11.23.

GTF2I

Deletions of the area of chromosome 7 include the gene *GTF2I*, officially known as the "general transcription factor IIi" gene. Normally, *GTF2I* instructs for making the two proteins BAP-135 and TFII-I. BAP135 is active in the B cells, a specialized part of the immune system. This protein is essential in a reaction that tells the B cells when to divide and when to produce the proteins necessary for fighting outside invaders. TFII-I protein is a transcription factor that binds to

DNA to regulate the activity of genes. This protein is active in the brain and other body tissues. It especially is essential in regulating cell access to calcium.

Loss of this gene may cause some of the unusual behavioral patterns and intellectual disabilities of people with the syndrome. It may also cause some of the dental abnormalities and problems with visual-spatial tasks. *GFT2I* appears as an autosomal dominant pattern and is located on the long arm (q) of chromosome 7 at position 11.23.

GTF2IRD1

Loss of a segment of chromosome 7 includes the *GTF2IRD1* gene or the "GTF2I repeat domain containing 1" gene. Normally, *GTF2IRD1* instructs for making a protein that is also a transcription factor, meaning that it binds to specific areas of DNA to turn off and on other genes. This protein is essential for brain, tissue, and facial development.

Researchers think that this gene may cause some of the distinctive facial characteristics and other behavioral disorders. The gene is found in the cells as an autosomal dominant pattern and is located on the long arm (q) of chromosome 7 at position 11.23.

LIMK1

Deletions in part of chromosome 7 include the *LIMK1* gene or the "LIM domain kinase 1" gene. Normally, *LIMK1* instructs for an essential brain protein. This protein is important for areas of the brain where visual-spatial information is processed. Within the cells, the LIMK1 protein appears to regulate cell size, shape, and movement and to control the organization of actin filaments, structures that make up the cytoskeleton.

As a result of the loss of this gene, people may experience loss of visual-spatial ability and other impairments. The gene is found in the cells as an autosomal dominant pattern and is located on the long arm (q) of chromosome 7 at position 11.23.

NCF1

Deletion of areas of chromosome 7 includes the *NCF1* gene, or the "neutrophil cytosolic factor 1" gene. Normally, *NCF1* provides instructions for making the protein neutrophil cytosolic factor 1. This essential protein is part of a large enzyme complex called NADPH oxidase, a factor in the phagocytes, a type of white blood cell in the immune system. When bacteria or other foreign invaders are present, this enzyme converts oxygen to a toxic molecule called superoxide, which then generates the strong disinfectant hydrogen peroxide and hypochlorous acid, the same chemical as in bleach. These substances, known are reactive oxygen species, are available for the phagocytes to kill foreign invaders and prevent them from reproducing in the body.

Deletions of *NCF1* gene lead to only one copy in the cell. Scientists believe this loss leads to hypertension and other symptoms of Williams syndrome. *NCF1* is found in an autosomal dominant pattern and is located on the long arm (q) of chromosome 7 at position 11.23.

What Is the Treatment for Williams Syndrome?

No cure exists for this syndrome. Serious, life-threatening conditions such as heart and lung disorders should be treated first. Treatment is symptomatic depending on the individual manifestation of the disorder. Physical therapy can alleviate some of the joint stiffness. Developmental and speech therapy may help children develop verbal strengths to overcome spatial deficits.

Further Reading

"Williams Syndrome." 2011. Genetics Home Reference. National Library of Medicine (U.S.). http://ghr.nlm.nih.gov/condition/williams-syndrome. Accessed 5/31/12.

"Williams Syndrome." 2011. MedlinePlus. National Library of Medicine (U.S.). http://www.nlm.nih.gov/medlineplus/ency/article/001116.htm. Accessed 5/31/12.

The Williams Syndrome Foundation. 2011. http://www.wsf.org. Accessed 5/31/12.

Wilms Tumor and WAGR Syndrome

Prevalence of Wilms Tumor	Affects about 10 children and adolescents per 1 million before the age of 15; 450–500 new cases diagnosed each year
Prevalence of WAGR	Rare; prevalence unknown; about one-third of people with aniridia have WAGR; about 7 in 1,000 people with Wilms tumor have WAGR
Other Names	11p deletion syndrome; 11p partial monosomy syndrome; WAGR Complex; Wilms tumor-aniridia-genital anomalies-retardation syndrome; Wilms tumor-aniridia-genitourinary anomalies-MR syndrome

At the beginning of the twentieth century, Dr. Max Wilms, a German surgeon, first described a type of cancerous tumor of the kidneys that usually occurs in children. Over 75% of the children are otherwise normal. However, about 25% will have other developmental disorders along with the tumor. Dr. Wilms would be pleased to know that today the tumor is very responsive to treatment, and about 90% survive at least five years.

What Is Wilms Tumor, and What Is WAGR Syndrome?

Wilms Tumor Alone

Wilms tumor is called a nephroblastoma. The word "nephroblastoma" comes from three Greek words: *nephr*, meaning "kidney"; *blast*, meaning "germ"; and *oma*, meaning "tumor of." Wilms tumor is a malignant growth in the kidneys that develops during the embryonic stage. The condition is not inherited, but a gene mutation may predispose the person to developing the tumor. There are five stages of Wilms tumors, depending on how confined the tumor is and whether it has spread to other areas of the body through the lymph nodes.

WAGR Syndrome

WAGR is an acronym for several conditions that affect many body systems. The symptoms involve Wilms tumor, aniridia, genitourinary anomalies, and mental retardation.

- *W—Wilms Tumor*: The cancerous growth is present in about 25% of the cases of WAGR. Wilms tumor is seen mostly in children, but can be seen in adults.

- *A—Aniridia*: Aniridia is a condition in which the iris, the colored part of the eye, is completely or partly missing. The term "aniridia" comes from two Greek words: *an*, meaning "without," and *irid*, meaning "the colored part of the eye." Because of the missing iris, which is the diaphragm that opens and closes in response to light, the person will be extremely sensitive to light and lose visual acuity. Other eye problems such as cataracts, glaucoma, and involuntary eye movements may also be present.

- *G—Genitourinary problems*: These problems appear more frequently in males than in females. A male may have undescended testicles. The female may have nonfunctional ovaries and simple clumps of tissue called gonad streaks. Females may also have a heart-shaped uterus, which makes carrying a fetus to full term difficult.

- *R—Retardation*: Several intellectual disabilities and behavioral disabilities may be present. In addition to mental retardation, the person may have attention deficit hyperactivity disorder (ADHD), obsessive-compulsive disorder (OCD), autism, anxiety, or depression.

If the child has childhood-onset obesity, the condition may be referred to as *WAGRO* syndrome. Other disorders may include pancreatitis and kidney failure.

What Are the Genetic Causes of Wilms Tumor and WAGR Syndrome?

Chromosome 11 and two genes, *PAX6* and *WT1*, are involved in Wilms tumor and the syndrome WAGR.

Chromosome 11

Most of the cases of this disorder are not inherited, but the result of the loss of genetic material on the short arm of chromosome 11. WAGR appears to be the result of the loss of a number of contiguous genes and neighboring genes. The deletions happen as a random event during the formation of sperm and egg or during early fetal development. Most people with the disorder have no family history of these abnormalities. One instance of heredity may happen when an unaffected person has an arrangement of chromosome 11; the chromosome is rearranged with no genetic material lost of gained. This is called a balanced translocation. However, if an offspring inherits the balanced translocation, he or she can have an arrangement with a missing gene or genes.

PAX6

PAX6 gene, officially known as the "paired box 6" gene, is located on chromosome 11. Normally, *PAX6* is essential in the formation of organs and tissues during embryonic development. *PAX6* provides instructions for a protein that activates other genes responsible for eye, brain, spinal cord, and pancreas development. *PAX6* appears to be involved in sensory development, especially the sense of smell and vision.

More than 250 mutations in *PAX6* have been found to cause aniridia. A person with aniridia will have no iris in the eye, and the eye will appear black because there is only the pupil of the eye. There is no iris to open and shut in response to light. Most of the mutations are the result of a premature stop signal in the instructions for making the protein. *PAX6* is located on the short arm (p) of chromosome 11 at position 13.

WT1

The *WT1* gene, officially known as the "Wilms tumor 1" gene, is located in the same region of the short arm of chromosome 11. Normally, this gene instructs for a protein that is essential for the development of kidneys and reproductive organs. *WT1* protein is a regulator of many other genes by binding to specific areas of DNA.

WT1 is on the area of chromosome 11 that is often deleted in WAGR syndrome. The loss of this gene is the cause of many abnormalities associated with the syndrome. *WT1* is located on the short arm (p) at position 13.

What Is the Treatment for Wilms Tumor and WAGR Syndrome?

For Wilms tumor, the combination of surgery, chemotherapy, and radiation has been very successful in 90% of the cases. Children with an advanced stage may need dialysis or a kidney transplant. For the other symptoms of WAGR, the person may need glasses to shield from the sun and bright lights. Educational intervention is available to assist with intellectual and behavioral disabilities.

Further Reading

"Wilms Tumor." 2011. Mayo Clinic. http://www.mayoclinic.com/health/wilms-tumor/DS00436. Accessed 12/5/11.

"Wilms Tumor." 2011. Medscape. http://emedicine.medscape.com/article/989398-overview#showall. Accessed 12/5/11.

"Wilms Tumor, Aniridia, Genitourinary Anomalies, and Mental Retardation Syndrome." 2011. Genetics Home Reference. National Library of Medicine (U.S.). http://ghr.nlm.nih.gov/condition/wilms-tumor-aniridia-genitourinary-anomalies-and-mental-retardation-syndrome. Accessed 12/5/11.

Wilson Disease (WD)

Prevalence   Rare; affects about 1 in 30,000

Other Names   copper storage disease; hepatolenticular degeneration syndrome; WD; Wilson's disease

In 1912, Samuel Alexander Kinnier Wilson, a British neurologist, noted a condition in which people displayed yellowing of the skin and whites of the eyes, fatigue, and abdominal swelling. He related it to the accumulation of copper in the body. The disorder was named after this physician—Wilson disease.

What Is Wilson Disease (WD)?

Wilson disease is a condition in which large amounts of copper collect in the brain, liver, and eyes, causing a multitude of symptoms. The signs begin most often during the teenage years but can appear any time between the ages of 6 and 45. Symptoms affect the following major organs of the body:

- Liver: This organ displays the first signs and symptoms of Wilson disease. The person has yellowing of the skin and eyes, a condition called jaundice. At first the person will be very tired, lose appetite, and possibly feel abdominal pain because of an enlarged spleen and fluid that has collected in the area. Small distended blood vessels may be seen on the chest.

- Motor problems: The person may have difficulty walking, trembling, and clumsiness.

- Behavioral symptoms: At first speech and learning problems may occur, and then the individual may develop depression, anxiety, and mood swings.

- Eye issues: Copper deposits may form green-to-brownish rings in the cornea. In addition, eye movement may become erratic, including the inability to gaze upward.

- Psychiatric problems: Frontal lobe problems in the brain may present impulsivity, impaired judgment, promiscuity, apathy, and dementia.
- Kidneys: Calcium may accumulate in the kidneys, causing weakening of the bones and loss of amino acids necessary for protein synthesis.
- Heart: This symptom is somewhat rare but can occur due to accumulation of fluid.
- Low hormone level: Failure of the parathyroid glands may lead to infertility and abortion.

What Is the Genetic Cause of Wilson Disease (WD)?

Mutations in the *ATP7B* gene and the *PRNP* gene are involved in Wilson disease.

ATP7B

Mutations in the *ATP7B* gene, known officially as the "ATPase, Cu++ transporting, beta polypeptide" gene, cause Wilson disease. Normally, *ATP7B* provides instructions for making the copper-transporting ATPase 2 protein. A large family of enzymes called ATPase move metals in and out of cells using the energy stored in the adenosine triphosphate molecule. The copper-transporting ATPase 2 is part of the "P" of the ATPase family, which is responsible for moving copper. These enzymes are located mostly in the liver, with small amounts in the kidneys and brain. Here it moves copper from the liver to other parts of the body.

The element copper is essential in keeping normal cell function. The enzyme maintains the balance of copper and is also important for the removal of excess copper, which can be toxic. The process is accomplished in the Golgi bodies of the liver cells, where ATPase 2 works to supply copper to a protein called ceruloplasmin. This protein moves copper into the bloodstream. If copper levels are excessive, the enzyme transfers copper to small sacs for elimination in the bile.

Over 260 *ATP7B* gene mutations cause Wilson disease. Most of the mutations are the result of the change of only one amino acid building block that makes the copper-transporting ATPase 2 enzyme. The mutations disrupt the structure of the protein, preventing the copper-carrying function from working properly. Excess copper is not removed and accumulates to toxic levels that damage organs, including the liver and brain. *ATP7B* is inherited in an autosomal recessive pattern and is found on the long arm (q) of chromosome 13 at position 14.3.

PRNP

Mutations in a second gene called *PRNP*, officially called the "prion protein" gene, cause Wilson disease. Normally, *PRNP* instructs for making the prion protein (PrP), which is especially located in the brain. Researchers think that PrP is involved in the transport of copper into the cells and that it also has a role in cell signaling and communication.

Changes in this gene may alter the course of those who have the mutated *ATP7B* gene. It may delay or may affect some of the symptoms. One mutation appears to affect the nervous system. *PRNP* is inherited in an autosomal recessive pattern and is located on the short arm (p) of chromosome 20 at position 13.

What Is the Treatment for Wilson Disease (WD)?

This genetic disorder is one that is very treatable. Early diagnosis and proper therapy can halt progress of the disease and improve symptoms. Most of the treatment is aimed at removing excess copper by a process called chelation therapy, which binds certain metals. A number of these drugs have been approved. However, taking the medications is a lifelong commitment, and not getting treatment can be fatal.

Further Reading

"Treatment of Wilson Disease." 2009. Wilson Disease Foundation. http://www.wilsonsdisease.org/wilson-disease/wilsondisease-treatment.php. Accessed 5/31/12.

"Wilson Disease." 2011. National Institute for Diabetes and Digestive and Kidney Diseases (U.S.). http://digestive.niddk.nih.gov/ddiseases/pubs/wilson. Accessed 5/31/12.

"Wilson's Disease." 2011. Mayo Clinic. http://www.mayoclinic.com/health/wilsons-disease/DS00411/DSECTION=treatments-and-drugs. Accessed 5/31/12.

Wiskott-Aldrich Syndrome (WAS)

Prevalence   Between 1 and 10 cases per million males worldwide
Other Names   Aldrich syndrome; eczema-thrombocytopenia-immunodeficiency syndrome; IMD2; immunodeficiency 2

In 1937, Dr. Alfred Wiskott, a German pediatrician, noted an abnormal blood-clotting issue in three brothers. The females in the family were unaffected. Later in 1954, Dr. Robert Anderson Aldrich, an American pediatrician, described a family of Dutch Americans who had abnormal blood-clotting issues, along with some other physical symptoms also present only in males. Members of the medical community named the syndrome Wiskott Aldrich syndrome after the two pioneering doctors.

What is Wiskott-Aldrich Syndrome (WAS)?

Wiskott-Aldrich syndrome (WAS) is a disorder of the cells of the blood and immune system. The disease is characterized by the term microthrombocytopenia, which is made up of four Greek root words: *micro*, meaning "small"; *thrombo*, meaning "clot"; *cyto*, meaning "cell"; and *penia*, meaning "lack of." The word then describes a decrease in the size and number of the blood involved in clotting.

The disorder, which is present at birth, is found exclusively in males and has the following symptoms:

- Easy bleeding and bruising.
- Prolonged bleeding following minor trauma.
- Skin disorders: The person will have patches of red, irritated skin, called eczema.
- Infections: A lack of the immune cells, such as T cells, B cells, dendritic cells, and natural killer cells, lead to infections of all kinds.
- Autoimmune disorders: When a person has an autoimmune disorder, the cells of the person's body attack other cells. People with WAS may develop any of several autoimmune disorders.
- Certain types of cancer: Persons with this immune system are prone to developing such cancers as lymphomas.

What Is the Genetic Cause of Wiskott-Aldrich Syndrome (WAS)?

Mutations in the *WAS* gene, officially known as the "Wiskott-Aldrich syndrome (eczema-thrombocytopenia)" gene, causes Wiskott-Aldrich syndrome. Normally, *WAS* instructs for making the protein WASP, which is located in the stem cells that make blood and some immune system cells, called the hemopoietic cells. The protein is essential for sending signals form the cell surface to the cytoskeleton, which is made up of the protein actin. Actin has several roles. It determines the shape of the cell, allows the cell to move, and assists with interaction between immune cells and outside invaders, such as bacteria.

Over 250 mutations in *WAS* cause Wiskott-Aldrich syndrome. Most of the variations produce very short, nonworking versions of WASP. The mutations that produce no WASP at all appear to have the most serious symptoms and the greatest risk of developing cancer. The impaired protein then cannot send signals from the cell's surface to the cytoskeleton and disrupts the functions of the bone marrow in producing immune cells. Thus, the person has infections, eczema, and autoimmune disorders. The impaired WASP also interferes with platelet development, causing the bleeding and bruising issues. *WAS* is inherited in an X-linked recessive pattern and is located on the short arm (p) of the X chromosome at position 11.4-11.21.

What Is the Treatment for Wiskott Aldrich Syndrome (WAS)?

The only current hope for cure is a cord blood or bone marrow transplant. Some studies are being done using gene therapy with a lentivirus. There are many things that the person must avoid. For example, aspirin and nonsteroidal anti-inflammatory drugs can interfere with blood platelet action. People with low platelet counts may require a blood transfusion.

Further Reading

"Wiskott Aldrich Syndrome." 2011. Genetics and Rare Diseases Information Center (GARD). National Library of Medicine (U.S.) http://rarediseases.info.nih.gov/GARD/Disease.aspx?PageID=4&DiseaseID=7895. Accessed 5/31/12.

"Wiskott-Aldrich Syndrome." 2011. Genetics Home Reference. National Library of Medicine (U.S.). http://ghr.nlm.nih.gov/condition/wiskott-aldrich-syndrome. Accessed 5/31/12.

"Wiskott-Aldrich Syndrome." 2011. Medscape. http://emedicine.medscape.com/article/137015-overview. Accessed 11/22/11.

WNT4 Müllerian Aplasia and Ovarian Dysfunction

Prevalence   Very rare; identified in only a few people worldwide

Other Names   Biason-Lauber syndrome; Mayer-Rokitansky-Küster-Hauser-like syndrome; Mullerian aplasia and hyperandrogenism; Müllerian duct failure; *WNT4* deficiency; *WNT4* Müllerian aplasia

The term "*WNT4* Müllerian aplasia and ovarian dysfunction" is a long name that can be understood by breaking down the individual terms. *WNT4* is the name of the gene that is associated with the condition. The term "aplasia" comes from two Greek words: *a*, meaning "without," and *plas*, meaning "form." The Müllerian ducts are structures named for the German scientist Johannes Müller, who lived in the early eighteenth century and first found the embryonic tubes that become the female reproductive organs. In Müllerian aplasia, these ducts do not form, leading to dysfunction of the ovaries.

What Is *WNT4* Müllerian Aplasia and Ovarian Dysfunction?

WNT4 Müllerian aplasia and ovarian dysfunction is a disorder of the reproductive system. In the embryo, a structure called the Müllerian duct develops into the uterus, fallopian tubes, cervix, and upper part of the vagina. Women with this condition have either a missing or undeveloped uterus, in addition to other abnormalities of the reproductive system. However, other parts of the female genitalia, as well as the secondary sex characteristics, are normal. These women may begin the menstrual cycle late, usually after the age of 16, and some may never begin the monthly period. In addition, women with *WNT4* Müllerian aplasia and ovarian dysfunction may have the following conditions:

- High levels of male sex hormones called androgens.
- Skin conditions: The high level of male sex hormones causes acne.
- Facial hair: The high level of male sex hormones causes excessive facial hair, a condition known as hirsutism.
- Kidney abnormalities.

What Is the Genetic Cause of *WNT4* Müllerian Aplasia and Ovarian Dysfunction?

Mutations in the *WNT4* gene, officially known as the "wingless-type MMTV integration site family, member 4," gene causes the condition. Normally, *WNT4* is a member of the *WNT* family that is essential during embryonic development. *WNT4* instructs for a protein that directs chemical signaling and the pathways of certain other genes. *WNT4* provides the information for making the female reproductive system, the sex hormones, and the kidneys. WNT4 protein is especially responsible for the formation of the Müllerian duct, the male sex hormones, and the development of the egg cells.

About three mutations in *WNT4* cause *WNT4* Müllerian aplasia and ovarian dysfunction. The mutations result in single amino acid exchanges, causing a dysfunctional protein. *WNT4* is inherited in an autosomal recessive pattern and is located on the short arm (p) of chromosome 1 at position 36.23-p35.1.

What Is the Treatment for *WNT4* Müllerian Aplasia and Ovarian Dysfunction?

Because this is a genetic condition occurring at a crucial time in embryonic development, no cure exists. The person learns to manage the symptoms. However, if kidneys are affected, treatment is essential because this can be life-threatening.

Further Reading

"*WNT4*." 2011. Genetics Home Reference. National Library of Medicine (U.S.). http://ghr.nlm.nih.gov/gene/WNT4. Accessed 12/5/11.

"WNT4 Müllerian Aplasia and Ovarian Dysfunction." 2011. Genetics Home Reference. National Library of Medicine (U.S.). http://ghr.nlm.nih.gov/condition/wnt4-mullerian -aplasia-and-ovarian-dysfunction. Accessed 12/5/11.

"WNT4 Müllerian Aplasia and Ovarian Dysfunction." Inherited Health. http:// www.inheritedhealth.com/condition/WNT4_M%C3%BCllerian_Aplasia_and_Ovarian _Dysfunction/726. Accessed 12/5/11.

Wolf-Hirschhorn Syndrome (WHS)

Prevalence   About 1 per 50,000 to 1 per 20,000 births; more prevalent in females with a ratio of 2 to 1.

Other Names   chromosome 4p deletion syndrome; chromosome 4p monosomy; del(4p) syndrome; 4p deletion syndrome; 4p- syndrome WHS; monosomy 4p; partial monosomy 4p; Pitt-Rogers-Danks syndrome (PRDS) or Pitt syndrome.

In 1961, Americans Herbert L. Cooper and Kurt Hirschhorn described a child with distinct brain and facial defects that were traced to a deletion in the short arm of chromosome 4. Four years later, the German scientist Ulrich Wolf, along with Hirschhorn, wrote several papers that brought the attention to the genetic professionals. Many other cases were later noted that included mental retardation and seizures. The condition was recognized by the profession as Wolf-Hirschhorn syndrome.

What Is Wolf-Hirschhorn Syndrome (WHS)?

Wolf-Hirschhorn syndrome is a condition with distinct facial appearance and head shape. Following are the major features of this disorder:

- Abnormally small head, a condition known as microcephaly.
- Distinct facial appearance: The person has a broad, flat nose with a high forehead. Protruding eyes are widely spaced. There is a short distance between nose and upper lip. The mouth is turned downward, with a small chin. Ears are poorly formed and may have small pits or holes, or ear opening, or just flaps of skin.
- Delayed growth: Growth disorders begin before birth. The infant has feeding problem and consequently does not gain weight.
- Short stature
- Weak muscle tone with underdeveloped muscles
- Poor motor skills: The child will be delayed in sitting, standing, and walking.
- Intellectual impairment: Disabilities range from mild to severe. Verbal and language skills are weak. However, the child may be outgoing and sociable.
- Seizures: These tend to disappear with age.
- Skin problems: Skin may acquire a mottled appearance with very dry skin.
- Skeletal abnormalities: These abnormalities include curvature of the spine and cleft palate or lip.
- Other issues: People with WHS may have eye, heart, and genitourinary tract disorders.

Death occurs at an estimated rate of 34% in the first two years of life. However, many children may die before WHS is suspected, so the condition may be underestimated. Life-threatening conditions usually occur from heart defects, pneumonia, infection, or seizures.

What Are the Genetic Causes of Wolf-Hirschhorn Syndrome (WHS)?

Changes involve deletions in a number of genes located near the short arm (p) of chromosome 4. The genes are *LETM1*, *MSX1*, and *WHSC1*.

Chromosome 4

This chromosome has about 1,300 to 1,600 genes that play a number of roles in the body. WHS is caused by deletion of genetic material at the end of the short arm. This area has many important genes, especially related to embryonic development. Loss of the genes causes many of the characteristic facial appearances, developmental delay, and symptoms of WHS.

Several problems with the chromosome may occur. Most of the cases of WHS are not inherited but occur as a random action during the formation of egg or sperm. About 85% to 90% of cases do appear to be random, and many of these happen in several complex chromosome arrangements, such as ring chromosomes. Ring chromosomes occur when the chromosomes break together in two places and then form a circle. In only a few cases, the rearranged chromosome may be inherited. This balanced transformation does not actually lose genetic material but simply be rearranged. The person with the translocation may not be affected but may pass the unbalanced gene, which will delete some of the end material of chromosome 4 and cause the symptoms of WHS.

LETM1

Deletion of the *LETM1* gene, officially called the "leucine zipper-EF-hand containing transmembrane protein 1" gene, causes some of the symptoms of WHS. Normally, *LETM1* instructs for a protein that is active in the mitochondria, the powerhouses of the cells. This protein may transport charged calcium atoms across the membranes of the mitochondria and possibly act to determine the shape of the mitochondria.

Because this gene is in the deleted area of chromosome 4, there will be only one gene in the cell, which may change the structure of the mitochondria. The loss of *LETM1* is associated with seizures and other impairments of the electrical activity in the brain. The gene appears as an autosomal dominant pattern in the cells because one gene is missing. It is located on the short arm (p) of chromosome 4 at position 16.3.

MSX1

Deletions of chromosome 4 include the *MSX1* gene, known officially as the "msh homeobox 1" gene. Normally, *MSX1* encodes for a protein that controls other genes. It is a member of a family called homeobox genes, which are essential in embryonic development of many body parts. *MSX1* appears to be critical for the development of the structures of the mouth, teeth, and nails. Loss of this gene affects many body parts and may be especially involved in the cleft palate and other facial and dental abnormalities. MSX1 appears as an autosomal dominant pattern in the cells and is located on the short arm (p) at position 16.2.

WHSC1

Another deletion in chromosome 4 is the *WHSC1* gene, officially known as the "Wolf-Hirschhorn syndrome candidate 1" gene. Normally, this gene instructs for three similar proteins called MMSET I, MMSET II, and RE-IIBP. These proteins appear before birth and are essential for normal development. MMSET I and MMSET II appear to function as enzymes called as histone methyltransferases. Histones are important proteins related to DNA. When the enzymes add a methyl group to the histones, they turn off the activity of certain genes. If the gene is lost and is not present in one of the cells, then the entire process is disrupted leading to the many characteristic features of WHS. *WHSC1* is inherited in an autosomal dominant pattern and is located on the short arm (p) of chromosome 4 at position 16.3.

What Is the Treatment for Wolf-Hirschhorn Syndrome (WHS)?

There is no cure for WHS. Certain symptoms can be treated. Treatments may include rehabilitation, speech and communication therapy, mediations for seizures, and tube feeding. Other symptoms may be treated according to individual needs.

Further Reading

"Wolf-Hirschhorn Syndrome." 2011. Genetics Home Reference. National Library of Medicine (U.S.). http://ghr.nlm.nih.gov/condition/wolf-hirschhorn-syndrome. Accessed 5/31/12.

"Wolf-Hirschhorn Syndrome." 2011. Medscape. http://emedicine.medscape.com/article/950480-overview#showall. Accessed 5/31/12.

"Wolf-Hirschhorn Syndrome." 2011. Wolf-Hirschhorn.org. http://www.wolfhirschhorn.org. Accessed 5/31/12.

Wolff-Parkinson-White Syndrome

Prevalence	Common cause of heart arrhythmia; affects 1 to 3 people in 1,000 worldwide; most frequent cause in Chinese arrhythmia patients, where it is responsible for more than 70% of cases.
Other Names	Ventricular pre-excitation with arrhythmia; WPW syndrome

In 1930, cardiologists Louis Wolff, Sir John Parkinson, and Paul Dudley White described a series of young patients who showed a special pattern on the electrocardiography (ECG) called a bundle branch block pattern. Several researchers had found the pattern earlier in 1915 and 1921, but the three doctors published extensively and were honored with the name in 1940. Two well-known performers have had the condition: the rock 'n' roll singer Meat Loaf and the shock rocker Marilyn Manson.

What Is Wolff-Parkinson-White Syndrome?

Wolff-Parkinson-White syndrome is a disorder of the electrical patterns of the heart. Normally, heartbeat is controlled by very well-organized electrical signals. An important group of cells called the atrioventricular node or AV node conducts the impulses from the upper chambers of the heart called the atria to the lower chambers called the ventricles. The electrical signals move through the AV node during each heartbeat, causing the ventricles to contract slightly later than the atria.

People with Wolff-Parkinson-White syndrome have an extra connection in the heart that lets the signal bypass the AV node and causes the ventricle to contract faster than normal. This additional pathway is called an accessory pathway. With the accessory pathway, the whole electrical pathway is disrupted, leading to an abnormally fast heartbeat, a condition known as tachycardia. Following are the symptoms of Wolff-Parkinson-White syndrome:

- A sensation of fluttering or pounding in the chest, known as palpitations
- Dizziness
- Shortness of breath
- Fainting
- Rarely, cardiac arrest and sudden death
- Few other genetic conditions may accompany the disorder

The most common arrhythmia associated with Wolff-Parkinson-White syndrome is called paroxysmal supraventricular tachycardia. The condition may occur at any age, but some people with the accessory pathway never experience any health problems.

What Is the Genetic Cause of Wolff-Parkinson-White Syndrome?

Mutations in the *PRKAG2* gene, officially called the "protein kinase, AMP-activated, gamma 2 non-catalytic subunit" gene, cause Wolff-Parkinson-White syndrome. Normally, *PRKAG2* instructs for making a part of a larger enzyme called AMP-activated protein kinase (AMPK) called the gamma-2 subunit. This enzyme, which responds to the energy demands of cells, is especially active in the tissue of the heart and skeletal muscles. The AMP-activated protein kinase (AMPK) regulates chemical pathways that are related to adenosine triphosphate (ATP), the cell's main energy source. It is the breakdown of ATP that releases energy for the body. If the level of ATP is too low, the enzyme restores the balance of energy.

About seven mutations in *PRKAG2* cause Wolff-Parkinson-White syndrome. Studies have shown that mutations cause the enzyme to allow the complex sugar glycogen to build up within the cardiac muscles, causing the cells to enlarge and possibly lead to the problems of the electrical signaling. Although most of the cases are sporadic, the familial type is of the gene inherited in an autosomal dominant pattern and is located on the long arm (q) of chromosome 7 at position 36.1.

What Is the Treatment for Wolff-Parkinson-White Syndrome?

The person must be under the care of a competent cardiologist. Medications may be prescribed depending upon the individual.

Further Reading

"What Is Wolff-Parkinson-White Syndrome? What Causes Wolff-Parkinson-White Syndrome?" 2011. *Medical News Today*. http://www.medicalnewstoday.com/articles/220163.php. Accessed 12/6/11.

"Wolff-Parkinson-White (WPW) Syndrome." 2011. Mayo Clinic. http://www.mayoclinic.com/health/wolff-parkinson-white-syndrome/DS00923. Accessed 12/6/11.

"Wolff-Parkinson-White Syndrome." 2011. Medscape. http://emedicine.medscape.com/article/159222-overview. Accessed 12/6/11.

Wolman Disease

Prevalence   Occurs in about 1 in 350,000 newborns

Other Names   acid lipase deficiency; familial xanthomatosis; LAL deficiency; LIPA deficiency; liposomal acid lipase deficiency, Wolman type; lysosomal acid lipase deficiency

In 1956, Abramov, Schorr, and Wolman noted a condition in which an infant was healthy at birth but soon became listless and lost weight. The child was a member of a family of Persian Jews, whose members were closely related to one another. Moshe Wolman studied the condition in several other cases and wrote about his observations in a journal. The disorder was named after him—Wolman disease.

What Is Wolman Disease?

Wolman disease is a metabolic condition in which the body cannot break down fats or lipids, and the accumulation of these lipids causes harmful effects. Because the fats are not broken down, the lipids collect in the spleen, liver, bone marrow, small intestines, adrenal glands, and lymph nodes. In addition, the person may have heavy calcium deposits in the adrenal glands. Infants are healthy at birth but develop symptoms within the first few weeks of life.

Affecting so many body systems, the disorder produces a variety of the following symptoms:

- Feeding difficulties with frequent explosive vomiting.
- Diarrhea: Stools are oily and foul-smelling.
- Poor weight gain: Because of the vomiting and diarrhea, the child becomes malnourished. Fats are beginning to accumulate in the walls of the intestine, causing nutrients not to be absorbed.
- Poor muscle tone.
- A distended abdomen because of the enlarged liver and spleen.
- Anemia.
- Yellowing of skin and eyes: This symptom occurs as the disease progresses and fat accumulates in the liver.
- Low-grade fever.
- Chalky deposits in the adrenal gland seen in X-rays.

The child seldom lives past the first year of life.

What Is the Genetic Cause of Wolman Disease?

Mutations in the *LIPA* gene, known officially as the "lipase A, lysosomal acid, cholesterol esterase" gene, cause Wolman disease. Normally, *LIPA* instructs for the lysosomal acid lipase, an enzyme found in the lysosomes. Lysosomes are located in the cells and act to digest and recycle materials such as fats, cholesterol, and triglycerides.

In the body, high-density lipoproteins or HDL carry cholesterol to the liver for removal. Cholesterol that is attached to a fatty acid is called a cholesteryl ester. The lysosomal acid lipase breaks down the esters into cholesterol and the fatty acid that is then excreted or used for nutrients.

Over 10 mutations in *LIPA* cause Wolman disease. Most of the mutations produce a very short version of the lysosomal acid lipase, which cannot function. When this enzyme is missing or inactive, cholesteryl esters and triglycerides accumulate in the cells of many tissues and organs, leading to malnutrition and the other symptoms of Wolman disease. *LIPA* is inherited in an autosomal recessive pattern and is located on the long arm (q) of chromosome 10 at position 23.3-q23.3.

What Is the Treatment for Wolman Disease?

No treatment exists for this disease; thus treatment must focus on management of symptoms. To receive nourishment, the child may be fed through the veins. Medications can be given to replace the hormones of the adrenal glands.

Further Reading

"Wolman Disease." 2011. About.com Rare Diseases. http://rarediseases.about.com/od/lysosomalstoragediseases/a/wolman.htm. Accessed 5/31/12.

"Wolman Disease." 2011. National Organization for Rare Diseases. http://www.cigna.com/individualandfamilies/health-and-well-being/hw/medical-topics/wolman-disease-nord1213.html. Accessed 5/31/12.

"Wolman Disease." 2012. Genetics Home Reference. National Library of Medicine (U.S.). http://ghr.nlm.nih.gov/condition/wolman-disease. Accessed 5/31/12.

"Wolman Disease—Facts and Information." Disabled World Towards Tomorrow. http://www.disabled-world.com/disability/types/wolman-disease.php. Accessed 5/31/12.

X

X Chromosome: A Special Topic

The Big Sex Chromosome

Throughout history, people have wondered why some children were male and others were female. Early physicians had to concoct various ideas about the origins. The liveliest debate originated in the eighteenth century, with the most celebrated philosophers and scientists of the time taking strong sides. Most scientists were convinced that all embryos existed since creation as preformed miniatures, held within their parents, ready to be born. This group believed in *preformation*. The other side argued the embryo was formed structure by structure starting with undifferentiated materials in the eggs. They called their position *epigenesis*. These eighteenth-century thinkers laid the foundation for prevailing today's thought. The DNA of sperm and egg have preformed instructions, and this information instructs proteins to create an embryo from undifferentiated stem cells. The instructions lie on the X and Y chromosomes.

Scientists are still looking for the precise chemical and structural detail of the X and Y chromosomes. Looking in unsophisticated microscopes, early geneticists located a large chromosome that baffled in many ways. The large structure held itself apart from other non-sex chromosome pairs, and so they called it X for unknown. When one looks at the chromosome, it does resemble the letter "X" during phases of cell division, but other chromosomes also look like "X's" during division stages. Although many think the "X" shape gave the chromosome the name, the name actually came from the unknown and baffling activity of the chromosome. Soon, when the tinier male counterpart was found, the researchers went down the alphabet and called it Y. These researchers realized that much information was on the unknown "X" chromosome, and some of the information could lead to some serious health disorders.

Humans have two sex chromosomes. One is the Y chromosome, which is small and carries little information except the determination of sex. On the other hand, the X chromosome is large, spanning about 155 million DNA building blocks or base pairs and representing about 5% of all DNA found in cells. Females have two X chromosomes, and males have one X and one Y chromosome.

Early in embryonic development in females, one of the two X chromosomes becomes inactivated in cells other than X cells. This X-inactivation, called lyonization, ensures that females will have only one working copy of the X chromosome in each one of the cells. This X activation is random so that in normal females, some of the X chromosomes are active in some cells, and the X chromosome from the father is active in the other.

The X chromosome is unique. It carries about 900 to 1,400 of the approximately 25,000 genes. It has arms structure from the centromere (where two sides are joined) that are designated as "p" for the short arm and "q" for the long arm. A few genes located at the end of the chromosome escape X-inactivation and occupy areas called pseudoautosomal regions, which means they act like somatic chromosomes and are present in both the X and Y chromosomes. Many of these genes are important for normal development. However, if the mother contributes a mutated gene on the on the arm of the X chromosome, the offspring may have problems with health and development because there is no area on the small Y chromosome to mask it.

Finding genes on each chromosome has been an area of exciting research. Mutations in the genes on the X chromosome cause many genetic disorders. Following are a list and brief description of some of the conditions related to the X chromosome:

- *Intestinal pseudo-obstruction*: Mutations, duplications, and deletions of the *FLNA* gene cause a condition that keeps food from moving though the digestive tract. Changes in nearby genes can cause neurological problems and a distinct facial appearance.

- *Klinefelter syndrome*: One or more extra copies of X are in the male's cells. The extra material disrupts normal sexual and intellectual development. The chromosome may be 47,XXY, 48,XXXY, or 49,XXXXY.

- *Microphthalmia with linear skin defects syndrome*: A deletion of a region called Xp22, which also has the *HCCS* gene, causes a lethal condition in which the infant dies before or soon after birth.

- *Triple X syndrome*: An extra copy of the X chromosome or 47,XXX may result in women with tall stature and learning problems. Sometimes there will be more X chromosomes: 48,XXXX or 49,XXXXX.

- *Turner syndrome*: The syndrome occurs when a normal X chromosome is present, but the other one is missing or altered. These girls have distinct features, such as short stature and skeletal abnormalites.

- *46,XX testicular disorder of sex development*: Some material from the Y chromosome with the *SRY* gene will translocate onto an X chromosome. The fetus will be a male without a Y chromosome.

- *48,XXYY syndrome*: The presence of two X and two Y chromosomes interferes with male sexual development.

- *Immune X-linked recessive disorders*: Chronic granulomatous disease (CYBB); Wiskott-Aldrich syndrome; X-linked severe combined immunodeficiency; X-linked agammaglobulinemia; Hyper-IgM syndrome type 1; IPEX; X-linked lymphoproliferative disease; Properdin deficiency

- *X-linked blood disorders*: Hemophilia A; Hemophilia B; X-linked sideroblastic anemia

- *X-linked endocrine disorders*: Androgen insensitivity syndrome/Kennedy disease; KAL1 Kallmann syndrome; X-linked adrenal hypoplasia congenita

- *X-linked recessive metabolic disorders*: Ornithine transcarbamylase deficiency; Oculocerebrorenal syndrome; dyslipidemia; Adrenoleukodystrophy; carbohydrate metabolism; Glucose-6-phosphate dehydrogenase deficiency; Pyruvate dehydrogenase deficiency; Fabry's disease; mucopolysaccharidosis; Hunter syndrome; Lesch-Nyhan syndrome; Menkes disease/Occipital horn syndrome

- *Nervous system recessive disorders*: Coffin-Lowry syndrome; MASA syndrome; X-linked alpha thalassemia mental retardation syndrome; Siderius X-linked mental retardation syndrome; Color blindness (red and green, but not blue); Ocular albinism type 1; Norrie disease; Charcot-Marie-Tooth disease (CMTX2-3); Pelizaeus-Merzbacher disease; SMAX2

- *Skin issues*: Dyskeratosis congenita; Hypohidrotic ectodermal dysplasia (EDA); X-linked ichthyosis; X-linked endothelial corneal dystrophy

- *Neuromuscular*: Becker's muscular dystrophy/Duchenne; Centronuclear myopathy (MTM1); Conradi-Hünermann syndrome; Emery-Dreifuss muscular dystrophy 1

- *Urologic*: Alport syndrome; Dent's disease; X-linked nephrogenic diabetes insipidus

- *Bone/teeth*: AMELX Amelogenesis imperfecta

- *No system*: Barth syndrome; McLeod syndrome; Smith-Fineman-Myers syndrome; Simpson-Golabi-Behmel syndrome; Mohr-Tranebjærg syndrome; Nasodigitoacoustic syndrome

See also Y Chromosome: A Special Topic

Further Reading

Bailey, Regina. 2012. "Chromosomes and Sex." About.com. http://www.biology.about.com/od/basicgenetics/p/chromosgender.htm. Accessed 2/10/12.

"Chromosome Map." 2011. *Genes and Disease*. National Center for Biotechnology Information (U.S.). http://www.ncbi.nlm.nih.gov/books/NBK22266. Accessed 12/14/11.

"X Chromosome." 2012. Genetics Home Reference. National Library of Medicine (U.S.). http://www.ghr.nlm.nih.gov/chromosome/X. Accessed 2/10/12.

Xeroderma Pigmentosum (XP)

Prevalence   About 1 in 1 million in the United States and Europe; more common in Japan, North Africa, and Middle East

Other Names   DeSanctis-Cacchione syndrome; XP

Although the condition is rare, several pieces of literature and film have had characters with xeroderma pigmentosum. In 1988, a film called *Dark Side of the Sun* featured Brad Pitt as the main character suffering from XP. In 2003, Jodi Picoult, the novelist who tackles ethical issues in medicine, wrote *Second Glance*, the story of a eugenics project in Vermont. The nine-year-old nephew of the main protagonist, Ross Wakeman, had XP. A 2003 novel, *A Cool Moonlight* by Angela Johnson, centers on a girl who has XP and can never be out in the sun. The story tells of the family struggles to make her life normal.

What Is Xeroderma Pigmentosum (XP)?

Xeroderma pigmentosum is a condition in which the person is extremely sensitive to ultraviolet or UV rays from sun. Most affected are areas that are exposed to the sun, although some people may have additional neurological problems. The term "xeroderma" comes from the Greek terms *xero*, meaning "dry," and *derm*, meaning "skin." The term pigmentosum relates to the pigment, the colored part of the skin.

Following are the most common symptoms of XP:

- Severe sunburn: Children with the condition will develop a serious sunburn when exposed for only a few minutes in the sun. Redness and blistering can last for weeks.

- Freckles: These pigmented freckles develop early in life. By age two, the freckles will appear on exposed skin areas. This kind of freckling is not normal for children without the disorder.

- Very dry skin: Exposure to the sun causes dry and colored, irregular dark spots on the skin. These symptoms give the condition its name, xeroderma pigmentosum. The dry skin may become scaly, crack, and ooze fluid.

- Sensitive eyes: When the person is in the sun, eyes become painful and become easily irritated. Vision may be impaired.
- Eyelashes: Eyelashes may be thin or fall out. Eyelids may turn in or out.
- Spidery blood vessels.
- Limited hair on chest and legs.
- Skin cancers: Children may develop the first cancer on face, lips, and eyelids by age 10. Cancers may also be found on scalp, eyes, tip of tongue. Multiple cancers are common, including brain and lung cancer.
- Neurological conditions: About 30% of people with XP develop serious problems with walking, coordination, hearing, swallowing and talking, and intellectual development. These problems are progressive and worsen over time.

At least eight forms exist, labeled from A (XP-A) to XP-G. There is also a variant type called XP-V. Almost 40% of people with XP die before the age of 20. Some people with less severe cases may live well into the 40s.

What Are the Genetic Causes of Xeroderma Pigmentosum (XP)?

Several genes cause XP. All the genes are associated with repairing damaged DNA and with an increase in skin cancer risk. The genes that are most connected with XP are the *ERCC2*, *ERCC3*, *POLH*, *XPA*, and *XPC* genes.

ERCC2

Mutations in the *ERCC2* gene, officially named the "excision repair cross-complementing rodent repair deficiency, complementation group 2" gene, cause XP and appear to be associated with neurological problems. Normally, *ERCC2* instructs for a protein XPD. This protein is part of a group known as the TFIIH complex, a transcription factor that helps repair DNA. XPD works with another protein called XPB in the transcription process. This complex is active in repairing damaged DNA, possibly using a mechanism called the nucleotide excision repair (NER).

Gene transcription is the first step in protein production. By controlling gene transcription, the TFIIH complex helps regulate the activity of many different genes. The XPD protein appears to stabilize the TFIIH complex. Studies suggest that the XPD protein works together with XPB, another protein in the TFIIH complex that is produced from the *ERCC3* gene, to start (initiate) gene transcription.

The TFIIH complex also plays an important role in repairing damaged DNA. DNA can be damaged by ultraviolet (UV) rays from the sun and by toxic chemicals, radiation, and unstable molecules called free radicals. DNA damage occurs frequently, but normal cells are usually able to fix it before it can cause problems. One of the major mechanisms that cells use to fix DNA is known as nucleotide excision repair (NER). When the region has been repaired, the enzyme complex attaches to DNA, temporarily unwinding the spiral. Other proteins snip out the damaged section and replace with the correct DNA.

Over two dozen mutations in *ERCC2* are the second-most cause of XP. These mutations disrupt the working of the TFIIH complex. As a result, the DNA goes unrepaired and cells die. If these are the genes that control growth, the UV rays cause the cells to grow too fast, causing cancer. These mutations are also correlated with neurological problems. *ERCC2* is inherited in an autosomal recessive pattern and is located on the long arm (q) of chromosome 19 at position 13.3.

ERCC3

Mutations in *ERCC3*, officially known as the "excision repair cross-complementing rodent repair deficiency, complementation group 3 (xeroderma pigmentosum group B complementing)" gene, cause a rare form of XP. Normally, *ERCC3* instructs for the protein XPB, which is an essential part of the TFIIH complex. This complex is active in transcription of many other genes and helps repair DNA by using the excision and repair process.

Mutations in the *ERCC3* gene usually involve a single exchange of one amino acid building block. One such mutation occurs when serine replaces phenylalanine. This exchange inhibits the general ability of the TRIIH complex to repair cell damage, causing the problems of XP. *ERCC3* is inherited in an autosomal recessive pattern and is located on the long arm (q) of chromosome 2 at position 21.

POLH

The *POLH* gene is officially known as the "polymerase (DNA directed), eta" gene. Normally, *POLH* instructs for the protein DNA polymerase eta, which is active in reading the sequences of DNA and using them to produce templates for replicating the cell's genetic material. The DNA polymerases also are essential for DNA repair, especially of damage caused from UV rays. However, this polymerase is relatively error-prone and can insert the wrong information that can result in mutations.

About 30 mutations in *POLH* cause the rare variant XP-V. The mutations prevent the production of DNA polymerase eta and keep the cells from repairing the damage caused by UV rays. The cells grow rapidly and increase the likelihood of skin cancer. *POLH* is inherited in an autosomal recessive pattern and is located on the short arm (p) of chromosome 6 at position 21.1.

XPA

Mutations in the *XPA* gene, officially named the "xeroderma pigmentosum, complementation group A" gene, cause XP. Normally, *XPA* instructs for a protein that repairs damaged DNA. UV rays, toxic chemicals, radiation, and free radical molecules may cause DNA damage. The XPA protein is part of the nucleotide excision repair (NER) that is seen in the other proteins that are associated with XP. The molecule repairs and stabilizes damaged areas of DNA.

About 25 mutations in *XPA* are found in XP, especially in the Japanese population. The mutation keeps the cell from producing a functional version of the XPA protein. As a result, abnormalities occur and cancers develop in areas exposed to the sun. This type is also prone to neurological disorders. *XPA* is inherited in an autosomal recessive pattern and is located on the long arm (q) of chromosome 9 at position 22.3.

XPC

Mutations in the *XPC* gene, officially called the xeroderma pigmentosum, complementation group C" gene, cause XP. The gene works in a similar way to the other genes to repair damaged DNA, but it may have other roles in DNA repair and in the functions of the XPC protein.

More than 40 mutations have been found in people with XP, especially those who reside in the United States and Europe. This form is generally not associated with neurological disorders. *XPC* is inherited in an autosomal recessive pattern and is located on the short arm (p) of chromosome 3 at position 25.

What Is the Treatment for Xeroderma Pigmentosum (XP)?

The obvious treatment is avoiding exposure to sunlight. The person should have checkups about every three to six months with a dermatologist, preferably one who is skilled in Mohs micrographic surgery. Small premalignant lesions known as actinic keratoses can be treated by freezing with liquid nitrogen or fluorouracil. Other lesions may need to be surgically removed.

Further Reading

Kraemer, Kenneth H., and John J. DiGiovanna. 2011. "Xeroderma Pigmentosum." *GeneReviews*. http://www.ncbi.nlm.nih.gov/books/NBK1397. Accessed 5/31/12.

"Xeroderma Pigmentosa." 2011. MedlinePlus. National Library of Medicine (U.S.). http://www.nlm.nih.gov/medlineplus/ency/article/001467.htm. Accessed 5/31/12.

Xeroderma Pigmentosum (XP) Society. 2005. http://www.xps.org. Accessed 5/31/12.

X-Linked Adrenal Hypoplasia Congenital

Prevalence   About 1 in 12,500 newborns

Other Names   adrenal hypoplasia congenital; X-linked AHC

X-linked adrenal hypoplasia congenital is one of those recessive conditions carried on the X chromosome and found only in males. (For more information on the X chromosome, see the entry "X Chromosome: A Special Topic.")

What Is X-Linked Adrenal Hypoplasia Congenital?

X-linked adrenal hypoplasia congenita is a disorder of the endocrine glands, especially the adrenal glands that sit on top of each kidney. The word "hypoplasia" comes from the Greek roots: *hypo* meaning "under," and *plas* meaning "form." The term "congenital" means "born with." (Occasionally, the Latin form "congentia" will be used in the literature.) Thus the condition X-linked adrenal hypoplasia congenital results in an underperforming adrenal gland that the child has at birth. The work of the adrenal glands it to produce hormones, and for them not to function is called adrenal insufficiency. Following are the symptoms of X-linked adrenal hypoplasia congenital, which begin in infancy:

- Vomiting
- Difficulty feeding
- Dehydration
- Very low blood sugar, a condition called hypoglycemia
- Shock
- Underdeveloped reproductive tissue—the person has a shortage of male hormones that cause undescended testicles, delayed puberty, and infertility

Severity and age of onset may vary, even among members of the same family.

What Is the Genetic Cause of X-Linked Adrenal Hypoplasia Congenital?

Mutations in the *NROB1* gene, officially known as the "nuclear receptor subfamily 0, group B, member 1" gene, cause X-linked adrenal hypoplasia congenital. Normally, *NROB1* instructs for making the protein DAX1, which is important in many of the following endocrine glands:

- Adrenal glands on top of each kidney
- Pituitary at the base of the brain
- Hypothalamus in the brain
- Reproductive organs—the ovaries in the female and testes in male

In addition to forming the glands, DAX1 helps regulate the hormones after the tissues are formed.

More than 110 mutations in *NROB1* cause X-linked adrenal hypoplasia congenita. Some of the mutations are from deletions in all or part of the genes. Other cause very short proteins, with others the result of one amino acid exchange. Whatever the cause, the production of DAX1 protein is disrupted, resulting in the abnormal conditions of the endocrine glands. *NROB1* is located on the short arm (p) of the X chromosome at position 21.3.

What Is the Treatment for X-Linked Adrenal Hypoplasia Congenital?

When a person has an attack of adrenal insufficiency, usually he must be admitted to the hospital for intensive care. Dehydration must be treated with glucose and electrolytes. Other hormones may need to be replaced depending upon the individual situation. Stress should be avoided.

Further Reading

Vilain, Eric J. 2009. "X-Linked Adrenal Hypoplasia Congenita." *GeneReviews*. http://www.ncbi.nlm.nih.gov/books/NBK1431. Accessed 12/8/11.

"X-Linked Adrenal Hypoplasia Congenital." 2011. Medscape. http://www.medscape.com/viewarticle/474575_10. Accessed 12/8/11.

X-Linked Adrenoleukodystrophy

Prevalence   Occurs about 1 in 17,000 worldwide; most frequent inherited disorder of white matter

Other Names adrenoleukodystrophy; adrenomyeloneuropathy; ALD (adrenoleukodystrophy); Schilder-Addison complex; X-ALD

X-linked adrenoleukodystrophy is one of those conditions related to the X chromosome that is found mostly in males. (For more information on the X chromosome, see the entry "X Chromosome: A Special Topic.")

What Is X-Linked Adrenoleukodystrophy?

X-linked adrenoleukodystrophy is a condition that affects the adrenal glands and the nervous system. The adrenal glands are small glands that sit on top of the kidneys and make hormones that are essential for the body. The term "leukodystrophy" comes from three Greek terms: *leuko*, meaning "white"; *dys*, meaning "with difficulty"; and *troph*, meaning "nourish." A leukodystrophy relates to the white matter or the fatty myelin sheath insulating the nerves of the brain and spinal cord. These nerve coverings do not get the nourishment that they need to survive.

People with X-linked adrenoleukodystrophy have a shortage of hormones because of damage to the outer layer of the adrenal glands called the adrenal cortex. The myelin sheath is progressively damaged, causing both physical and intellectual problems. There are three types of X-linked adrenoleukodystrophy: a childhood form, an adrenomyeloneuropathy type, and a type called Addison disease only.

Pattern of X-linked inheritance

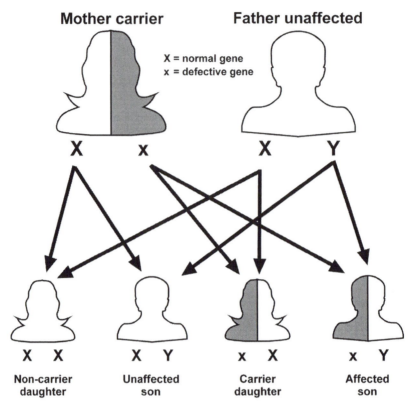

Diagram showing the pattern of X-linked inheritance, which is the case in x-linked adrenoleukodystrophy. It is a disorder found mainly in males and inherited from the mother. (ABC-CLIO)

Childhood Cerebral Form

These children have learning and behavioral issues beginning about the age of 10. As they get older, the conditions worsen. They may display aggressive behavior and experience vision problems. They may become totally disabled in a few years.

Adrenomyeloneuropathy Type (AMP)

This type occurs later in life. In early adulthood or middle age, the person may begin to experience stiffness that worsens over time. He may also show urinary and genital tract disorders, some brain dysfunction, and some adrenal gland deficiencies.

Addison Disease Only

When the person has only the deficiency of the adrenal glands without other symptoms, he has Addison disease. The adrenocortical insufficiency causes the person to lose weight and experience weakness, vomiting, and coma. Color in the skin may also change.

What Is the Genetic Cause of X-Linked Adrenoleukodystrophy?

Mutations in the *ABCD1* gene, officially known are the "ATP-binding cassette, sub-family D (ALD), member 1" gene, causes this disorder. Normally, *ABCD1* instructs for the adrenoleukodystrophy protein (ALDP). ALDP is called a half-transporter because it must join with another protein to work. The work is done in a membrane that surrounds peroxisomes, the small sacs in cells that process many types of molecules. ALDP and its coworker may transport a molecule that breaks down very long-chain fatty acids in the peroxisomes.

Over 480 mutations in *ABCD1* cause X-linked adrenoleukodystrophy. Most of the mutations stop the production of ALDP completely. The others may produce the protein, but it is nonfunctional. With none of the functioning protein, toxic fats build up and damage the adrenal gland and the fatty white layer of the myelin sheath. The destruction of these tissues causes the disorder. *ABCD1* is located on the long arm (q) of the X chromosome at position 28.

What Is the Treatment for X-Linked Adrenoleukodystrophy?

For the adrenal insufficiency, corticosteroids replacement therapy is essential. Children in school need psychological and educational support. Men who develop the AMP type will benefit from physical and occupational therapy.

Further Reading

Kemp, Stephan. 2010. "X-Linked Adrenoleukodystrophy." *SciTopics*. August 27. http://www.scitopics.com/X_linked_adrenoleukodystrophy.html. Accessed 2/12/12.

Moser, Ann B.; Steven J. Steinberg; and Gerald V. Raymond. 2009. "X-Linked Adrenoleukodystrophy." *GeneReviews*. http://www.ncbi.nlm.nih.gov/books/NBK1315. Accessed 12/8/11.

X-Linked Agammaglobulinemia (XLA)

Prevalence   Occurs in 1 in 200,000 newborns

Other Names   agammaglobulinemia; Bruton's agammaglobulinemia; congenital agammaglobulinemia; hypogammaglobulinemia; XLA

X-linked agammaglobulinemia (XLA) is one of those conditions and is found only in males. (For more information on the X chromosome, see the entry "X Chromosome: A Special Topic.")

What Is X-Linked Agammaglobulinemia (XLA)?

In 1952, Dr. Ogden Bruton described a condition in which his patients had severe infections and seemingly no ability to produce antibodies. Antibodies are proteins that make up the gamma globulin or immunoglobulin part of blood plasma. He noted that the condition was mainly in boys. He had found the first inherited immunodeficiency inherited by means of the X chromosome. Sometimes, the condition is called Bruton's agammaglobulinemia, but mostly it is known as X-linked agammaglobulinemia (XLA)

XLA affects the immune system of males. The people with XLA are missing B cells, a group of specialized white cells that fight infection. B cells produce antibodies that ferret out foreign invaders and mark them for destruction. If the antibodies are not working properly, the body does not fight off germs.

Infants with XLA appear healthy at birth, but after a month or two, the mother's antibodies that have protected them disappear, and they begin to get one infection after another. Following are the most common infections:

- Ear infections
- Pink eye or conjunctivitis
- Sinus infections
- Pneumonia
- Chromic diarrhea
- Serious bacterial infection; however, they do not seem to get viruses

What Is the Genetic Cause of X-Linked Agammaglobulinemia (XLA)?

Mutations in the *BTK* gene, officially known as the "Bruton agammaglobulinemia tyrosine kinase" gene, cause XLA. Normally, *BTK* instructs for the protein Bruton tyrosine kinase (BTK), which is active in the development of B cells.

Over 600 mutations in *BTK* are associated with X-linked agammaglobulinemia. Most of the mutations result in no BTK protein or an abnormal version. When the B cells do not develop, then antibodies are missing, causing an increase in the number of infections. *BTK* is inherited in a recessive pattern and is located on the long arm (q) of the X chromosome at position 21.33-q33.

What Is the Treatment for X-Linked Agammaglobulinemia (XLA)?

Individuals with XLA may be treated with antibiotics to prevent infection and with weekly or monthly injections with gamma globulin to replace the missing protein.

Further Reading

Conley, Mary Ellen, and Vanessa C. Howard. 2011. "X-Linked Agammaglobulinemia." *GeneReviews*. http://www.ncbi.nlm.nih.gov/books/NBK1453. Accessed 12/8/11.

"X-Linked Agammaglobulinemia." 2011. Immune Deficiency Foundation. http://primaryimmune.org/about-primary-immunodeficiency-diseases/types-of-pidd/x-linked-agammaglobulinemia. Accessed 12/8/11.

"X-Linked Agammaglobulinemia." 2011. Mayo Clinic. http://www.mayoclinic.org/x-linked-agammaglobulinemia. Accessed 12/8/11.

X-Linked Congenital Stationary Night Blindness

Prevalence   Unknown; appears more common in Dutch-German Mennonite communities

Other Names   XLCSNB; X-linked CSNB

X-linked congenital stationary night blindness is one of those recessive conditions that is carried on the X chromosome and found only in males. (For more information on the X chromosome, see the entry "X Chromosome: A Special Topic.")

What Is X-Linked Congenital Stationary Night Blindness?

X-linked congenital stationary night blindness is a disease of the retina of the eye. In the structure of the eye, the retina is a screen at the back of the eye where images are received and then translated to electrical impulses and sent via the optic nerve to the brain. The retina also houses the rods and cones, the structures that enable one to see light and color. In this congenital disorder, the person has difficulty seeing in low light, a condition known as night blindness. Other vision problems, such as loss of sharpness, severe nearsightedness, involuntary movements of the eye, and eyes that do not look in the same direction, can also exist. The disorder affects light vision only; color vision is normal. In addition, the disease is not progressive and does not worsen over time.

Two types of X-linked congenital stationary night blindness are known: complete and incomplete. The symptoms are generally the same and are only distinguished by genetic tests and a special test of the retina, called an electroretinogram. The complete type is more common than the incomplete.

What Are the Genetic Causes of X-Linked Congenital Stationary Night Blindness?

Two genes—*CACNA1F* and *NYX*— are related to X-linked congenital stationary night blindness.

CACNA1F

Mutations in the *CACNA1F* gene, officially known as the "calcium channel, voltage-dependent, L type, alpha 1F subunit" gene, cause X-linked congenital stationary night blindness. Normally *CACNA1F* provides instructions for calcium channels, which transport calcium atoms across the membranes of the cells. The gene makes one part, called the alpha-1 subunit, of the channel called CaVI.4, which enables the atoms to flow across the membrane. The channels are found in many cells but are essential in the photoreceptors in the retina, which receive light. The receptors that receive light are called rods; the receptors that determine color are called cones. The channels created by *CACNA1F* relay information from the rods and cones to cells called bipolar cells that then transmit the signals to the brain.

Over 70 mutations in the *CACNA1F* gene cause X-linked congenital stationary night blindness. This gene is related to the incomplete form of the disorder and causes night blindness in addition to other eye problems. The mutations change the structure of the alpha-1 subunit, which changes the creation of the proper calcium channels. The atoms cannot cross the cell membranes, disrupting the signals of the rods and cones of the retina. *CACNA1F* is inherited in a recessive pattern and is found on the short arm (p) of the X chromosome at position 11.23.

NYX

Mutations in the *NYX* gene, officially known as the "nyctalopin" gene, cause a type of X-linked congenital stationary night blindness. Normally, *NYX* provides instructions for making the protein nyctalopin. This protein is located on the surface of the photoreceptors, the rods and cones. This protein is essential for normal vision because it plays a critical role in relaying the impulses to the retinal bipolar cells.

Over 50 mutations in *NYX* are related to the complete form of X-linked congenital stationary night blindness. Only one change in the protein building blocks alters the size and shape of nyctalopin, which especially affects the function of the rods without disrupting the function of the cones. *NYX* is inherited in a recessive pattern and is located on the short arm (x) of the X chromosome at position 11.4.

What Is the Treatment for X-Linked Congenital Stationary Night Blindness?

Glasses or contact lenses can treat the nearsightedness and other eye problems. However, the problems with night vision have no real treatment. The person may be restricted from driving at night.

Further Reading

Boycott, Kym M; Yves Sauvé; and Ian M. MacDonald. 2008. "X-linked Congenital Stationary Night Blindness." *GeneReviews*. http://www.ncbi.nlm.nih.gov/books/ NBK1245. Accessed 12/8/11.

"X-Linked Congenital Stationary Night Blindness." 2011. Genetics and Rare Diseases Information Center (GARD). National Institutes of Health (U.S.). http://rarediseases .info.nih.gov/GARD/Condition/3995/Night_blindness_congenital_stationary.aspx. Accessed 12/8/11.

"X-Linked Congenital Stationary Night Blindness." 2011. Genetics Home Reference. National Library of Medicine (U.S.). http://ghr.nlm.nih.gov/condition/x-linked-congenital-stationary-night-blindness. Accessed 12/8/11.

X-Linked Copper Deficiency. *See* Menkes Syndrome

X-Linked Corpus Callosum Agenesis. *See* L1 Syndrome

X-Linked Creatine Deficiency

Prevalence Exact number unknown, but may account for between 1% and 2% of males with intellectual disability; more than 150 males identified

Other Names creatine transporter defect; creatine transporter deficiency; SLC6A8 deficiency; SLC6A8-related creatine transporter deficiency; X-linked creatine deficiency syndrome

X-linked creatine deficiency is one of those recessive characteristics located on the X chromosome. (For more information on the X chromosome, see the entry "X Chromosome: A Special Topic.")

What Is X-Linked Creatine Deficiency?

X-linked creatine deficiency is a disorder that especially affects the brain and muscles. The word "creatine" comes from the Greek word *kreas*, meaning "flesh." In biochemistry, creatine is a colorless organic crystalline substance that is found in many organs and body fluids. In the body, creatine combines with phosphate to make phosphocreatine, which is an important source of energy. If a deficiency of creatine exists, several body systems are affected, especially the brain.

Following are symptoms of X-linked creatine deficiency:

- Delayed growth: Children with the deficiency experience slow growth.
- Mild to severe intellectual disability.

- Behavioral disorders: These symptoms may appear as attention deficit hyperactivity disorder or autistic signs of poor communication and social interaction.
- Delayed motor skills: The child may be late in sitting, standing, or walking.
- Delayed speech development.
- Tiredness.
- Seizures.
- Abnormal heart rhythms: Some may experience cardiac abnormalities.
- Distinct facial features: Some, but not all, will have a very small head, a broad forehead, and sunken, flat part of the middle of the face.

What Is the Genetic Cause of X-Linked Creatine Deficiency?

Mutations in the *SLC6A8* gene, officially known as the "solute carrier family 6 (neurotransmitter transporter, creatine), member 8" gene, cause X-linked creatine deficiency. Normally, *SLC6A8* instructs for the protein called sodium- and chloride-dependent creatine transporter 1, which is a transport agent to get the compound creatine in the cells. Creatine is essential for body energy.

About 20 mutations in the *SLC6A8* gene are liked to this disorder. The mutations disrupt the ability of the protein to transport creatine to the cells, which results in a damaging shortage to tissues that require lots of energy, especially the brain and muscles. *SLC6A8* is inherited in a recessive pattern and is found on the long arm (q) of the X chromosome at position 28.

What Is the Treatment for X-Linked Creatine Deficiency?

Treatment for the deficiency is to give the person oral creatine monohydrate to increase cerebral creatine levels. Supplementation with ornithine and dietary restriction of the amino acid arginine may also improve the outcome. Special education may be required to assist in correcting behavioral issues.

Further Reading

"X-Linked Creatine Deficiency." 2011. Genetics Home Reference. National Library of Medicine (U.S.). http://ghr.nlm.nih.gov/condition/x-linked-creatine-deficiency. Accessed 12/9/11.

"X-Linked Creatine Deficiency Syndrome." 2011. RightDiagnosis.com. http://www.rightdiagnosis.com/medical/x_linked_creatine_deficiency_syndrome.htm. Accessed 12/9/11.

X-Linked Dystonia-Parkinsonism

Prevalence   Linked to people of Filipino descent where more than 500 cases have been reported; traced to the ancestry of a particular woman on island of Panay, with 5.24 per 100,000

Other Names   dystonia musculorum deformans; dystonia-parkinsonism, X-linked; dystonia 3, torsion, X-linked; DYT3; Lubag; torsion dystonia-parkinsonism, Filipino type; XDP; X-linked dystonia-parkinsonism syndrome; X-linked torsion dystonia-parkinsonism syndrome

X-linked dystonia-parkinsonism is one of the recessive conditions found on the X chromosome but identified with one particular geographical group—Filipinos, whose ancestors came from the island of Panay in the Philippines.(For more information on the X chromosome, see the entry "X Chromosome: A Special Topic.")

What Is X-Linked Dystonia-Parkinsonism?

X-linked dystonia-parkinsonism is a disorder that affects the nerves and the ability to move. This neurodegenerative movement disorder is found only in people of Filipino descent. The Filipino name is Lubag. Around the age of 35, the person begins to have symptoms. However, one person, as young as 14, was reported in the literature. The symptoms include both parkinsonism and dystonia.

Parkinsonism

These movement disorders are usually the first sign of the disorder and may include the following progressive symptoms:

- Slow movement, a condition known as bradykinesia
- Tremors
- Shuffling gait with recurrent falls
- Inability to hold the body upright
- Poor balance with many falls.

Dystonia

The Parkinson symptoms continue, but later in life, the symptoms of dystonia appear. Dystonia is a condition in which muscle contractions that are involuntary may cause unusual contortions of the body. The following symptoms of dystonia may occur:

- Progressive pattern: The symptoms usually start in one area, such as the eyes, jaw, tongue, or neck, and then spread to other parts of the body.
- Muscle cramping and spasms.

- Difficulty speaking and swallowing.
- Difficulty with coordination and walking.

Scans of the brain may show a loss of neurons or nerve cells within the caudate nucleus and putamen areas of the brain. The symptoms vary widely, with both mild and severe cares reported. Those with severe dystonia-parkinsonism may become fully dependent in just a few years and die prematurely from breathing difficulties and complication.

What Is the Genetic Cause of X-Linked Dystonia-Parkinsonism?

Mutations in the TAF1 gene, known also as the "TAF1 RNA polymerase II, TATA box binding protein (TBP)-associated factor, 250kDa" gene, cause X-linked dystonia-parkinsonism. Normally, *TAF1* provides instructions for part of the protein transcription factor IID (TFIID). Because it is a transcription factor that binds to parts of DNA, *TAF1* is essential in controlling the action of many genes. The gene is part of a complex called the *TAF1/DYT3* system, which acts in multiple areas. Researchers have found this area of genetic material can be combined to form various sets of proteins that make instructions. Some of the variations are essential for the function of neurons in the brain.

Many mutations in the *TAF1/DYT3* multiple transcript system cause X-linked dystonia-parkinsonism. Some of the changes are the result of a single exchange of an amino acid building block. Other changes involve deletions or addition of extra genetic material in several nucleotides. The extra material is called a retroposon, a small piece of DNA that moves around and disrupts the function of a gene, or in this case the complete *TAF1/DYT* transcript system. In the case of X-linked dystonia-parkinsonism, the disruption is in the brain, leading to the death of neurons especially in an area called the caudate nucleus and putamen. These areas are critical for normal movement, learning, and memory. *TAF1* is inherited in a recessive pattern and is located on the long arm (q) of the X chromosome at position 13.1.

What Is the Treatment for X-Linked Dystonia-Parkinsonism?

Treating the patient with the right drugs or combination of drugs may slow some of the progress and aid some of the symptoms, but it does not cure the disorder. The Parkinson symptoms may improve with L-dopa or dopamine agonist therapy. Dystonic features may respond to anticholinergics or benzodiazepines. However, these powerful drugs may be only partially effective and can have severe side effects when used over a period of time.

Further Reading
Evidente, Virgilio Gerald H. 2010. "X-linked Dystonia-Parkinsonism." *GeneReviews*. http://www.ncbi.nlm.nih.gov/books/NBK1489. Accessed 12/10/11.

"X-Linked Dystonia-Parkinsonism." 2010. Dystonia Medical Research Foundation. http://www.dystonia-foundation.org/pages/more_info___x_linked_dystonia_parkinsonism/72.php. Accessed 12/10/11.

"X-linked Dystonia-Parkinsonism (Lubag)." 2011. Worldwide Education and Awareness for Movement Disorders. http://www.wemove.org/dys/dys_dxlink.html. Accessed 12/10/11.

X-Linked Hyper IgM Syndrome

Prevalence   Occurs in about 2 per million newborn boys

Other Names   HIGM1; hyper-IgM syndrome 1; immunodeficiency with hyper-IgM, type 1

In 1961, Rosen and a team first described X-linked hyper IgM syndrome. It is one of those disorders found on the X chromosome and found only in males. (For more information on the X chromosome, see the entry "X Chromosome: A Special Topic.")

What Is X-Linked Hyper IgM Syndrome?

X-linked hyper IgM syndrome is a disorder of the immune system. As the name "hyper" implies, people with this disorder have an excess level of antibodies called immunoglobulins. Antibodies help protect against germs and other foreign invaders. Several classes of antibodies make up the immune system:

- Immunoglobulin G, or IgG
- Immunoglobulin A, or IgA
- Immunoglobulin E, or IgE
- Immunoglobulin M, or IgM

As the name implies, people with hyper IgM syndrome have abnormal levels of this antibody and low levels of the other three. The lack of any balance in the various immunoglobulins affects the function of the immune system.

Some of the following symptoms may develop in people with X-linked hyper IgM syndrome:

- Frequent infections in infancy and early childhood, which involve pneumonia, ear infections, and sinus infections
- Failure to thrive, which means poor weight gain and growth
- Low levels of white blood cells, a condition called neutropenia
- Autoimmune disorders

- Brain and nervous system disorders
- Liver disease
- Gastrointestinal tumors
- Lymphoma, a cancer of the immune system cells

The symptoms may vary even in the same family.

What Is the Genetic Cause of X-Linked Hyper IgM Syndrome?

Mutations in the *CD40LG* gene, officially known as the "CD40 ligand" gene, cause X-linked hyper IgM syndrome. Normally *CD40LG* instructs for the protein called the CD40 ligand. A ligand is a protein that attaches like a key in a lock to a protein receptor. The CD40 ligand is found of the surface of B cells, which are involved in producing immunoglobulins to help the body fight infection. Of the classes of immunoglobulins, each has a different function. B cells mature into the cells that produce IgM without signals from other cells. The CD40 receptor must interact with the CD40 ligand in order for it to produce the other antibodies. When the process works correctly, a series of chemical signals tell the B cells to make the other three cells: IgG, IgA, and IgE. The CD40 ligand allows T cells to interact with other cells of the immune system.

Over 150 mutations in the *CD40LG* gene are connected with the hyper IgM syndrome. The mutations lead to an abnormal CD40 ligand or the absence of one completely. When this ligand does not attach to its receptor on the B cells, no IgG, IgA, or IgE is produced. The ability of the T cells is also impaired. The IgM is present in abundance while the other factors are not functioning properly, causing the symptoms of the syndrome. *CD40LG* is inherited in a recessive pattern and is located on the long arm (q) of the X chromosome at position 26.

What Is the Treatment for X-Linked Hyper IgM Syndrome?

Without treatment, this condition can result in death during childhood or adolescence. A blood transplant of hematopoietic cells performed before organ damage can avoid the life-threatening complications. Treating the symptoms, such as diarrhea and other infections, is very individual and performed on a case-by-case basis.

Further Reading

Immune Deficiency Foundation. 2007. *2007 Patient and Family Handbook for Primary Immunodeficiency Diseases*. 4th ed. http://www.primaryimmune.org/publications/ book_pats/ book_pats.htm. Accessed 2/12/12.

Johnson, Judith; Alexandra H. Filipovich; and Kejian Zhang. 2010. "X-linked Hyper IgM Syndrome." *GeneReviews*. http://www.ncbi.nlm.nih.gov/books/NBK1402. Accessed 12/10/11.

"X-linked Immunodeficiency with Hyper IgM." 2011. Medscape. http://emedicine .medscape.com/article/889104-overview. Accessed 12/10/11.

X-Linked Hyperuricemia. *See* Lesch-Nyhan Syndrome (LNS)

X-Linked Infantile Nystagmus

Prevalence   About 1 in 5,000 newborns; exact number not known

Other Names   congenital motor nystagmus; FRMD7-related infantile nystagmus; idiopathic infantile nystagmus; NYS1; X-linked congenital nystagmus; X-linked idiopathic infantile nystagmus

X-linked infantile nystagmus is one of those recessive disorders located on the X chromosome and found mostly in males. (For more information on the X chromosome, see the entry "X Chromosome: A Special Topic.")

What Is X-Linked Infantile Nystagmus?

X-linked infantile nystagmus is a disorder of the eye. The Greek word *nystagmos*, meaning "nod," describes the involuntary movements of the eyes. The movement can be in any direction. Typically the motion is horizontal, and the movement is "to-and-fro" like one is nodding. As the person gets older, the range of the movement may decrease; but the frequency may increase, especially if he tries to stare directly at an object. The movement may also worsen when the person is tired or sick. The person is born with the condition, or it may develop within the first few months of life.

Other symptoms may also accompany the movement. The young children may nod, shake, or bobble the head, although this may disappear over time. Vision may be reduced; the person may find an area where vision is best, and position the head to maximize this vision. Balance may also be affected. However, symptoms may vary, even among members of the same family.

What Is the Genetic Cause of X-Linked Infantile Nystagmus?

Mutations in the *FRMD7* gene, officially known as the "FERM domain containing 7" gene, cause X-linked infantile nystagmus. Normally, *FRMD7* provides instructions for a protein that is found in midbrain and cerebellum, regions that control eye movement. Although the exact function of the protein is not known, it probably is essential in the development of the nerve cells in the areas of the brain and retina.

Over 35 mutations in *FRMD7* have been linked to X-linked infantile nystagmus. A single change in one of the amino acid building blocks can disrupt the function of the protein. This change produces an unstable protein that interrupts the normal development of the nerve cells in the retina and areas that control eye movement. *FRMD7* is inherited in a recessive pattern and is located on the long arm (q) of the X chromosome at position 26.2.

What Is the Treatment for X-Linked Infantile Nystagmus?

The condition cannot be cured. Glasses or contact lenses may be beneficial. Surgery on the muscles of the eye may be helpful, especially if the person adopts a head position that may put strain on the neck muscles. Low-vision aids and special education are available.

Further Reading

"Congenital Nystagmus." 2010. University of Arizona Hereditary Ocular Disease. http://disorders.eyes.arizona.edu/disorders/congenital-nystagmus-0. Accessed 12/10/11.

"X-Linked Infantile Nystagmus." 2011. Genetics Home Reference. National Library of Medicine (U.S.). http://ghr.nlm.nih.gov/condition/x-linked-infantile-nystagmus. Accessed 12/10/11.

"X-Linked Infantile Nystagmus." 2011. Medscape. http://www.medscape.org/viewarticle/707024_3. Accessed 12/10/11.

X-Linked Juvenile Retinoschisis

Prevalence   Estimated 1 in 5,000 to 25,000 males worldwide

Other Names   congenital X-linked retinoschisis; retinoschisis; retinoschisis, degenerative; retinoschisis, juvenile, X-linked

In 1898, Haas first described some male patients in which the retina of the eye appeared damaged. He called the condition vitreous veils because of the appearance when he looked into the eyes. Later, in 1953, Jaeger coined the term "retinoschisis." The word "retinoschisis" combines the term retina, the screen at the back of the eye, with the Greek work *schisis*, meaning "splitting." The condition was linked to a gene on the X chromosome. (For more information on the X chromosome, see the entry "X Chromosome: A Special Topic.")

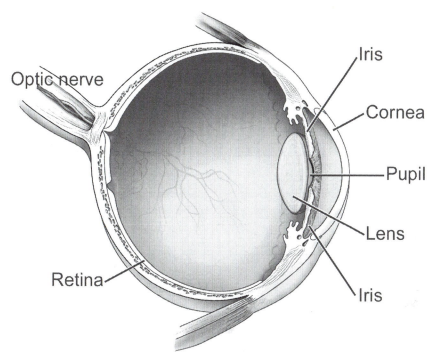

A diagram of the human eyeball. In juvenile retinoschisis, part of the retina in the back of the eye is damaged, impairing vision. This genetic disorder is almost always found only in boys. (CDC)

What Is X-Linked Juvenile Retinoschisis?

Juvenile X-linked retinoschisis is a disorder of the nerve tissue of the eye, which occurs in young boys. The disease affects the cells in the part of the retina of the eye called the macula, where the central point of sharp vision occurs. The macula enables the vision necessary for reading, recognizing faces, and driving. Occasionally, in retinoschisis, the retina splits into two layers, with a cyst or growth forming between the two layers. The young boys will first be noticed when they start to school and obviously have poor vision. In addition, the boys may squint and have involuntary eye movements or nystagmus.

What Is the Genetic Cause of X-Linked Juvenile Retinoschisis?

Mutations in the *RS1* gene, officially called the "retinoschisin 1" gene, causes X-linked juvenile retinoschisis. Normally, *RS1* provides information for making the protein retinoschisin, which plays an essential role in developing and maintaining the light sensitive photoreceptors in the retina. Retinoschisin is probably necessary for the retinal cells to organize properly and for them to adhere to one another.

Over 160 mutations in *RS1* are connected to X-linked juvenile retinoschisis. Most mutations result from the exchange of just one amino acid building block in the protein. The mutations affect the three-dimensional structure of the protein, disrupting the organization of the cells and the cell adhesion process. This disruption can cause the splitting or tearing of the retina and result in the symptoms of the disorder. *RS1* is inherited in a recessive pattern and is located on the short arm (p) of the X chromosome at position 22.13.

What Is the Treatment for X-Linked Juvenile Retinoschisis?

The person needs an early diagnosis for the earliest intervention possible. Using many low-vision aids such as large-print textbooks, preferential seating, and handouts with high contrast may help. Surgery may be needed to address the complications of the vitreous bleeding and retinal detachment. The person should avoid contact sports.

Further Reading

"Retinoschisis." 2011. University of Michigan Kellogg Eye Center. http://www.kellogg.umich.edu/patientcare/conditions/retinoschisis.html. Accessed 12/12/11.

"Retinoschisis, Juvenile." 2011. Medscape. http://emedicine.medscape.com/article/1225857-overview. Accessed 12/12/11.

Sieving, Paul A.; Ian M. MacDonald; Meira Rina Meltzer; and Nizar Smaoui. 2009. "X-Linked Juvenile Retinoschisis." GeneReviews. http://www.ncbi.nlm.nih.gov/books/NBK1222. Accessed 12/12/11.

X-Linked Lissencephaly. *See* Lissencephaly

X-Linked Lymphoproliferative Disease (XLP)

Prevalence	About a million males worldwide
Other Names	Duncan disease; Epstein-Barr virus-induced lymphoproliferative disease in males; familial fatal Epstein-Barr infection; Purtilo syndrome; severe susceptibility to EBV infection; severe susceptibility to infectious mononucleosis

X-linked lymphoproliferative disease (XLP) is one of the recessive disorders found in males located on the X chromosome. (For more information on the X chromosome, see the entry "X Chromosome: A Special Topic.")

In 1964, M. A. Epstein and Y. M. Barr discovered a virus that showed different disorders in different geographical areas of the country. No explanation was given for the reason. It is known that the condition XLP is also associated with a virus known as Epstein-Barr virus (EBV).

What Is X-Linked Lymphoproliferative Disease (XLP)?

X-linked lymphoproliferative disease is a serious disorder of the immune system. Most of the males who have this condition are especially reactive to the Epstein-Barr virus or EBV. This very common virus eventually affects most people. Normally, after the initial infection, EBV remains in the person's immune cells but is inactive because the person's T cells are there to fight the virus off. In some people it can erupt to cause "mono" or infectious mononucleosis.

With XLP, the person overreacts to the EBV virus by producing large amounts of immune cells. Large numbers of T cells, B cells, and other lymphocytes called macrophages cause a condition known as hemophagocytic lymphohistiocytosis. In this condition, blood producing cells in the bone marrow are broken down, causing fever and damage to the liver, spleen, heart, kidneys, and other organs and tissues.

In addition to the destruction of blood-forming cells, some of the people with XLP have very high levels of some kinds of antibodies, a condition known as dysgammaglobulinema. The gamma globulins in the blood overshadow the other antibodies, causing the immune system to fail to function properly and making the person prone to infection. The person may eventually get lymphoma or cancer of the immune system.

What Are the Genetic Causes of X-Linked Lymphoproliferative Disease (XLP)?

Two genes—*SH2D1A* and *XIAP*—are related to XLP.

SH2D1A

Mutations in the *SH2D1A* gene, officially known as the "SH2 domain containing 1A" gene, cause XLP. Normally, *SH2D1A* instructs for making a protein called the signaling lymphocyte activation molecule (SLAM) associated protein (SAP). The immune system is a complex one. The SAP interacts with other proteins in the SLAM family to initiate signaling pathways to control immune cells. This action causes certain lymphocytes to destroy certain toxic lymphocytes and assist in the production of cells called killer T cells. When the lymphocytes are no longer needed, it aids in their self-destruction.

Over 70 mutations in *SH2D1A* cause XLP. Most of the mutations destroy the SAP functions or result in a short or unstable SAP. This loss of function disrupts the whole pathway signals of the immune system, allowing the life-threatening EBV to take hold. Lymphomas may also develop. *SH2D1A* is inherited in a recessive pattern and is located on the long arm (q) of the X chromosome at position 25.

XIAP

The second gene involved with XLP is the *XIAP* gene, officially known as the "X-linked inhibitor of apoptosis" gene. Normally, *XIAP* instructs for a protein that is found especially in the immune cells. Its mission is to protect these cells from self-destructing by keeping them from certain enzymes called caspases, which are related to apoptosis. *XIAP* keeps caspase 3, 7, and 9 from acting on the immune cells.

Several gene mutations in *XIAP* are found in people with XLP. The mutations reduce or completely eliminate the protein, which results in the increase if the lymphocytes that destroy the blood-forming cells and damage liver and other organs. *XIAP* is inherited in a recessive pattern and is located on the long arm (q) of the X chromosome at position 25.

What Is the Treatment for X-Linked Lymphoproliferative Disease (XLP)?

This condition is life-threatening. In order to life past childhood, the child must have a stem cell transplant called an allogeneic stem cell transplantation. A healthy donor must give the cells for the transplant. If not, the person usually dies from complications associated with the EB virus.

Further Reading

Filipovich, Alexandra; Judith Johnson; Kejian Zhang; and Rebecca Marsh. 2011. "Lymphoproliferative Disease, X-Linked." *GeneReviews*. http://www.ncbi.nlm.nih.gov/books/NBK1406. Accessed 12/12/11.

"X-Linked Lymphoproliferative Disease." 2011. Genetics Home Reference. National Library of Medicine (U.S.). http://ghr.nlm.nih.gov/condition/x-linked-lymphoproliferative-disease. Accessed 12/12/11.

"X-Linked Lymphoproliferative Disease." 2011. Medscape. http://emedicine.medscape.com/article/203780-overview. Accessed 12/12/11.

X-Linked Mental Retardation and Macroorchidism. *See* Fragile X Syndrome

X-Linked Myotubular Myopathy

Prevalence　　About 1 in 50,000 newborns

Other Names　CNM; MTMX; X-linked centronuclear myopathy; XLMTM; XMTM

X-linked myotubular myopathy is one of the serious muscular conditions that are found on the X chromosome that is found only in males. (For more information on the X chromosome, see the entry "X Chromosome: A Special Topic.")

What Is X-Linked Myotubular Myopathy?

X-linked myotubular myopathy is a condition that primarily affects muscles used for movement (skeletal muscles) and occurs almost exclusively in males. This disorder is a member of a family called the centronuclear myopathies. The myotube is a stage in development of the skeletal muscles in which the center of the cell occupies most of the cells. In a functioning muscle cell, the center is found at either end.

Following are the symptoms of X-linked myotubular myopathy:

- Muscle weakness, a condition known as myopathy, evident at birth
- Poor muscle tone, a condition known as hypotonia, also evident at birth
- Poor motor skills—the muscle problems affect learning to sit, stand, or walk
- Eating difficulties—due to muscle weakness, infants may not be able to feed
- Breathing difficulties—individuals may need mechanical assistance to breathe
- Eye movements
- Weakness in muscles of the face
- Poor reflexes
- Curvature of the spine
- Joint deformities especially of the hips and knees
- Large head
- Narrow elongated face

The breathing disorders usually cause an early death, although some with X-linked myotubular myopathy have lived into adulthood.

What Is the Genetic Cause of X-Linked Myotubular Myopathy?

Mutations in the *MTM1* gene, officially known as the "myotubularin 1" gene, cause X-linked myotubular myopathy. Normally, *MTM1* provides instruction for the enzyme myotubularin, which acts as a phosphatase enzyme. This enzyme removes oxygen and phosphorous from two molecules, phosphatidylinositol 3-phosphate and phosphatidylinositol 3,5-biphosphate. These two molecules transport molecules within cells.

Over 200 mutations in *MTM1* are related to X-linked myotubular myopathy. Most of the mutations are a change in only one building block, and others result in an abnormally short, nonworking enzyme. All the mutations act to prevent myotubularin from properly removing the oxygen and phosphorous from the important two transport molecules. The result is that the development and maintenance role is disrupted, leading to the symptoms of X-linked myotubular myopathy. *MTM1* is inherited in a recessive pattern and is located on the long arm (q) of the X chromosome at position 28.

What Is the Treatment for X-Linked Myotubular Myopathy?

Treatment is symptomatic and usually involves long-term care that includes a pulmonologist, neurologist, physical therapist, and rehabilitation specialist. A tracheostomy, G-tube feeding, and assistive communication devices are often needed.

Further Reading

Das, Soma; James Dowling; and Christopher R. Pierson. 2011. "X-Linked Centronuclear Myopathy." *GeneReviews*. http://www.ncbi.nlm.nih.gov/books/NBK1432. Accessed 12/12/11.

"Myotubular Myopathy (MTM)/Centronuclear Myopathy (CNM)." Muscular Dystrophy Association. 2011. http://www.mda.org/disease/mm.html. Accessed 12/12/11.

Myotubular Myopathy Resource Group. 2001. http://www.mtmrg.org. Accessed 12/12/11.

X-Linked Opitz Syndrome (XLOS). *See* Opitz G/BBB Syndrome

X-Linked Primary Hyperuricemia. *See* Lesch-Nyhan Syndrome (LNS)

X-Linked Severe Combined Immunodeficiency

Prevalence	Most common form of SCID; exact incidence unknown; affects about 1 in 50,000 to 100,000 newborns
Other Names	SCIDX1; XSCID; X-SCID

X-linked severe combined immunodeficiency or SCID is one of those conditions that received wide publicity when David Vetter, the Bubble Boy, lived in a plastic bubble for over 12 years. (For more information on the X chromosome, see the entry "X Chromosome: A Special Topic.")

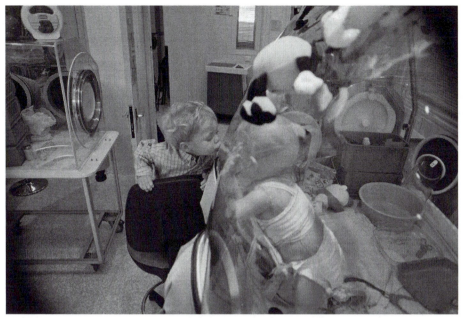

A cured child gives a kiss through the plastic sheet of the sterile "bubble" to another child affected with severe combined immunodeficiency (SCID) X1 who had received a bone marrow transplant at Necker Hospital for Sick Children, in Paris,in 1999. (Gamma-Rapho via Getty Images)

What Is X-Linked Severe Combined Immunodeficiency?

X-linked severe combined immunodeficiency (SCID) is a disorder of the immune system. The immune system is a body-wide network of organs, tissues, and blood factors that work together to fight off bacteria, viruses, fungi, or other foreign invaders, called antigens. In the bone marrow, the cells develop that will fight these foreign antigens and destroy them. The cells that are part of the immune system are the following:

- *T cells*: Three forms of T-lymphocytes are Helper cells, which direct and assist all immune cells in attacking foreign antigens; cytotoxic T cells commonly called killer T cells, which kill the unwanted antigens; and regulatory T cells, called T-suppressor cells, which control the off-on switch to attack.

- *B cells*: When the outsider gets into the body, the helper T cells tell the B cells to make antibodies against it. Antibodies bind to the antigen and neutralize it allowing the phagocytes to destroy it completely.

- *Phagocytes*: The Greek word *phage* means "eat." These cells literally devour the foreign invaders that the B cells have neutralized.

- *NK cells*: These are the natural killer cells that seek out invaders, such as tumor cells and viruses, to destroy them.

If the body is strong, the person with normal immune system cells is able to fight off the invaders and win most of the time. However, boys with X-linked severe combined immunodeficiency (SCID) are missing some major components of the system. They have a severe defect in the T-cell production that keeps the B-lymphocytes and other immune system cells from functioning. Therefore, they are prone to recurrent and persistent infections. These infections can be life-threatening, and without medical intervention, the children usually die in infancy or early childhood.

Boys with SCID get all types of opportunistic infections, which would not cause illness in a person with a normal immune system. As infants, they experience chronic diarrhea, skin rashes, and slow growth. At present, researchers know of 10 different varieties of SCID that have various effects. All of them result in the same critical, deadly results if not treated.

What Is the Genetic Cause of X-Linked Severe Combined Immunodeficiency?

Mutations in the *IL2RG* gene, officially known as the "interleukin 2 receptor, gamma" gene, cause X-linked SCID. Normally, *IL2RG* gives the instructions for making the protein called the common gamma chain. The protein, which is a transcription factor with one end inside the cell and the other end outside the cell, functions to transmit signals to the nucleus of the cell. These receptors are on the surface of immature blood-forming cells in the bone marrow and partner with other cells to regulate T cells, B cells, and NK cells.

Over 300 mutations in the *IL2RG* gene cause SCID. Most of the mutations result from a change in only one building block. The changes either lead to a non-working common gamma chain or prevent the protein from functioning. Without the gamma chain, major chemical signals are not relayed to the nucleus, and the cells of the immune system do not develop properly. This glitch in the immune system leads to the severe life-threatening SCID. *IL2RG* is inherited in a recessive pattern and is located on the long arm (q) of the X chromosome at position 13.1.

What Is the Treatment for X-Linked Severe Combined Immunodeficiency?

Children with this disorder cannot survive without medical intervention. They must receive a bone marrow transplant (BMT) from a relative with matching bone marrow. Interim management includes treatment of infection with immunoglobulin infusions and antibiotics. Gene therapy has been successful in some individuals.

The Boy in the Bubble

In the early 1980s, David Vetter, a young boy from Shenandoah, Texas, became known as "David, the Bubble Boy." He was born with the rare genetic disease X-linked severe combined immunodeficiency. Because he had no immune responses, he was forced to live his life inside a special large plastic bubble at the Texas Children's Hospital in Houston.

When an older son died at seven months of SCID, the parents were told that each son they conceived would have a 50/50 chance of having the condition. However, according to the doctors, there was hope. The parents had a daughter who could donate bone marrow for a transplant. Within minutes after birth, David was placed in a little plastic cocoon bed, to wait for the transplant. But, some bad news emerged; the daughter was not a match, and David had no cure. The doctors had not planned for this but said they would keep him in the bubble until a cure was found.

Everything the child needed had to be disinfected in a chamber filled with ethylene oxide before it could go into the bubble. David was handled with special plastic gloves that were built into the walls of the bubble. As he grew, he had tutors, television, and a playroom inside the bubble. He was set up with a special bubble so he could go home for two or three weeks at a time. The parents let friends visit him and even arranged a special showing at a movie theater for David in his traveling bubble. At about the age of four, he learned that he could poke holes in the bubble, leading the doctor to tell him about germs and his special condition.

David's story became a media sensation, and VIPs from everywhere came to visit him. He appeared as a calm and happy child, but actually he was quite depressed about the isolation and prospects for the future. In 1977, NASA designed for him a special $50,000 suit so he could get out of the bubble and walk in the outside. On the day of his debut to the outside world, David refused to get into the suit for the press. Later, he used the suit only seven times before outgrowing it.

David's life was not pleasant. Doctors feared that as he got older, he would become more difficult. In addition, people were beginning to question the ethics of the research, and the government was threatening to cut funding, claiming they had spent more than $1.3 million for research that was showing no results. The doctors then convinced the parents to let them try a bone marrow transplant using his sister, even though the bone marrow donor was not an exact match.

David did fine for several months and hoped that soon he would be out of the bubble. But he became very ill with diarrhea, vomiting, and high fever. He was taken out of the bubble for treatment. He died on February 22, 1984, at the age of 12. His sister's marrow had a dormant Epstein-Barr virus, which was activated when transplanted to David. David's medical and personal artifacts, including the space suit, are on display at the National Museum of American History in Washington.

Further Reading

Bonilla, Francisco A. 2011. "X-Linked Severe Combined Immunodeficiency (SCID)." UpToDate. http://www.uptodate.com/contents/x-linked-severe-combined-immunodeficiency-scid. Accessed 12/11/11.

Davis, Joie, and Jennifer M. Puck. 2005. "X-Linked Severe Combined Immunodeficiency." *GeneReviews*. http://www.ncbi.nlm.nih.gov/books/NBK1410. Accessed 12/11/11.

"Missing Body Defense Systems." 2010. SCID.Net. http://www.scid.net/about.htm. Accessed 12/11/11.

X-Linked Sideroblastic Anemia

Prevalence   Unknown; may be more common than once thought

Other Names   Anemia, sex-linked hypochromic sideroblastic; ANH1;
congenital sideroblastic anaemia; erythroid 5-aminolevulinate
synthase deficiency; hereditary iron-loading anemia;
X chromosome–linked sideroblastic anemia; X-linked
pyridoxine-responsive sideroblastic anemia; XLSA

X-linked sideroblastic anemia is one of those X-linked conditions found only in males. (For more information on the X chromosome, see the entry "X Chromosome: A Special Topic.") *Note that this condition is different from sideroblastic anemia and ataxia.* This order has a serious anemia component and none of the ataxia or movement disorders. Also, the two disorders have different genetic causes.

What Is X-Linked Sideroblastic Anemia?

X-linked sideroblastic anemia is a disorder of the red blood cells. The word "sideroblastic" comes from two Greek words: *sidero*, meaning "iron," and *blast*, meaning the "original germ cells." This term refers to the blood cells called sideroblasts that, in the bone marrow, are related to several disease disorders. A sideroblast is a red blood cell that has heavy deposits of iron in the cytoplasm, and these abundant deposits from a ring around the nucleus. Normally, the iron would have been used to make heme, but because the mature erythrocytes lack mitochondria, the sideroblasts do not do so.

People with X-linked sideroblastic anemia have mature blood cells that are smaller than normal and appear small and pale because the hemoglobin does not work normally. The red blood cells that are produced in the bone marrow are called ring sideroblasts, and this abnormal shape leads to the accumulation of iron in the red blood cells.

Following are the signs of X-linked sideroblastic anemia:

• Pale skin

• Fatigue

• Dizziness

• Rapid heartbeat

• Enlarged liver and spleen

The symptoms of this disorder result from the combination of an overload of iron and reduced hemoglobin. Although the symptoms range from mild to severe and appear only in young adulthood, the person over time will develop serious heart disease and damage to the liver because of excess iron buildup.

What Are the Genetic Causes of X-Linked Sideroblastic Anemia?

Two genes are involved in X-linked sideroblastic anemia: *ALAS2* and *HFE*.

ALAS2

Mutations in the *ALAS2* gene, officially known as the "aminolevulinate, delta-, synthase 2" gene, cause X-linked sideroblastic anemia. Normally, *ALAS2* provides instructions for the enzyme 5-aminolevulinate synthase 2 or erythroid ALA-synthase. Two genes, *ALAS1* and *ALAS2*, make this protein. *ALAS1* is active throughout the body, but *ALAS2* is active only in developing red blood cells.

ALA-synthase is one of eight enzymes in a multistep process that produces heme, the iron-containing substance in the red blood cells and component that is essential in the oxygen–carbon dioxide exchange in blood vessels called capillaries. ALA-synthase is responsible for the first step in the process.

At least 50 mutations in *ALAS2* cause X-linked sideroblastic anemia. Most of the mutations result in only a single change in only one amino acid building block in the ALA-synthase enzyme. These changes disrupt the entire normal production of heme and prevent the cells from making enough hemoglobin. Normally, most of the iron in the erythroblasts is converted into heme. However, if this process does not take place, excess iron builds up in the cells, causing damage to the body organs and the features of X-linked sideroblastic anemia. *ALAS2* is inherited in a recessive pattern and is located on the short arm (p) of the X chromosome at position 11.21.

HFE

The second gene involved in X-linked sideroblastic anemia is *HFE*, officially known as the "hemochromatosis" gene. Normally, *HFE* instructs for a protein that is on the surface of cells, primarily liver, intestinal, and immune system cells. This protein acts as a regulator of the amount of iron in the body. It works with another protein called hepcidin, which is considered the master iron regulator hormone. Produced in the liver, hepcidin determines how much iron is absorbed from food and how it is stored in the body. A person may absorb about 10% of the iron eaten in foods. HFE protein also interacts with the transferring receptor, whose role in iron regulation in unknown.

Only one mutation of this gene may interact with a mutation in *ALAS2* to cause more severe symptoms of X-linked sideroblastic anemia. The mutation may increase the severity of the disorder because the amount of iron from food may lead to a greater overload of iron in the cells and eventually the organs. *HFE* is inherited in an autosomal recessive pattern and is located on the short arm of chromosome 6 at position 21.3.

What Is the Treatment for X-Linked Sideroblastic Anemia?

Because this disorder is potentially life-threatening, treatment is essential. It may involve removing toxic agents and administering pyridoxine, thiamine, or folic acid and other antidotes for iron overload. Several medical measures could include chemotherapy and bone marrow or liver transplant.

Further Reading

"Sideroblastic Anemias." 2011. Medscape. http://emedicine.medscape.com/article/ 1389794-overview. Accessed 12/13/11.

"X-Linked Sideroblastic Anemia." 2011. Genetics Home Reference. National Library of Medicine (U.S.). http://ghr.nlm.nih.gov/condition/x-linked-sideroblastic-anemia. Accessed 12/13/11.

X-Linked Sideroblastic Anemia and Ataxia

Prevalence   Rare; only a few families affected

Other Names   X-linked sideroblastic anemia with ataxia; XLSA/A

X-linked sideroblastic anemia and ataxia is one of those X-linked conditions found only in males. (For more information on the X chromosome, see the entry "X Chromosome: A Special Topic.") *Note that this condition is different from sideroblastic anemia. It has different symptoms as well as different genetic causes.*

What Is X-Linked Sideroblastic Anemia and Ataxia?

X-linked sideroblastic anemia and ataxia is a condition involving a blood disorder called anemia, combined with movement issues. The term "sideroblast" comes from Greek words: *sidero*, meaning "iron," and *blast*, meaning "germ cell." This term refers to the blood cells called sideroblasts that, in the bone marrow, related to several disease disorders. A sideroblast is a red blood cell that has heavy deposits of iron in the cytoplasm, and these abundant deposits from a ring around the nucleus. Normally, the iron would have been used to make heme, but because the mature erythrocytes lack mitochondria, the sideroblasts do not do so. However, this structure does not appear to cause a serious toxic buildup in X-linked sideroblastic anemia and ataxia, and the anemia is usually mild and treatable.

However, major problems do arise with ataxia. Early in life, the young boy has movement and balance problems. The following issues are related to ataxia:

- Movement issues: First, the trunk of the person is affected, making sitting, standing, and walking impossible.
- Coordination of movements: The person cannot judge distance or scale and cannot make rapid, alternating movements.
- Speech difficulties.
- Tremors.
- Rapid eye movements.

What Is the Genetic Cause of X-Linked Sideroblastic Anemia and Ataxia?

Mutations in the *ABCB7* gene, officially known as the "ATP-binding cassette, subfamily B (MDR/TAP), member 7" gene, causes X-linked sideroblastic anemia and ataxia. Normally, *ABCB7* makes instructions for the protein called an ATP-binding cassette (ABC) transporter. A transporter protein carries many types of molecules across cell membranes. ABCB7 protein is found in the mitochondria, the bean-shaped powerhouse of the cell. In the developing red blood cells, ABCB7 protein, which is located in the cell membrane of the mitochondria, is essential for developing heme, the compound that contains iron and is an important part of hemoglobin that performs the oxygen–carbon dioxide exchange in the blood capillaries. The protein is probably involved in maintaining the balance of iron in the developing red blood cells.

About three mutations in the *ABDB7* are associated with X-linked sideroblastic anemia and ataxia. Most of the changes involve a single building block in the protein, altering its structure. The small changes disrupt the normal role in heme production and balance of iron. Anemia results; however, scientists are not clear about the role in ataxia. *ABCB7* is inherited in a recessive pattern and is located on the long arm (q) of the X chromosome at position 13.3.

What Is the Treatment for X-Linked Sideroblastic Anemia and Ataxia?

Males with the disorder can benefit from early physical therapy to attain gross motor skills. Adaptive devices may help. Speech therapy may improve articulation of words. The anemia is usually easily managed with supplements and medication.

Further Reading

Pagon, Roberta A., and Thomas D. Bird. "X-Linked Sideroblastic Anemia with Ataxia." *GeneReviews*. http://www.ncbi.nlm.nih.gov/books/NBK1321. Accessed 12/12/11.

"X-Linked Sideroblastic Anemia and Ataxia." Genetics Home Reference. National Library of Medicine (U.S.). http://www.ghr.nlm.nih.gov/condition/x-linked-sideroblastic-anemia-and-ataxia. Accessed 12/12/11.

"X-Linked Sideroblastic Anemia with Ataxia: Another Mitochondrial Disease?" http://www.ncbi.nlm.nih.gov/pmc/articles/PMC1763461. Accessed 12/12/11.

X-Linked Uric Aciduria Enzyme Defect. *See* Lesch-Nyhan Syndrome (LNS)

XX Male Syndrome. *See* **46,XX Testicular Disorder of Sex Development**

XX Sex Reversal. *See* **46,XX Testicular Disorder of Sex Development**

XXX Syndrome. *See* **Triple X Syndrome**

XXY Syndrome. *See* **Klinefelter Syndrome**

XY Trisomy. *See* **Klinefelter Syndrome**

XXYY Syndrome. *See* **48,XXYY Syndrome**

XYY Karyotype. *See* **47,XYY Syndrome**

XYY Syndrome. *See* **47,XYY Syndrome**

Y

Y Chromosome: A Special Topic

The Small Chromosome

Human beings have two sex chromosomes; one is X, and the other Y. Females have two X chromosomes, but no Y chromosome. Males have one X and one Y chromosome. Thus, it is up to the male to contribute a Y chromosome to create a male embryo. There are 22 pairs of body or somatic chromosomes and one set of sex chromosomes, making a total of 23 pairs. These are found in each cell of the body.

The Y chromosome is the smallest of all chromosomes. It spans about 58 million amino acid building blocks and is only about 2% of the total DNA. The Y chromosome has between 70 and 200 genes. The large region of the chromosome is the one that has the *SRY* gene, which determines the development of the fetus into a male. The other regions of the Y chromosome have very little information and determine little else.

Like its companion the X chromosome, the Y also has a few genes located at the end of the chromosome pseudoautosomal regions, which means they act like somatic chromosomes. Many of these genes are important for normal development. However, if the mother contributes a mutated recessive gene on the one arm of the X chromosome, the offspring may have problems with health and development because there is no area on the small Y chromosome to mask it.

Following are the disorder related to the Y chromosome:

- 46,XX testicular disorder of sex development: Some genetic material, containing the *SRY* gene, from the Y chromosome translocates to the X chromosome. The fetus will be a male without a Y chromosome.

- 48,XXYY syndrome: An extra X chromosome and an extra Y chromosome in each of the male's cells interferes with make sexual development and the level of the sex hormone testosterone.

- 47,XYY syndrome: The extra Y chromosome is associated with tall stature, learning disorders, and other features in boys and men.

- Y chromosome infertility: Deletions of genetic material in the Y chromosome result in abnormal sperm development.

See also X Chromosome: A Special Topic

Further Reading

Johnson, George B. 2003. "Y Chromosome: Men Really Are Different." TxtWriter. http://
www.txtwriter.com/onscience/articles/ychromosome.html. Accessed 12/14/11.

"Y Chromosome." 2012. Genetics Home Reference. National Library of Medicine (U.S.).
http://ghr.nlm.nih.gov/chromosome/Y. Accessed 2/10/12.

"The Y Chromosome." 2003. Nature.com. http://www.nature.com/nature/focus/
ychromosome. Accessed 12/14/11.

Y Chromosome Infertility

Prevalence   About 1 in 2,000 to 1 in 3,000 males of all groups.

Other Names   Y chromosome-related azoospermia

What Is Y Chromosome Infertility?

Y chromosome infertility is a condition that affects sperm. The man cannot father
children for the following reasons:

- No sperm is produced, a condition called azoospermia
- A smaller than usual number of sperm is produced, a condition called oligo-
  spermia
- Sperm cells are abnormally shaped
- Sperm cells do not move properly

Some men may have mild to moderate infertility and can possibly father a child by
using assisted reproductive technologies. In this procedure, sperm may be harvested from the testes for fertilization in vitro, or outside the body. Most men with
Y chromosome infertility have no other physical or health symptoms.

What Are the Genetic Causes of Y Chromosome Infertility?

Changes in the Y chromosome cause Y chromosome infertility. The Y chromosome is very small compared to the X chromosome and carries very little information except male sex determination and development. Two basic factors are
involved—the azoospermia factor and the *USP9Y* gene.

Azoospermia Factor or AZF

The small Y chromosome has only a few regions. Deletion of any genetic material
can disrupt the normal male factors. An area called the azoospermia factor A, B, or

C houses genes that instruct for proteins related to sperm development. Any missing genetic material will disrupt the proteins needed for normal sperm development and result in Y chromosome infertility.

This condition is caused by new deletions in men with no family history. If men do father children through assisted reproductive technologies, they will pass the genetic condition called Y-linked to all sons. Daughters who do not have a Y chromosome will not be affected.

USP9Y

Mutations in the *USP9Y* gene, officially known as the "ubiquitin specific peptidase 9, Y-linked" gene, cause Y chromosome infertility. Normally, *USP9Y* instructs for the protein ubiquitin-specific protease 9. Found on the Y chromosome in the region known as azoospermia factor A (AZFA), the protease is believed to be involved in sperm development.

Mutations in this gene are rare. However, a small number of men may have mutations or deletions in all or part of the gene. The changes disrupt the production of ubiquitin-specific protease 9 or result in an abnormal, nonworking protein. The person will not be able to father children. *USP9Y* is located on the long arm (q) of the Y chromosome at position 11.2.

What Is the Treatment for Y Chromosome Infertility?

Recent research into the field of assisted reproductive technologies has enabled some men with Y chromosome infertility to father children. A technique called in vitro fertilization retrieves sperm from semen or from testicular biopsies. The harvested sperm can then fertilize the egg in vitro or outside the body. The fertilized embryo is then implanted in the uterus for development. The risk of birth defects are not increased in children conceived in this manner.

Further Reading

Silver, Sherman T. 2011. "The Y Chromosome and Male Infertility." The Infertility Center of St. Louis. St. Luke's, Hospital, St. Louis, Missouri. http://www.infertile .com/infertility-treatments/y-chromosome.htm. Accessed 12/12/11.

"Y Chromosome Infertility." 2011. Genetics Home Reference. National Library of Medicine (U.S.). http://ghr.nlm.nih.gov/condition/y-chromosome-infertility. Accessed 12/12/11.

Yakut Short Stature Syndrome. *See* 3-M Syndrome

YY Syndrome. *See* 47,XYY Syndrome

Z

ZAP70-Related Severe Combined Immunodeficiency (ZAP70-Related SCID)

Prevalence Rare; only about 15 affected individuals identified

Other Names Selective T-cell defect; ZAP70-related SCID; zeta-associated protein 70 deficiency

What Is ZAP70-Related Severe Combined Immunodeficiency (ZAP70-Related SCID)?

ZAP70-related severe combined immunodeficiency (SCID) is a rare disorder of the immune system. The prevalence of severe combined immunodeficiency disorder or SCID from all genetic causes is 1 in 50,000, but only 15 cases of this type of condition have been identified. This disorder is one of the most serious because children with SCID have absolutely no immunity to anything. Bacteria, viruses, fungi, or any other outside invader cause very serious infections that can be life-threatening. Many other organisms, called opportunistic infections, cause problems that would not otherwise be a threat to healthy people. Following are some of the symptoms of ZAP70-related SCID:

- Pneumonia
- Chronic diarrhea
- Skin rashes
- Oral moniliasis, also known as candidiasis
- Failure to grow and thrive

If the children are not treated with a stem cell transplant, they will not live past their second year.

What Is the Genetic Cause of ZAP70-Related Severe Combined Immunodeficiency (ZAP70-Related SCID)?

Mutations in the *ZAP70* gene, officially known as the "zeta-chain (TCR) associated protein kinase 70kDa" gene, causes ZAP70-related SCID. Normally, *ZAP70* instructs for the protein zeta-chain-associated protein kinase, which is part of a signaling pathway that turns on the T cells of the immune system. The T cells search out and defend the body against foreign invaders. *ZAP70* is responsible for the activation of several types of T cells, including the cytotoxic cells that destroy cells infected by viruses and helper T cells that direct the activity of other immune cells.

Over 12 mutations in *ZAP70* have been related to ZAP70-related SCID. The mutations result in a single change in one amino acid building block. This one change disrupts the zeta-chain-associated protein kinase, leading to an unstable, nonworking protein that cannot produce the proper T cells. Without T cells, the person is susceptible to infection. *ZAP70* is inherited in an autosomal recessive pattern and is located on the long arm (q) of chromosome 2 at position 12.

What Is the Treatment for ZAP70-Related Severe Combined Immunodeficiency (ZAP70-Related SCID)?

The immediate short-term symptoms of diarrhea and infections must be treated with intravenous immunoglobulin and other antibiotic procedures. However, the child should be given a hematopoietic stem cell transplant within the first few months of life for survival.

Further Reading

Capece, Tara, and David Nash. 2009. "*ZAP70*-Related Severe Combined Immunodeficiency." *GeneReviews*. http://www.ncbi.nlm.nih.gov/books/NBK20221. Accessed 12/14/11.

"ZAP70-Related Severe Combined Immunodeficiency." 2011. Genetics Home Reference. National Library of Medicine (U.S.). http://www.ghr.nlm.nih.gov/condition/zap70 -related-severe-combined. Accessed 12/14/11.

Zellweger Syndrome

Prevalence   Occurs in about 1 in 50,000 individuals

Other Names   cerebrohepatorenal syndrome; PBD, ZSS; peroxisome biogenesis disorders, Zellweger syndrome spectrum

Hans Ulrich Zellweger was a Swiss American pediatrician at the University of Iowa who specialized in a family of disorders called leukodystrophies. The word "leukodystrophy" comes from three Greek words: *leuko*, meaning "white"; *dys*, meaning "with difficulty"; and *troph*, meaning "nourish" or "feed." A leukodystrophy is a condition in which the white matter, the myelin sheath that covers the neurons in the brain and spinal cord, does not receive the needed substances for proper development. Zellweger spectrum syndrome is a leukodystrophy named after the Swiss doctor.

What is Zellweger Spectrum?

Zellweger spectrum is a rare group of conditions that affect many parts of the body. The conditions at one time were considered separate diseases but now are considered a spectrum disorder because of the complete absence of working peroxisomes in cells. Peroxisomes are cell structures that break down toxic substances and synthesize lipids for the cells to function. They are essential for normal brain development because they act to form the myelin sheaths of the neurons. They are also important in normal function of eyes, liver, kidney, and bone.

Sometimes Zellweger syndrome is called cerebrohepatorenal syndrome, which describe the clinical symptoms of cerebro (brain), hepato (liver), and renal (kidney) functions. If the peroxisomes are not working properly, problems in lipid metabolism cause an accumulation of very long-chain fatty acids, phytanic acid, and other materials, which are toxic to many systems. Following are the symptoms of Zellweger syndrome:

- Distinct facial characteristics: Child will have high forehead, wide-set eyes, and underdeveloped eyebrow ridges.
- Skeletal abnormalities, including a large fontanelle.
- Mental retardation.
- Seizures.
- Lack of muscle tone: Sometimes the condition is so severe that the child will not be able to move, suck, or swallow.
- Eye disorders: The child may be born with glaucoma and retinal degeneration.
- Impaired hearing.
- Gastrointestinal bleeding.
- Jaundice.

Three disorders comprise the Zellweger spectrum, which is part of the peroxisome biogenetic disorders group. Following are the three disorders in the spectrum:

- Zellweger syndrome or ZS: This is the most severe form of the disorder, present at birth. Most of the symptoms listed above are present.

- Neonatal adrenoleukodystrophy or NALD: Some of the symptoms may be present but may not be as severe and may develop later in childhood.

- Infantile Refsum disease or IRD: This is the least severe form. Some children may live into adulthood.

The symptoms of each of these forms may overlap and be very difficult to distinguish between the three.

What Are the Genetic Causes of Zellweger Spectrum?

Mutations in the *PEX1* gene, officially known as the "peroxisomal biogenesis factor 1" gene, cause the spectrum disorder. Normally, *PEX1* instructs for the protein peroxisomal biogenesis factor 1, which is part of the peroxins family. Peroxins are critical for the development of the cell structures known as peroxisomes, tiny sacs that contain enzymes that break down certain fats and toxic compounds. The peroxisomes are also essential for the production of fats used in the nervous system. The myelin sheath of neurons is made of a fatty material.

About 67 mutations in the *PEX1* gene are related to Zellweger spectrum. This gene is responsible for about 70% of cases. Most mutations are the result of the change on one amino acid building block. The mutations either completely eliminate the activity of the protein or reduce its function. Thus, the cell has empty peroxisomes that cannot work. The less severe mutations allow some of the peroxisomes to form. *PEX1* is inherited in an autosomal recessive pattern and is located on the long arm (q) of chromosome 7 at position 21.2.

What Is the Treatment for Zellweger Spectrum?

Because this disorder has such severe metabolic and neurological symptoms that develop during the fetal stage, it has no cure. Nor is there a standard protocol for treatment. Most of the treatments deal with the abnormalities that occur after birth and are symptomatic and supportive.

Further Reading

"NINDS Zellweger Syndrome Information Page." 2011. National Institute for Neurological Disorders and Stroke (U.S.). http://www.ninds.nih.gov/disorders/zellweger/zellweger.htm. Accessed 12/14/11.

"Zellweger Spectrum." 2011. Genetics Home Reference. National Library of Medicine (U.S.). http://www.ghr.nlm.nih.gov/condition/zellweger-spectrum. Accessed 12/14/11.

Helpful Resources about Genetic Disorders

Most Viewed Genetic Disorders

The following list of genetic disorders includes the top 10 disorders that had the most views in the online Genetics Home Reference resource from the National Library of Medicine of the National Institutes of Health (http://ghr.nlm.nih.gov/BrowseConditions). These change continually, but as of February 2012, the conditions, which are all included in this encyclopedia, are listed in order:

The most often viewed:

1. Sickle-cell disease
2. Hemophilia
3. Cystic fibrosis
4. Down syndrome
5. Tay-Sachs disease
6. Klinefelter syndrome
7. Angelman syndrome
8. Cri-du-chat syndrome
9. Triple X syndrome
10. Fragile X syndrome

Helpful Resources for Human Genetic Disorders

The following resources are provided in an order roughly based on accessibility, relevance, and authority of information, for those seeking to find information on genetic disorders and human genetics.

Websites

American Society of Human Genetics

http://www.ashg.org

The official site of the American Society of Human Genetics, this is mostly for a technical and research audience. A section called "educate" may be of interest to students. Abstracts of research presentations at meetings are available but users have to pay for full articles.

GeneReviews

http://www.ncbi.nlm.nih.gov/books/NBK1116

This website, produced by the National Institutes of Health, is for more in-depth research on a genetic conditions topic. Each subject in "GeneReviews" offers several authors who are experts in the particular topic. For many of the rare genetic disorders, this type of review is excellent because those who focus and research the disorder write the articles. The topics are updated every two to three years as needed. The articles are peer reviewed for accuracy and pertinence. An alphabetical listing of all diseases is given.

Genetic and Rare Conditions Site

http://www.kumc.edu/gec/support

This website, from the Medical Genetics Department of the University of Kansas Medical Center, supports lay advocacy and support groups and presents information on genetic conditions, birth defects for professionals, educators, and individuals. A link is given to national and international organizations and support groups.

Genetic and Rare Diseases Information Center (GARD)

http://rarediseases.info.nih.gov/GARD

This site is a collaborative effort of two agencies of the National Institutes of Health, The Office of Rare Diseases Research (ORDR) and the National Human Genome Research Institute (NHGRI), to help people find useful information about genetic conditions and rare diseases.

Genetics Home Reference

http://ghr.nlm.nih.gov

This publication from the National Library of Medicine, part of the National Institutes of Health, is an excellent resource for students and consumers who want a good beginning reference for human genetics. There are more than 650 disorders and 900 genes. The information is reviewed and updated every three years. Included also is a handbook for understanding genetics and resource references for each genetic condition.

Genomics and Health Impact Update
http://www.cdc.gov/genomics/update/current.htm
This website, produced by the Centers for Disease Control and Prevention, and published by the Office of Public Health Genomics (OPHG), offers news updates on genomics.

Human Genome Project
http://www.genome.gov/10001772
As this site proclaims, "The Human Genome Project (HGP) was one of the great feats of exploration in history—an inward voyage of discovery rather than an outward exploration of the planet or the cosmos; an international research effort to sequence and map all of the genes—together known as the genome—of members of our species, *Homo sapiens*. Completed in April 2003, the HGP gave us the ability to, for the first time, to read nature's complete genetic blueprint for building a human being."

Medline Plus
http://www.nlm.nih.gov/medlineplus/healthtopics.html
This large website, also from the National Library of Medicine and the National Institutes of Health, gives general information about "causes, treatment and prevention for over 900 diseases, illnesses, health conditions and wellness topics," including genetic disorders. Current news about trials is presented. It is updated daily.

Medscape Reference
http://reference.medscape.com/Medscape Reference
Aimed at the advanced student or professional, this commercial site provides comprehensive coverage across more than 30 medical specialties and is composed of different areas of medicine. Medscape reference articles represent the expertise of physicians and pharmacists from leading academic medical centers in the United States and worldwide. Users may have to register, but it is free.

My Family Health Portrait
https://familyhistory.hhs.gov
On this website, the surgeon general of the United States is promoting a Family Health Initiative. In association with the Department of Health and Human Services (HHS), individuals are encouraged to trace the medical history of their families, looking for specific conditions present. A screening tool is included on the website.

Nemours Children's Health Systems
http://www.nemours.org/service/medical/genetic.html
This website is from Nemours Health System. Nemours has hospital locations throughout the country and focuses on certain genetic disorders, such as Down syndrome.

National Human Genome Resource Information
http://www.genome.gov
This site provides a wealth of educational materials about genetics and genomics. Browse the navigation bar on the left to find animated presentations, a download-able CD entitled "Understanding the Human Genome Project," a series of teaching modules and more. Consider the following lists: NHGRI Resources for Students, which include National DNA day, NHGRI webinar series, and Educational re-sources that include summer workshops, and news about research.

OMIM
http://www.ncbi.nlm.nih.gov/omim
OMIM stands for Online Mendelian Inheritance in Man. This is a database of human genes and genetic disorders developed by staff at Johns Hopkins University and hosted on the web by the university. It is, however, intended for use by physi-cians, scientists, and advanced students, not the general public.

Orphanet
http://www.orpha.net/consor/cgi-bin/index.php
This British website is a portal for rare diseases and orphan drugs. The main menu includes current research on rare diseases and drugs that target them. These drugs are called orphan drugs because they are for conditions that are relatively rare and affect only a few people. The theme on the website is: Rare diseases are rare, but rare-disease patients are numerous. There is information about the diseases arranged in an alphabetical list and a searchable database for people who are look-ing for answers using symptoms.

Websites for General Information about Genetics

Biology in Motion
http://www.biologyinmotion.com
This site, developed by a scientist, features entertaining, interactive biology learn-ing activities, including genetics, primarily aimed at younger students.

The Biology Project
http://www.biology.arizona.edu
Online interactive site for learning biology from the University of Arizona. It features a meaningful section of Mendelian genetics and molecular biology.

Dolan DNA Learning Center
http://www.dnalc.org
Dolan DNA Learning Center is an operating unit of the Cold Spring Harbor Laboratory and is a science center devoted to helping students understand genet-ics. It features an interactive feature that explores the nature of genes, DNA, as well as news.

Howard Hughes Medical Institute
http://www.hhmi.org
This website offers lectures, tutorials, and interactive demonstrations on many topics in biology, including genetics.

Nova: Body + Brain (Public Broadcasting System)
http://www.pbs.org/wgbh/nova/body/
This is the section on the body and brain from the website for the television show *Nova*. Several programs relate to biology, including "Cracking Your Genetic Code," and "How Cells Divide: Mitosis and Meiosis." The viewer can watch excerpts from the programs.

Understanding Genetics: The Tech Museum
http://www.thetech.org/genetics/index.php
A project supported by the Department of Genetics at the Stanford Medical School, and in association with the Tech Museum of Innovation in San Jose, California, "Understanding Genetics" provides basic information on genetics, news, an "Ask the Geneticist" feature, surveys on ethical issues, and other topics.

University of Utah
http://learn.genetics.utah.edu
This website presents genetics in an understandable way. One can take a tour of the basics, DNA to protein, heredity and traits, and cells.

Wellcome Trust Sanger Institute
http://www.yourgenome.org
This website helps people understand genetics and genomic science and the implications for us all. This is a popular course developed by Wellcome Trust Sanger Institute, Europe's largest genome center.

Dictionaries

Many medical dictionaries are on the market. Medical dictionaries give basic information about diseases and disorders.
Taber's Cyclopedic Medical Dictionary. 2001. Edited by Clayton Thomas, MD. Philadelphia: F. A. Davis Company. This dictionary is an example of one that is excellent for looking up background. For genetic disorders, for example, the dictionary gives a brief history of the person that a disease might have been named, such as Drs. Down and Alzheimer.

MedlinePlus Dictionary Feature   http://www.nlm.nih.gov/medlineplus/mplusdictionary.html
This Merriam-Webster online medical dictionary, provided through MedlinePlus (see above) gives a brief description of the disorder and, in the case of many genetic disorders, gives a background of the disorder.

Merck Manual Online
http://www.merckmanuals.com/home/index.html
http://www.merckmanuals.com/home/resources/anatomical_drawings/brain_and
_spinal_cord.html
The *Merck Manual* has been published since 1899 as a medical reference book of diseases and health conditions. It is now available for free online from the Merck Company, a pharmaceutical manufacturer.

Books and Textbooks

Lewis, Ricki. 2006. *Human Genetics: Concepts and Applications*. 10th ed. Dubuque, IA: William Brown.

Books on genetics can provide a foundation for understanding; however, many books do not contain information on specific human genetic disorders., especially many of the rare disorders. This is one textbook that presents complex information in a clear and interesting format. Many illustrations and graphics make this a clear and well-written book for anyone interested in human genetics. The author, Ricki Lewis, who has a PhD in molecular genetics, decided that the laboratory was not right for her. She studied journalism and became a science writer.

Books and Resources on Specific Disorders   Books, written generally to help people cope and understand a specific disorder, are available on the more familiar genetic disorders. For example, books on Down syndrome, Alzheimer disease, and cystic fibrosis may be in your public library or available for purchase at a bookstore or online. In fact, the children's section of the library is a good source for finding books that understandably explain the more common disorders.

Rare disorders seldom have entire books dedicated to them. The best way to search is to use the websites listed above or put the name of the disorder in a search engine. Many of the disorders have foundations for support, which is often the best source for the rarer disorders.

About the Author

Evelyn B. Kelly, PhD, is an independent scholar in Ocala, Florida. A college teacher by profession, Kelly has extensively studied genetics and biotechnology since its emergence in the 1990s. She has attended three institutes on biotechnology at the Center for Biotechnology Study at the University of Florida. She had a TRUE scholarship to study in the summer of 1995 at a laboratory of a molecular geneticist at the University of Florida, searching for the white-eyed gene in *Plodia interpunctella*, the Indian meal worm. In the summer of 1996, she received a fellowship to study from the American Council of Biology at Shands Teaching Hospital, University of Florida Medical School. This fellowship included a visit to the National Institutes of Health and the National Library of Medicine in Bethesda, Maryland. She has covered numerous conferences on genetic disorders and has written frequently for *Genetic Engineering News*, *Drug and Market Development*, and *Modern Drug Discovery*. Kelly has written the *Encyclopedia of Attention Deficit Hyperactivity Disorders*, *Gene Therapy*, *Stem Cells*, *Obesity*, and *The Skeletal System* for ABC-CLIO/Greenwood Press. She has written more than 400 articles and 16 books.

Index

Note: Main entries have bold page numbers